Computer Vision and AI for Indoor/Outdoor Thermal Environment, Construction Safety, and HVAC Equipment Monitoring

This book introduces the application of computer vision (CV) and AI technologies to indoor personnel information monitoring, outdoor thermal environment monitoring, construction safety monitoring, and HVAC equipment monitoring.

Computer Vision and AI for Indoor/Outdoor Thermal Environment, Construction Safety, and HVAC Equipment Monitoring explores non-contact measurement techniques relevant to the field of indoor human information and their applications in on-demand regulation of HVAC systems in sleep and office scenarios. It explores the application of CV technologies and machine learning algorithms in the field of outdoor environment monitoring, such as urban thermal environment improvement and building construction safety management. The book also covers HVAC equipment monitoring with CV technologies, including the use of visible and infrared images to detect heat pump frost and heat exchanger condensation. Finally, the authors address emerging technologies in indoor human information monitoring, outdoor environment monitoring, and HVAC equipment monitoring as well as their applications.

This book is intended for researchers in computer science and image processing. It will also be useful to researchers in building environment as well as HVAC equipment maintenance.

Computer Vision and AI for Indoor/Outdoor Thermal Environment, Construction Safety, and HVAC Equipment Monitoring

Edited by

Bin Yang, Zhihan Lyu, Xiaogang Cheng, and Yihang Liu

CRC Press
Taylor & Francis Group
Boca Raton London New York

CRC Press is an imprint of the
Taylor & Francis Group, an **informa** business

Designed cover image: Shutterstock

First edition published 2026
by CRC Press
2385 NW Executive Center Drive, Suite 320, Boca Raton FL 33431

and by CRC Press
4 Park Square, Milton Park, Abingdon, Oxon, OX14 4RN

CRC Press is an imprint of Taylor & Francis Group, LLC

ISBN: 978-1-041-00747-0 (hbk)
ISBN: 978-1-041-00748-7 (pbk)
ISBN: 978-1-003-61145-5 (ebk)

DOI: 10.1201/9781003611455

Typeset in Times
by KnowledgeWorks Global Ltd.

Contents

Preface

The integration of computer vision (CV) into environmental monitoring and HVAC systems is transforming traditional approaches to indoor comfort, building efficiency, and safety management. This book provides a comprehensive overview of how non-contact vision-based techniques enable the detection of human skin temperature, posture, and indoor distribution, offering occupant-centric control strategies for optimizing HVAC operation in residential, office, and even sleep environments. Beyond the indoor environment, it introduces in detail the application of CV and geospatial imaging to monitor the urban thermal environment, including the monitoring of outdoor thermal comfort, the assessment of microclimate variations, and the quantification of urban heat island effects.

Another key focus is construction safety, where vision-based monitoring methods contribute to accident prevention and improved site management. In parallel, the book highlights how to intelligently monitor the operation of HVAC equipment by analyzing infrared images and visible images, allowing for non-contact fault detection and diagnosis as well as predictive maintenance strategies that improve reliability and operational efficiency.

By combining technical foundations, various application scenarios, and perspectives on future development, this book provides both breadth and depth in exploring the convergence of CV and the built environment. It will serve as an essential resource for students, researchers, and industry professionals engaged in smart building, thermal environmental monitoring, and the advancement of sustainable and human-centered built environments.

Editor Biographies

Dr. Bin Yang is full professor and dean of Energy and Safety Engineering School in Tianjin Chengjian University, China. He got his joint PhD from Technical University of Denmark (DTU) and National University of Singapore (NUS). He got his bachelor and master degree from Tianjin University (China). He is a researcher with more than 20 years research experience in indoor environment quality, energy efficient and smart buildings, and HVAC. He has worked in different countries with world famous research groups such as DTU and University of California Berkeley (UCB). He was a tenured professor in Umeå University, Sweden. He has obtained several awards such as Building and Environment Best Paper Award, REHVA Young Scientist Award, ASHRAE Ralph Nevins Physiology and Human Environment Award, ASHRAE Graduate Grant-in-Aid Award, Chinese Government Award for Outstanding Self-financed Students Abroad, and so on. He has given presentations as one invited speaker several times in globally famous Universities and research institutes such as University of Southern California, Alberta University, Tsinghua University, REHVA, etc. He has more than 200 high quality publications and 40 patents. His published books include "Personal Comfort Systems for Improving Indoor Thermal Comfort and Air Quality" (Springer Publisher) and two Chinese monographs. He is the Stanford/Elsevier Top 2% Scientists (2020-2026).

Dr. Zhihan Lyu (senior member, IEEE) received the joint Ph.D. degree in computer science from Ocean University of China in 2012. He authored or coauthored more than 300 papers, including more than 120 papers on IEEE/ACM Transactions. He is British Computer Society Fellow, ACM Distinguished Speaker, Career-long Scientific Influence Rankings of Stanford's Top 2% Scientists, Marie Skłodowska-Curie Fellow, Clarivate Highly Cited Researcher, and Elsevier Highly Cited Chinese Researcher. He is Associate Editor for 22 journals, including *ACM TOMM*, *IEEE Transactions on Intelligent Transportation Systems*, *IEEE Transactions on Network and Service Management*, *IEEE Transactions on Network Science and Engineering*, and *IEEE Transactions on Computational Social Systems*. He is Co-Chair or TPC of 50 conferences, including ACM MM 2021–2025 and ACM IUI 2015–2022. He has reviewed 400 papers. He was the recipient of more than 20 awards from China, Europe, and IEEE. He has given more than 80 invited talks for universities and companies in Europe and China. He has given 20 keynote speeches for international conferences.

Dr. Xiaogang Cheng graduated from Southeast University, Nanjing Tech University, and Nanjing University successively and received Ph.D. degree. He studied computer vision (CV) at the Royal Institute of Technology in Sweden and ETH Zurich. He is currently an associate professor at the School of Communication and Information Engineering of Nanjing University of Posts and Telecommunications, and the director of the Institute of Flexible Intelligence, mainly engaged in the research of CV and its applications. He has been selected as one of the "Hundred Talents Program" at the provincial level and “Nanjing Public Security Science and Technology Innovation Leaders”. He has published more than 30 academic papers and 34 authorized patents and was awarded the first prize of Jiangsu Provincial Public Security Department’s Science and Technology Award and the Third Prize for Science and Technology by the Ministry of Public Security of China. He has been selected for the China Postdoctoral International Exchange Program and has been awarded as an excellent supervisor for the open project of Nanjing University of Posts and Telecommunications.

Yihang Liu is a Ph.D. candidate at Tianjin University. Her research interest is the study of occupant-centric control strategies for ventilation and air conditioning systems based on computer vision, including sleep and office environments. The research topic focuses on dynamically adjusting the operational parameters of ventilation and air conditioning systems based on occupants’ thermal demand to reduce energy consumption. She has won two national scholarships and five national competition awards, published six journal papers and two conference papers, participated in the writing of two English academic monographs, and participated in one project supported by the National Natural Science Foundation of China and one project supported by the Tianjin Natural Science Foundation.

Contributors

Luting Bai
School of Energy and Safety Engineering
Tianjin Chengjian University
Tianjin, China

Xiaogang Cheng
School of Communications and Information Engineering
Nanjing University of Posts and Telecommunications
Nanjing, China

Ruiqi Guo
Department of Architecture
School of Design
Shanghai Jiao Tong University
Shanghai, China

Xinrou Han
School of Energy and Safety Engineering
Tianjin Chengjian University
Tianjin, China

He Li
School of Energy and Safety Engineering
Tianjin Chengjian University
Tianjin, China

Yihang Liu
School of Architecture
Tianjin University
Tianjin, China

Zhihan Lyu
School of Computer Science and Technology
Xidian University
Xi'an, China

Min Qi
School of Energy and Safety Engineering
Tianjin Chengjian University
Tianjin, China

Bo'an Wei
School of Energy and Safety Engineering
Tianjin Chengjian University
Tianjin, China

Bin Yang
School of Energy and Safety Engineering
Tianjin Chengjian University
Tianjin, China

Shuang Yang
School of Energy and Safety Engineering
Tianjin Chengjian University
Tianjin, China

Weiling Zhang
School of Energy and Safety Engineering
Tianjin Chengjian University
Tianjin, China

Haichuan Zhao
School of Energy and Safety Engineering
Tianjin Chengjian University
Tianjin, China

Zeyu Zhao
School of Energy and Safety Engineering
Tianjin Chengjian University
Tianjin, China

Acknowledgments

The editors and contributors of this book wish to acknowledge the following financial support and invaluable contributions that made this publication possible.

Chapters 2 and 4: The work in these chapters was supported by the National Natural Science Foundation of China (No. 52278119). Chapter 3: The work in this chapter was supported by Key Program of the Tianjin Natural Science Foundation Project (No. 25JCZDJC00060). Chapters 5–7: The work in these chapters was supported by the National Key Research and Development Program of China (No. 2023YFC3806101-03). We are deeply indebted to the funding agencies for their support during the writing process of this book.

We extend our special thanks to the authors of the various research cases cited throughout this book. Their work and insights were essential in illustrating key concepts and describing practical application scenarios.

Finally, acknowledgment is also extended to the various publishers and copyright holders who graciously granted permission to reproduce figures within this book.

1 Introduction

Zeyu Zhao, Xinrou Han, Shuang Yang, Yihang Liu, and Bin Yang

1.1 RESEARCH BACKGROUND

The comfort of indoor and outdoor environments and the efficient operation of heating, ventilation, and air conditioning (HVAC) systems are crucial for ensuring human health, improving productivity, and reducing energy consumption. Traditional monitoring methods for indoor/outdoor environmental conditions and HVAC equipment status face limitations due to spatiotemporal technical constraints.

With the rapid development of artificial intelligence (AI) technologies, particularly breakthroughs in deep learning, computer vision (CV), and natural language processing (NLP), new opportunities have emerged for intelligent monitoring of indoor/outdoor environments and HVAC systems. Through big data analytics, pattern recognition, and predictive modeling, AI can optimize environmental control strategies, enhance system response speed, identify outdoor risks, and accurately monitor equipment operation status. For instance, deep learning algorithms enable more precise predictions of indoor and outdoor temperature fluctuations, facilitating accurate thermal environment control and minimizing energy waste.

Advancements in AI not only drive environmental monitoring technologies and intelligent operation of HVAC systems but also significantly enhance operational efficiency and resource utilization. These innovations provide a scientific foundation for boosting productivity and optimizing energy consumption. Such technological breakthroughs are poised to overcome existing barriers in intelligent environmental monitoring and HVAC systems, unlocking new developmental opportunities. The research background for monitoring indoor/outdoor environments and HVAC equipment is elaborated in next sections.

1.1.1 Indoor Thermal Environment Monitoring and Regulation

In recent years, with the rise of the "occupant-centric" concept in building design, indoor environmental control has shifted toward prioritizing occupants' actual needs. This approach emphasizes real-time monitoring of occupant-related information tied to thermal comfort to enable automated control of HVAC systems, balancing thermal comfort requirements with energy efficiency. Occupant-centric control (OCC) strategies are primarily applied to public buildings such as offices and transportation hubs, as well as sleep environments, which will be detailed in Chapters 3 and 4.

In the 1970s, Fanger established classical thermal comfort theory [1], prompting subsequent research into methods for assessing human thermal comfort. Traditional measurement approaches include three primary methods: questionnaire surveys, environmental parameter measurements, and physiological parameter

DOI: 10.1201/9781003611455-1

measurements. Questionnaire-based methods heavily rely on continuous occupant cooperation, suffer from the Hawthorne effect, and lack operational practicality. Environmental parameter measurements utilize sensors to detect indoor temperature, humidity, air velocity, and other variables to evaluate environmental comfort. While feasible, fixed sensor placements fail to represent actual comfort conditions in occupied zones or address individual thermal preferences. Consequently, physiological parameter measurement methods have gained further attention.

Physiological parameter measurement employs sensors to assess physiological signals and thermal adaptive behaviors to infer thermal sensations. Studies indicate that physiological signals such as heart rate, pulse, blood perfusion rate, skin temperature, metabolic rate, electroencephalography (EEG), and surface electromyography (sEMG) correlate with human thermal comfort [2–5]. Thermal adaptive behaviors, including posture adjustments, also serve as predictors of thermal comfort [6]. Based on contact requirements, physiological measurement techniques are categorized into contact-based, semi-contact, and non-contact methods. The first two require wearable devices or sensors, where factors like sensor placement, angle, and user discomfort may introduce experimental errors, limiting real-world applicability. Thus, vision-based non-contact technologies have been widely adopted to detect physiological parameters and thermal states.

Implementing OCC strategies requires not only predicting thermal comfort but also analyzing occupant distribution. Vision-based deep learning algorithms can predict occupant count and location, which are critical for HVAC system operation, energy consumption, and control strategies. Adjusting operational parameters based on spatial occupancy enables energy-efficient performance while ensuring comfort in occupied zones. Chapter 2 will elaborate on non-contact detection methods for skin temperature, posture, and occupancy. Integrating HVAC system operations with occupancy data and actual thermal comfort metrics can reduce energy consumption, enhance comfort, and maintain air quality [7]. Chapter 3 focuses on sleep environments, while Chapter 4 discusses the thermal environment when occupants are in the waking state.

1.1.2 Outdoor Environmental Monitoring

The accelerating pace of global urbanization and intensifying climate change have exacerbated urban overheating [8]. Urban heat island (UHI) effects and heatwave disasters continuously degrade urban living environments, significantly reducing residents' thermal comfort and posing severe threats to public health, socioeconomic stability, and sustainable development. Creating comfortable outdoor environments, enhancing urban livability, and revitalizing cities have become urgent priorities [9, 10].

Outdoor thermal comfort is a complex outcome influenced by a variety of factors, including urban microclimate (air temperature, humidity, wind speed, solar radiation, etc.) and individual physiological (age, gender, physical activity) and psychological aspects. Pedestrian comfort is affected by the urban natural and artificial environments from multiple perspectives under various sensory stimuli, such as thermal sensation, vision, hearing, air quality [11, 12], and human activities [13–15].

Consequently, monitoring outdoor environmental parameters to inform rational urban planning has emerged as a critical strategy for improving pedestrian thermal comfort. In the era of big data, the proliferation of geolocation devices, mobile technologies, and mapping services has generated novel geospatial image datasets. These images capture urban physical spaces from diverse perspectives, enabling quantitative studies of urban environments. Outdoor thermal environment monitoring will be discussed in detail in Chapter 5.

In recent years, construction safety issues in China have become increasingly prominent. Statistics reveal 85 construction safety incidents in 2023, resulting in 320 fatalities (or missing persons), with collapses, falls from heights, and crane-related accidents accounting for 82.35%–83.12% of the total. For instance, a construction accident in Jiulongpo Metro occurred due to workers' non-compliance with safety protocols, leading to equipment failure and multiple injuries. Construction safety in China not only impacts public safety but also jeopardizes social stability. However, the vast scale of ongoing projects lacks comprehensive monitoring data and centralized digital management systems, hindering effective information accessibility. To address this, CV-based construction safety monitoring methods are being developed, which will be elaborated in Chapter 6.

1.1.3 HVAC Equipment Monitoring

In the late 1980s, fault detection and diagnosis (FDD) technologies began to be applied in the HVAC and building engineering fields. Most of the research was focused on the core equipment and piping in the refrigeration machine room. The core equipment includes chillers, chilled water pumps, cooling water pumps, collectors, headers, heat pumps, coolers, etc., while the piping includes chilled water loop, cooling water loop, and so on. Typically, FDD technology detects and diagnoses common faults by measuring the temperature or pressure at various locations in the system and analyzing the thermodynamic relationships. It can effectively extend the life of equipment and components, stabilize indoor temperature, and improve building energy efficiency. Traditional FDD methods in the HVAC field can be divided into three major categories: quantitative model-based methods, qualitative model-based methods, and process history-based methods. Quantitative models are developed based on sound physical or engineering principles but are computationally complex. Qualitative models are easy to develop and apply but rely heavily on the expertise of the developers, and some rules may not be applicable when the system is complex. Process history models are black-box models established when the system's physical characteristics are not well understood. Although the development concept is not difficult, a large amount of operational data from HVAC systems is required for training.

Traditionally, HVAC operational data are structured datasets collected via building automation systems (BASs). The operational data, such as temperature and humidity of equipment or piping, is obtained by contact measurement methods using temperature and humidity sensors. The operation is simple, but it is inevitable that it will have some impact on the normal operation of the system, causing errors in the measurement data, which in turn affects the results of fault diagnosis. Therefore, the development of a non-contact measurement method based on image signals, which can quickly and accurately diagnose and resolve faults, has a breakthrough significance for FDD technology in the HVAC field. The relevant technology will be introduced in Chapter 7, mainly including equipment surface temperature monitoring, monitoring of frost formation on heat pump surfaces, and monitoring of surface condensation on heat exchangers.

1.2 CONTRIBUTIONS OF THIS BOOK

In recent years, the rapid advancement of AI-related technologies, including CV, machine learning, internet of things (IoT), and robotics, has significantly expanded their applications across diverse fields. These technologies have demonstrated remarkable potential in non-contact measurement research. Building on a comprehensive review of recent advancements, this book systematically consolidates the applications of non-contact measurement technologies in three critical domains: indoor thermal environment monitoring, outdoor environmental monitoring, and HVAC equipment monitoring. It further analyzes challenges encountered during their development, aiming to inspire future research directions. To achieve these objectives, the book is structured as follows.

Chapter 1 provides an overview of the research background. Chapters 2–4 focus on CV-based non-contact measurement in indoor thermal environments.

Chapter 2 details non-contact techniques for detecting occupant thermal states and occupancy. It discusses methodologies based on skin temperature and human posture, including advancements in Eulerian video magnification (EVM) for skin temperature monitoring, skeletal keypoint recognition for posture analysis, infrared imaging and photoplethysmography (PPG) for skin temperature monitoring, and object tracking and detection for occupancy monitoring. These technologies are evaluated across multiple dimensions, including accuracy, practicality, and scalability.

Chapters 3 and 4 explore practical applications in sleep environment monitoring and regulation, and on-demand control for indoor thermal environment, respectively. These chapters highlight how CV-driven non-contact measurements enable precise parameter monitoring and energy-efficient optimization of HVAC systems. By presenting these case studies, the book demonstrates the transformative potential of vision-based non-contact technologies in achieving high-precision environmental monitoring and energy-saving HVAC operations, offering a roadmap for researchers and practitioners in building science and smart infrastructure.

Chapters 5 and 6 address the application of non-contact measurement in outdoor environmental monitoring. These chapters introduce research efforts applying CV technologies and machine learning algorithms to outdoor environmental monitoring. Chapter 5 summarizes key non-contact measurement technologies in three aspects: microclimate monitoring, city-level urban environmental information, and street-level urban environmental information, providing novel insights for evaluating urban environments and pedestrian thermal comfort. In Chapter 6, focusing on construction safety, the book explores the integration of CV technologies with robotic automation, emphasizing their applications in construction safety monitoring. This includes sensing technologies, behavior recognition technologies,

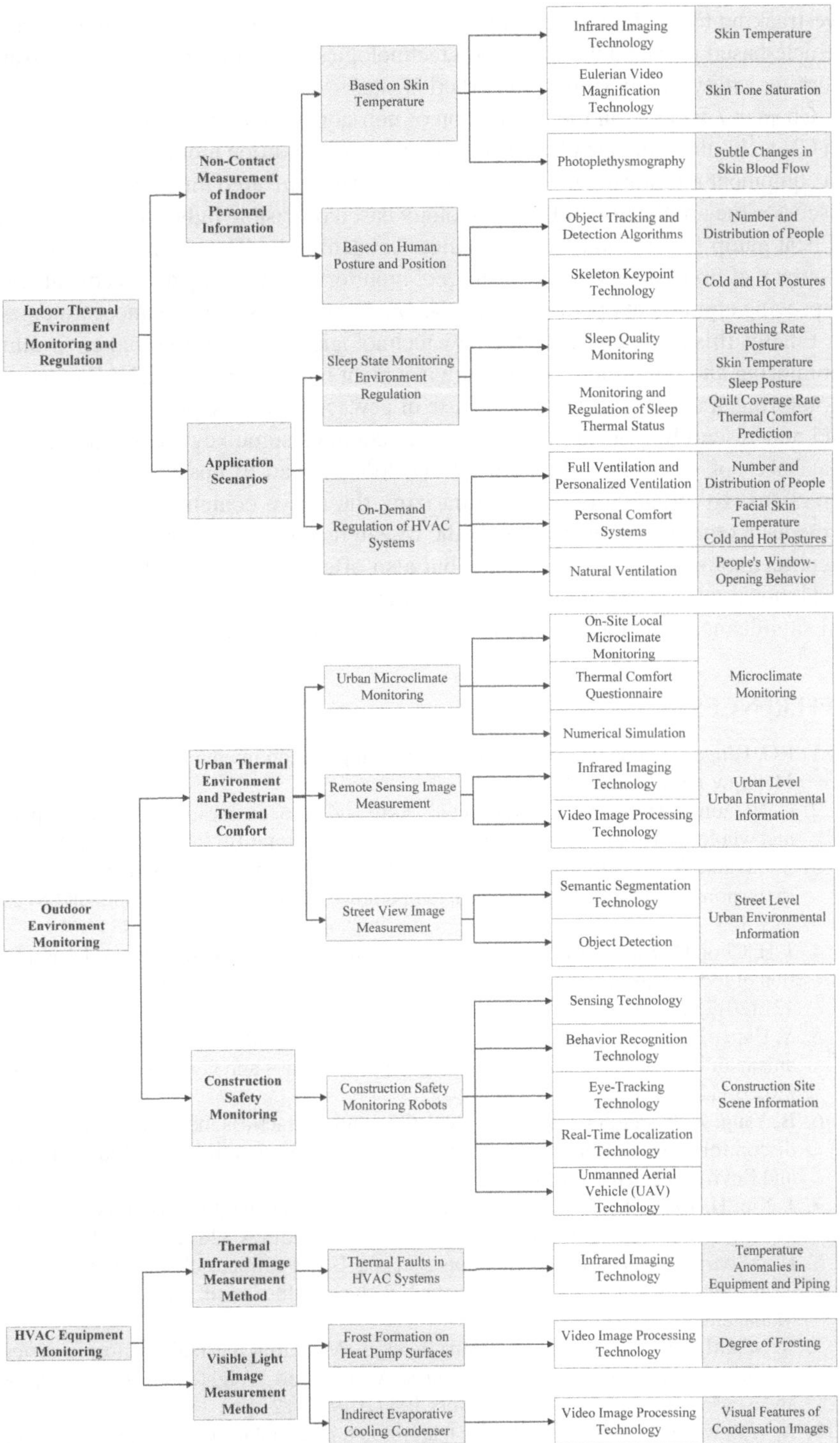

FIGURE 1.1 The logical framework of the book.

eye-tracking technologies, real-time localization technologies, and unmanned aerial vehicle-based on-site data acquisition technologies, offering a new technical framework for safety monitoring at construction sites.

Chapter 7 focuses on the application of non-contact measurement methods based on CV technology in the field of HVAC equipment condition monitoring. It elaborates on equipment monitoring methods based on thermal infrared images and particularly discusses the application of CV technology based on visible light images in the areas of heat pump frosting and heat exchanger condensation. The chapter introduces the accuracy of non-contact visual intelligent monitoring and analyzes its advantages in improving equipment operating efficiency and optimizing maintenance strategies.

Finally, this book discusses the new technologies in indoor thermal environment monitoring and regulation, outdoor environment monitoring, and HVAC equipment monitoring, and looks forward to the use of new technologies, possible intersections, and new research opportunities. The book identifies some key technical challenges that have not yet been fully resolved and looks ahead to possible future research directions. By organizing and summarizing the above content, the book not only provides a comprehensive review of the research progress of non-contact measurement technologies in different fields but also offers important technical references and innovative ideas for future research, which is of great academic value and practical significance. The logical framework of the book is shown in Figure 1.1.

REFERENCES

1. P.O. Fanger, Thermal comfort, analysis and application in environmental engineering, McGraw Hill, New York, 1972.
2. C. Huizenga, H. Zhang, E. Arens, et al., Skin and core temperature response to partial- and whole-body heating and cooling, Journal of Thermal Biology 29 (2004) 549–558.
3. S. Takada, S. Matsumoto, T. Matsushita, Prediction of whole-body thermal sensation in the non-steady state based on skin temperature, Building and Environment 68 (2013) 123–133.
4. J.-H. Choi, D. Yeom, Study of data-driven thermal sensation prediction model as a function of local body skin temperatures in a built environment, Building and Environment 121 (2017) 130–147.
5. Y. Dang, Z. Liu, X. Yang, et al., A fatigue assessment method based on attention mechanism and surface electromyography, Internet of Things and Cyber-Physical Systems 3 (2023) 112–120.
6. B. Yang, X. Cheng, D. Dai, et al., Real-time and contactless measurements of thermal discomfort based on human poses for energy efficient control of buildings, Building and Environment 162 (2018) 106284.
7. J. Xie, H. Li, C. Li, J. Zhang, M. Luo, Review on occupant-centric thermal comfort sensing, predicting, and controlling, Energy and Buildings 226 (2020) 110392.
8. H. Akbari, C. Cartalis, D. Kolokotsa, et al., Local climate change and urban heat island mitigation techniques – The state of the art, Journal of Civil Engineering and Management 22 (2016) 1–16.
9. N. Mijani, S.K. Alavipanah, S. Hamzeh, et al., Modeling thermal comfort in different condition of mind using satellite images: An Ordered Weighted Averaging approach and a case study, Ecological Indicators 104 (2019) 1–12.
10. A. Wibowo, K.O. Salleh, Landscape features and potential heat hazard threat: A spatial–temporal analysis of two urban universities, Natural Hazards 92 (2018) 1267–1286.

11. K. Pantavou, S. Lykoudis, B. Psiloglou, Air quality perception of pedestrians in an urban outdoor Mediterranean environment: A field survey approach, Science of the Total Environment 574 (2017) 663–670.
12. M. Zakaria, E. Ezani, N. Hassan, et al., Traffic-related air pollution (TRAP), air quality perception and respiratory health symptoms of active commuters in a university outdoor environment, IOP Publishing, Bristol, 2019.
13. W. Gao, Y. Qian, H. Chen, et al., Assessment of sidewalk walkability: Integrating objective and subjective measures of identical context-based sidewalk features, Sustainable Cities and Society 87 (2022) 104142.
14. X. Ma, C.K. Chau, J.H.K. Lai, Critical factors influencing the comfort evaluation for recreational walking in urban street environments, Cities 116 (2021) 103286.
15. D. Berkouk, T.A.K. Bouzir, S. Boucherit, et al., Exploring the multisensory interaction between luminous, thermal and auditory environments through the spatial promenade experience: A case study of a university campus in an oasis settlement, Sustainability 14 (2022) 4013.

2 Non-Contact Measurement Method of Indoor Personnel Information

Yihang Liu, Weiling Zhang, Bo'an Wei, and Xiaogang Cheng

2.1 INTRODUCTION

During the operation of building systems, various types of sensors and detection algorithms can be employed to capture environmental parameters, occupant distribution, and thermal states in building spaces. This data is used to predict system operating parameters, thereby enabling energy-efficient operation of heating, ventilation, and air conditioning (HVAC) systems and improving energy efficiency. Existing methods for detecting human thermal states can be classified into contact measurement methods, semi-contact measurement methods, and non-contact measurement methods [1].

Contact measurement methods involve questionnaire surveys and physiological parameter measurements. Questionnaire methods require interrupting ongoing activities for frequent survey completion and are limited by subjectivity, social desirability bias, dynamic environmental changes, and memory errors. Additionally, the Hawthorne effect causes discrepancies between survey results and actual thermal states [2]. Physiological measurements typically use devices such as polysomnography, mercury thermometers, and skin temperature sensors to monitor parameters, including heart rate, pulse, blood perfusion, skin temperature, metabolic rate, and electroencephalogram. While these physiological parameters correlate with human thermal states [3], the measurement devices can cause discomfort. Some of the devices are used under special conditions, which are not convenient for measurement, and will also cause measurement errors, and some measurement devices are not suitable for some special groups.

Semi-contact measurement methods integrate sensors into wearable devices such as infrared sensor-equipped glasses [4], smart wristbands [5], and heart rate monitors. These devices may still induce wearer discomfort. Non-contact detection methods eliminate both the discomfort from semi-contact devices and the interference with daily activities caused by contact methods. Non-contact measurement methods employ environmental sensors, motion sensors, radio frequency sensors, and video image processing technologies to obtain occupant information, including location, number, posture, and physiological parameters. The framework for non-contact measurement methods is shown in Figure 2.1.

 DOI: 10.1201/9781003611455-2

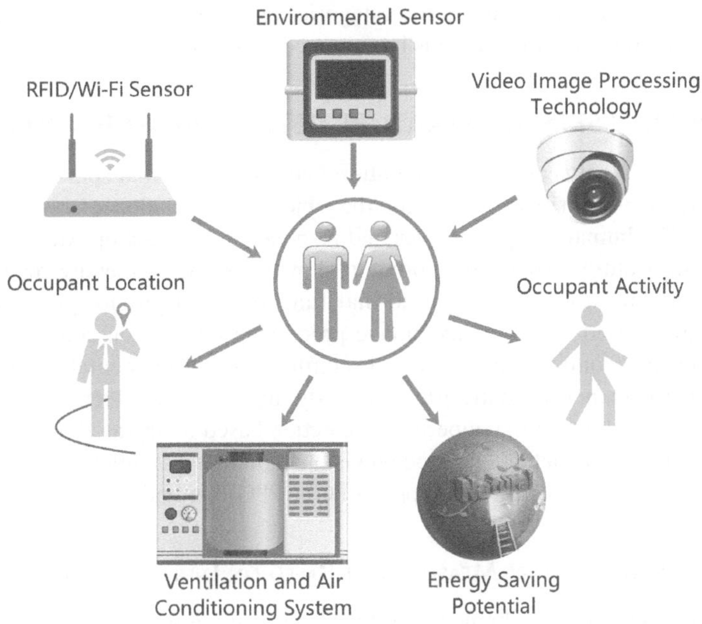

FIGURE 2.1 Framework for non-contact measurement methods.

Measuring indoor temperature and carbon dioxide concentration can help estimate occupant numbers and thermal perception [6]. However, carbon dioxide concentration and temperature sensors only provide rough occupancy estimates, with detection accuracy affected by factors like sensor placement relative to occupied zones and solar radiation. Moreover, individual differences in thermal comfort mean environmental parameters alone cannot accurately represent actual thermal sensations. Motion sensors like passive infrared (PIR) sensors, ultrasonic detectors, and pressure mats are commonly used for occupancy detection but cannot determine occupant numbers or precise locations [7]. With technological advancements, radio frequency identification (RFID) sensors can identify the spatial coordinates of occupants [8], though requiring all occupants to carry RFID tags interferes with normal activities and affects detection accuracy. Wi-Fi sensors estimate occupant numbers by detecting connected devices [9], but their accuracy is influenced by device usage patterns and occupant habits.

Given the aforementioned limitations, various camera-based occupant information detection technologies have been gradually developed. Computer vision, which combines cameras with deep learning (DL), can replace the human eye to automatically recognize and track targets in real time while predicting physiological parameters and thermal states [1]. Generally, camera-based deep learning algorithms can predict occupant numbers, locations, and postures with accuracy rates typically ranging between 80% and 97%, improving detection speed and precision without interfering with occupants' normal activities [10]. Additionally, captured images can be used to estimate information such as skin temperature, thereby deriving thermal state-related data to inform HVAC system control strategies [1]. Consequently, this chapter will focus on introducing the categories and principles of non-contact occupant information detection methods based on computer vision technologies.

2.2 HUMAN SKIN TEMPERATURE DETECTION TECHNOLOGY

The relationship between skin temperature and human thermal sensation is closely linked, which is a critical metric for the objective evaluation of body's thermal state [11]. The human body perceives and responds to changes in external environmental temperature through variations in skin temperature, thereby regulating its body temperature to maintain the normal functioning of physiological processes. Technologies based on video and image processing enable non-contact detection, addressing shortcomings such as the discomfort caused by wearable devices and interference with normal daily life. Currently, non-contact detection methods are primarily divided into skin temperature detection based on infrared thermal images and skin temperature detection based on visible light images (Euler video magnification and remote photoplethysmography (PPG) (rPPG) methods).

2.2.1 Skin Temperature Measurement Based on Infrared Thermal Image

Infrared thermography technology (IRT) can receive the infrared radiation emitted by an object through an infrared detector, generating an image that shows the continuous distribution of the temperature field. Using an infrared thermal imager can obtain a continuous temperature distribution across the entire surface of an object, allowing for direct imaging and comparison of temperature ranges in different areas without the need for contact or damage to the object. Thanks to these advantages, infrared thermography, as a non-contact and non-destructive technology, has been increasingly applied in various fields such as building inspections, metal crack detection, pipeline welding defect identification, electrical engineering, oil contamination detection, vehicle recognition, and fire detection in recent years [12].

With advancements in infrared imaging technology, an innovative method based on infrared thermography has been developed for automatically detecting human thermal discomfort. By collecting infrared thermal imaging data and other physiological signals of occupants in different thermal environments, researchers have constructed a multimodal model capable of analyzing these data and predicting human thermal comfort. This research highlights the advantages of infrared thermography as a non-contact and non-invasive technology, which can effectively capture the distribution of human surface temperatures, thereby helping to identify and improve environmental factors affecting thermal comfort. This, in turn, provides a scientific basis for designing energy-efficient and comfortable smart building systems [13]. ThermalSense system utilizes infrared thermography to dynamically detect and predict occupants' thermal preferences [14]. By analyzing changes in human skin temperature, ThermalSense can infer the physiological state of vasodilation or vasoconstriction, thereby determining an individual's comfort level under different temperature conditions. The application of this technology helps to more precisely control HVAC systems, aiming to reduce energy consumption while maintaining user comfort. Studies have shown that ThermalSense achieves an accuracy rate of 94%–95% in predicting heating or cooling needs, demonstrating the significant potential of infrared thermography in energy conservation and enhancing indoor environmental comfort [14].

The method of IRT monitors and evaluates an individual's thermoregulatory performance and thermal comfort by measuring the skin temperature at specific points on the face. This technology leverages the rich vascular network of the face and its typically unobstructed exposure, allowing for non-invasive, real-time capture of skin temperature changes, which in turn infers the body's thermal regulation state. By analyzing these data, researchers can define heuristic rules for assessing thermal comfort and explore predictive models based on these rules, offering new perspectives for designing smart building systems that enhance energy efficiency and occupant satisfaction [15]. Li et al. [16] proposed a non-invasive framework using infrared thermography for real-time monitoring and analysis of facial skin temperatures of building occupants to assess their thermal comfort. This technology utilizes low-cost infrared cameras to capture temperature data from different facial regions, which are then processed and analyzed using computer vision and machine learning algorithms. This enables accurate prediction of occupants' thermal comfort and dynamic control of indoor environment, aiming to improve energy efficiency and occupant satisfaction. Researchers have developed a system based on thermal imaging cameras, and experiments have shown that infrared thermography has the potential to be integrated into smart building systems for more efficient, occupant-centric thermal environment control, thereby reducing energy consumption without compromising comfort [17]. Cosma et al. [18] utilized infrared thermography to establish individual thermal comfort models by extracting temperature information from multiple localized body areas. Compared to traditional skin temperature–based measurement methods, this technology uses low-cost infrared cameras to capture skin temperatures from different facial regions, thereby minimizing disruption to building occupants. The results indicate that temperature measurements of ears, nose, and cheeks are most critical for predicting thermal comfort, and the framework achieves an average accuracy of 85% in effectively assessing occupants' thermal comfort. This work demonstrates the potential of infrared thermography in building environment control and automation systems, offering a novel approach for personalized HVAC control. Researchers also employed a low-cost infrared webcam system to predict individual thermal preferences by capturing and analyzing facial skin temperatures [19], as shown in Figure 2.2. This method not only eliminates the need

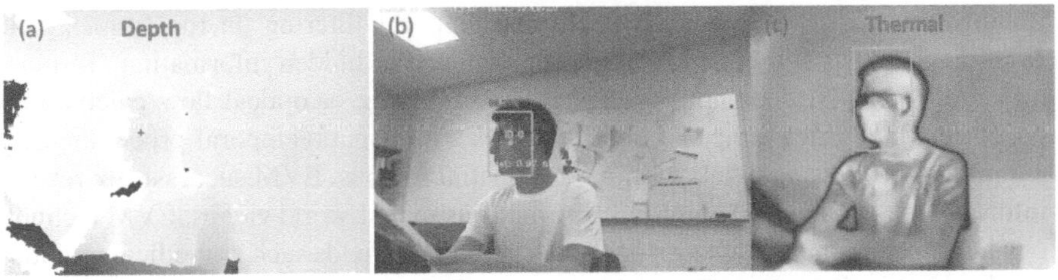

FIGURE 2.2 Dual camera face detection: (a) depth data from the Kinect; (b) RGB images from the Kinect (for face detection); (c) thermal images from the thermal camera (bounding box is mapped from the RGB image). (Source: Reference [19], Copyright 2019 by Elsevier.)

for occupant participation or wearable devices but also adapts to various environmental settings, providing high flexibility and scalability. Experimental validation shows that the system can accurately measure and reflect individual thermal comfort in multi-occupant environments, offering an effective solution for thermal comfort control and energy management in smart building HVAC systems.

Tejedor et al. [20] proposed a novel method using IRT to assess the thermal comfort of elderly individuals in indoor environments. IRT technology was employed to monitor and quantify the skin temperature at four specific points on the face (nose, forehead, cheekbones, and chin) as well as clothing temperature, using these data to evaluate the response of the human body's thermal regulation system to the indoor environment. The results indicate that the IRT model can estimate the heat exchange between three body parts (head, back-pelvis, chest-limbs) and the surrounding environment based on user characteristics and environmental parameters. This provides a reliable assessment tool for indoor thermal comfort and helps improve the control strategies of HVAC systems to meet the thermal environment needs of elderly individuals during winter.

He et al. [21] introduced a new approach that combines infrared thermography, computer vision, and machine learning to intelligently detect human thermal state. By analyzing skin temperatures of facial areas (such as cheeks and nose) and hands, the method predicts human thermal comfort. Infrared thermography cameras and computer vision techniques were used to collect skin temperature data, which, combined with subjective thermal sensation data, were used to build a random forest (RF) classification model. These models can predict human thermal state with high accuracy, offering potential improvements to air conditioning control systems to make indoor environments more comfortable and energy-efficient. The findings suggest that temperatures of specific body parts measured by infrared thermography can be effectively used to predict and automatically regulate indoor thermal comfort, highlighting its significance for the development of smart building environmental control systems.

2.2.2 Skin Temperature Measurement Based on Euler Video Magnification Technology

Eulerian video magnification (EVM) is an advanced video processing technique designed to reveal subtle changes in videos that are difficult or impossible to observe with the naked eye. This technique takes a standard video sequence as input, applies spatial decomposition, and then performs temporal filtering on the frames. The resulting signal is subsequently amplified to uncover hidden information. To magnify motion, EVM does not perform feature tracking or optical flow calculations but instead amplifies temporal color changes using spatiotemporal processing. By temporally processing pixels within fixed spatial regions, EVM successfully reveals informational signals and amplifies tiny motions in real-world videos. EVM technology holds broad application prospects, particularly in fields such as medical monitoring, health diagnostics, video enhancement, and security surveillance. By amplifying subtle changes in videos, EVM can help doctors and researchers observe and analyze physiological signals such as heart rate, respiration, and blood flow in real time without physical contact. Additionally, this technology can be used to improve video quality, enhance hard-to-notice details, and thus play a role in video editing, special effects production, and virtual reality (VR) [22].

Jazizadeh et al. [23] explored whether computers could quantify human thermal comfort by analyzing subtle changes in digital videos. The researchers developed an image processing framework based on EVM, which can amplify tiny changes in physiological signals related to human thermoregulation. Experimental results with subjects showed that the framework effectively identified the thermoregulatory states of the subjects, providing a new method for non-contact and non-invasive thermal comfort assessment. Jung et al. [24] utilized a non-invasive Doppler radar sensing system to detect human respiratory activity, using it as an indicator to assess human thermal comfort and thereby intelligently control HVAC systems. The study found that under different indoor temperature conditions, most subjects measured by the system showed changes in respiratory intensity, demonstrating the feasibility and potential of this method in quantifying human thermoregulation and assessing thermal comfort. Alghoul et al. [25] introduced two methods for extracting heart rate variability (HRV) from video signals: independent component analysis (ICA) and EVM. The researchers compared the effectiveness of the two methods in extracting HRV features. The ICA method performed better in high-frequency (HF) and low-frequency to HF ratio (LF/HF) aspects, while the EVM method excelled in LF and correlation with real data. The results indicated that the ICA method improved HF estimation, but the EVM method might be more suitable when motion is involved in the video or HF parameters are not critical. This study provides new insights into non-contact heart rate and HRV measurement technologies, which have potential value in emotion estimation, psychological stress monitoring, and other applications.

Jazizadeh et al. [26] proposed a method using red, green, blue (RGB) video images to infer individual thermal comfort for distributed control of HVAC systems. They developed a framework that leverages human thermoregulation mechanisms and EVM to infer thermoregulatory states by analyzing subtle changes in facial skin blood flow, as shown in Figure 2.3. The experiment

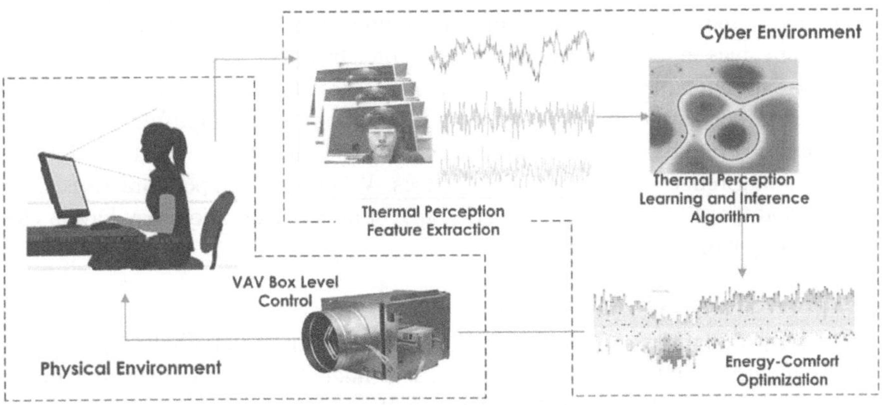

FIGURE 2.3 Conceptual framework of thermal comfort prediction based on video image and its integration with HVAC operation. (Source: Reference [26], Copyright 2018 by Elsevier.)

involved 21 subjects in low-temperature (20°C) and high-temperature (30°C) environments. The results demonstrated that the framework successfully inferred thermoregulatory states in 89% of statistically significant cases, proving the feasibility of non-invasive, cost-effective thermal comfort assessment through video images.

Jung et al. [27] introduced a vision-based and non-invasive method that utilizes RGB video images to infer human thermoregulatory states in response to changes in indoor thermal environments, aiming to achieve more energy-efficient HVAC system control. Cheng et al. [28] presented a non-contact measuring method based on subtleness magnification and deep learning (NIDL), designed to provide real-time feedback on human thermal comfort for smart buildings. By leveraging skin feature data (such as subtle motions and texture changes) and a 315-layer deep neural network (DNN), the relationship between skin features and skin temperature was established. Experimental data collection involved 1.44 million frames of feature data for algorithm validation. Compared to non-contact measurement algorithm based on a partly-personalized saturation temperature model (NIPST), the NIDL algorithm demonstrated significant improvements in measurement accuracy, with average and median errors reduced by 39.07% and 38.76%, respectively. The main contributions of this study include proposing a novel non-contact skin temperature measurement method, being the first to apply deep learning to feature extraction and relationship construction combined with EVM, and creating a large-scale image-based dataset for human thermal comfort [28]. The principles underlying this technology will be further elaborated in detail.

2.2.2.1 Skin Tone Saturation

Artificially defined color spaces use three or four independent variables to describe colors. Commonly used color spaces include RGB, CMYK (cyan, magenta, yellow, black), YUV (Y: luminance, UV: chrominance), YIQ (luminance, in-phase, quadrature), and HSV (hue, saturation, value). Different color spaces have their advantages and limitations, and some can be converted into one another. The HSV color space has axes that form an inverted hexagonal cone [29]. In the counterclockwise direction, hue is the dimension that represents colors such as red, yellow, green, blue, etc. Hue values measure the deviation of a color from black, which represents zero energy. Saturation measures the deviation of a color from achromaticity, with a range of 0–1. The deeper the color, the higher the saturation value. Amplifying and extracting the RGB signals of facial skin color can be used to analyze heart rate [30], but RGB cannot reflect saturation as HSV does. When skin temperature rises, pores dilate, and the skin turns red, indicating that skin tone saturation may have a close relationship with skin temperature.

2.2.2.2 Saturation Temperature Model

After identifying the user, identify the queue for each user. Record the ST models of different queues in the computer database. The model first samples the skin temperature and represents the sampled frames as vectors.

$$V = \{f_1, f_2, \ldots, f_n\} \tag{2.1}$$

where $\boldsymbol{V}$ is the video sampled from hand skin, f is a frame of the video, and n is the sampling time. As mentioned, saturation is a parameter in the HSV color space that represents color depth. This chapter proposes the research hypothesis that there is a linear relationship between skin color saturation and skin temperature.

$$T = aS + b \tag{2.2}$$

where $\boldsymbol{T}$ represents skin temperature, and $\boldsymbol{S}$ denotes skin color saturation. Specifically, $T = \{T_1, T_2, \ldots, T_n\}$, and similarly, $\boldsymbol{a}$, $\boldsymbol{b}$, and $\boldsymbol{S}$ are vectors. If the number of subjects is m, Equation (2.2) can be expressed as follows:

$$\begin{bmatrix} T_{11} & T_{12} & \cdots & T_{1m} \\ T_{21} & T_{22} & \cdots & T_{2m} \\ \vdots & \vdots & \cdots & \vdots \\ T_{n1} & T_{n2} & \cdots & T_{nm} \end{bmatrix} = \begin{bmatrix} a_{11} & a_{12} & \cdots & a_{1n} \\ a_{21} & a_{22} & \cdots & a_{2n} \\ \vdots & \vdots & \cdots & \vdots \\ a_{n1} & a_{n2} & \cdots & a_{nn} \end{bmatrix} \times \begin{bmatrix} S_{11} & S_{12} & \cdots & S_{1m} \\ S_{21} & S_{22} & \cdots & S_{2m} \\ \vdots & \vdots & \cdots & \vdots \\ S_{n1} & S_{n2} & \cdots & S_{nm} \end{bmatrix} + \begin{bmatrix} b_{11} & b_{12} & \cdots & b_{1m} \\ b_{21} & b_{22} & \cdots & b_{2m} \\ \vdots & \vdots & \cdots & \vdots \\ b_{n1} & b_{n2} & \cdots & b_{nm} \end{bmatrix} \tag{2.3}$$

The variation in skin color saturation caused by changes in skin temperature is extremely subtle. To characterize the saturation signal, this chapter introduces EVM technology [22].

$$\hat{S}(x, y, t) = (1+\alpha)S(x, y, t) + \varphi(x, y, t) \tag{2.4}$$

where $\hat{S}$ denotes the amplified skin color saturation, S represents the original saturation before amplification, (x, y) are spatial coordinates, t is time, α is the amplification factor, and φ represents Gaussian noise. $S(x, y, t)$ corresponds to the saturation value of the image at position (x, y) and time t. The algorithm assumes that $S(x, y, t)$ follows a Gaussian noise distribution, which is subsequently removed through filtering. A detailed explanation is provided below. The absolute error (E_{abs}) is defined as follows:

$$E_{abs} = T' - T \tag{2.5}$$

where T' represents the skin temperature calculated by the algorithm, while T denotes the measured skin temperature.

When capturing images with charge-coupled device cameras, HF noise, such as salt-and-pepper noise, may arise due to factors like illumination intensity and camera sensor temperature. To ensure accuracy, a median filtering method (a type of order-statistic filter) is employed. The median filter demonstrates excellent denoising performance for certain types of random noise, particularly salt-and-pepper noise [31]. As the effective information in the video is amplified, noise is also amplified.

To maintain accuracy, the median filter is reapplied. Six color parameters were analyzed in both the RGB and HSV color spaces. After extracting color parameters from one color space, the information is stored in a numerical matrix. Skin color saturation, an indicator describing color purity and signal intensity in images, exhibits a strong correlation with skin temperature.

2.2.2.3 Skin Sensitivity Index (SSI)

2.2.2.3.1 Definition of SSI

When the human body is exposed to thermal stimulation, blood circulation changes, which are also reflected in skin color and texture. Based on the HSV color space, the S-channel (saturation) is extracted to establish a linear ST model:

$$T = k_i \times S + b_i \tag{2.6}$$

where i denotes the number of test subjects, k varies with the rate of skin temperature change, b represents the intercept, and S and T correspond to skin saturation and temperature, respectively. In this chapter, k is defined as skin sensitivity index (SSI). SSI is a high-weight coefficient reflecting the change in skin temperature, indicating the skin's sensitivity to external thermal stimuli.

2.2.2.3.2 Calculation Steps of SSI

1. Extract each frame from the captured video.
2. Define the region of interest (ROI) for analysis.
3. Extract the S-channel from the ROI images and compute the average S-value for each ROI.
4. Determine the SSI value for each subject by correlating the measured skin temperature.

2.2.2.4 Skin Temperature Detection Algorithm

The NIDL algorithm first employs the EVM algorithm to amplify subtle skin changes [28] and selects the ROI. Based on the characteristics of human thermal comfort, a 315-layer DNN is optimized to extract skin image features, establishing a regression relationship between skin images and skin temperature. To account for individual differences, this chapter introduces the SSI, which is correlated with skin color saturation, and combines it with NIDL to propose the NISDL method [32], as shown in Table 2.1.

2.2.2.4.1 Principle of EVM Amplification

The EVM algorithm is based on the Eulerian perspective, proposing an EVM technique. It processes monocular video sequences in Eulerian spatiotemporal domains to amplify subtle changes imperceptible to the naked eye. The EVM algorithm can separately amplify spatial and temporal channels, making it particularly suitable for enhancing color variations in image pixels over time. This marks the first application of the Eulerian method to amplify both color and motion in videos [22, 33]. Wadhwa et al. [34] proposed a phase-based video amplification method,

TABLE 2.1
Non-Contact Skin Temperature Detection Method Based on Skin Sensitivity Index

Algorithm: NISDL Algorithm
Output: NISDL model (*.h5), Skin temperature (°C)
Steps:

1. Video Preprocessing
    1. Denoise the captured video and apply slight magnification.
    2. Magnification factor β=10 (Equation 2.8).
    3. Extract the ROI from each video frame, with a size of 150×150 pixels.
2. Label Preparation
    1. Perform numerical interpolation on the skin temperature data captured by iButton sensor.
    2. Apply uniform interpolation, adding 11 data points per minute.
    3. Establish a correspondence table between ROI images and interpolated skin temperature values.
3. Algorithm Training
    1. Common Settings for NISDL Methods I and II
        1. Training set to test set ratio: 12:4.
        2. Validation set: 500 images.
        3. Network training configuration: 32 images/batch, 8 epochs.
        4. Training process: ~30,000 images trained before each validation.
        5. Activation function: ReLU.
    2. NISDL Method I
        1. Concatenate SSI values with ROI images.
        2. Convolution kernel: 1×1.
    3. NISDL Method II
        1. Separately extract features from SSI values and ROI images.
        2. Concatenate these two features in the latter half of the network.
        3. Convolution kernel: 3×1.
4. Model Parameter Optimization

overcoming the limitation of small amplification factors at high spatial frequencies. Elgharib et al. [35] introduced a layered video magnification approach, enabling the amplification of small motions within large movement ranges. Due to its sensitivity to temporal color variations, this chapter's skin temperature prediction model utilizes EVM to correlate skin temperature (*T*) with skin color saturation (*S*). The study hypothesizes a linear relationship between these two variables. Cheng et al. were the first to apply video magnification for determining skin temperature and thermal sensation, enabling non-contact measurement technology in HVAC control systems [36].

When the human body undergoes thermal reactions (heat/cold stress), observable skin changes, such as pore contraction and color variations, can be amplified using the EVM method. f(*x*, *t*) represents the human skin image, where x = pixel position, k(*t*) = skin variations under different environmental conditions [22].

$$f(x,t) = Z(x + k(t)) \tag{2.7}$$

where Z denotes the relational function between f(x, t) and cutaneous variation k(t). β represents amplification factor. Through first-order Taylor expansion of Z [22], we obtain the following:

$$f(x,t) \approx Z(x) + (1+\beta)k(t)\frac{\partial f(x)}{\partial x} = Z\left(x + (1+\beta)k(t)\right) \tag{2.8}$$

where k(t) represents the spatial variation of skin texture at time t, which EVM amplifies to a magnitude of $1 + \beta$. As shown in Formula (2.8), imperceptible cutaneous changes can be amplified to visible levels. The captured video data undergo EVM processing to serve as input signals for the DNN.

2.2.2.4.2 Deep Neural Network (DNN) Architecture

This detection method employs deep learning for big data training and model generation. An optimized DNN architecture (Inception) [37] was implemented with the following modifications: retained core Inception functionality while removing the final fully connected layer. Rectified linear unit (ReLU) is used as the activation function.

2.2.2.4.3 Calibration Function

To enhance prediction accuracy, a calibration function was deployed for the DNN-predicted skin temperatures based on piecewise stationary theory. The calibration function is expressed as follows:

$$T_p' = T_p - \xi \tag{2.9}$$

where T_{p} is DNN-predicted skin temperature (uncalibrated); T_{p}' is calibrated skin temperature value; ξ is calibration coefficient derived from Equation (2.10).

$$\xi = \begin{cases} \dfrac{Error_{train_{mean}}}{\tau_1}, & if \ \dfrac{1}{n}\sum_{i=1}^{n} Error_{test}(i) > \eta \\ \dfrac{Error_{train_{mean}}}{\tau_2}, & if \ Error_{train\_mean} \geq \varepsilon \end{cases} \tag{2.10}$$

where $Error_{\mathrm{train\_mean}}$ is the average error of the entire training set, and $Error_{\mathrm{test}}$ represents the first few error values of the test set. The mean of the first n error values from the test set is used as prior knowledge to determine the calibration conditions. η and ε are controllable thresholds. Parameters τ_1 and τ_2 are precision adjustment coefficients. Based on data training and model generation, the parameters η, ε, τ_1, τ_2, and n are defined.

2.2.2.4.4 Evaluation Metrics

$$Error = \left|T_p(i) - T_r(i)\right| \qquad i = 1, 2, 3, \ldots, t \tag{2.11}$$

where $T_{\mathrm{p}}(i)$ is the skin temperature predicted by the algorithm; $T_{\mathrm{r}}(i)$ is the true skin temperature captured by iButton sensor; t represents sampling time.

2.2.2.4.5 Detection Method

2.2.2.4.6 Detection Results

A total of 16 subjects were recruited for subjective physiological experiments, resulting in the collection of 1.44 million images. Based on this dataset, the NISDL algorithm was validated and compared with the NIPST algorithm and DL algorithm. To verify the effectiveness of the NISDL algorithm (with SSI), the SSI component and its corresponding feature extraction hidden layer were removed from NISDL Method II, converting it into another deep learning network (without SSI), named the DL algorithm. The DL algorithm, NISDL Method I, and Method II are all nonlinear, data-driven approaches utilizing deep learning networks. For further validation, the NIPST algorithm (a linear, model-driven method) was also included in the comparative analysis.

1. Runtime Environment

   Image processing and algorithm validation were performed on a GPU-equipped computer with the following specifications: GPU: GeForce GTX TITAN X; CPU: Intel Core i5-4460 @ 3.2GHz×4; RAM: 16GB; System: 64-bit.
2. Training of NISDL Method I

   ROI image dimensions: n×150×150×3; expanded SSI values: n×150×150×1. The SSI matrix was treated as an additional channel and concatenated with ROI images, yielding n×150×150×4. Four convolutional layers used ReLU activation, with 1×1 kernels. Feature extraction employed DenseNet201, outputting a matrix of size n×4×4×1920. Two hidden layers were constructed, with the final fully connected layer sized 1920×n.
3. Training of NISDL Method II

   Expanded SSI matrix dimensions: n×1920×1 (different from Method I), processed with 3×1 kernels. SSI features were extracted via two hidden layers, while ROI image features were extracted using DenseNet201. In the latter half of the framework, both feature sets were concatenated. To balance SSI's influence (moderate, neither excessive nor negligible), ROI features were set to n×1920 and SSI features to n×640. Final three hidden layer sizes: 2560×1024, 1024×512, and 512×1.
4. Commonalities Between Methods I and II

   Both methods used the 1.44-million-image dataset with a 12:4 train–test split and 500-image validation set. Training parameters: 8 epochs, batch size of 32. Models (*.h5) were saved when validation error fell below 0.46°C or after processing 300,000 training images. Post-training, test set images were input to the models to predict skin temperatures.
5. Quantitative Comparison

   The set of values obtained from the iButton serves as the ground truth. The corresponding error statistics, including the mean and median values for NISPT, DL, NISDL Method I, and Method II, are 0.579, 0.359, 0.335, and 0.265 (mean) and 0.343, 0.309, 0.238, and 0.228 (median), respectively. The results indicate that the deep learning methods (DL, NISDL Method I, and II) outperform the nonlinear model (NIPST), and the methods incorporating SSI (NISDL Method I and II) surpass those without SSI (DL). The errors of the DL, NISDL I, and NISDL II methods are primarily concentrated within the range of 0°C–0.75°C. NISDL outperforms DL because the error rates of NISDL corresponding to the intervals [0, 0.25) are 52.25% and 55.62%, respectively. Moreover, the error rates of NISDL for the intervals [0.25, 0.5) and [0.5, 0.75) are both lower than those of DL. The error rate of NIPST increases in the [0.75, 1) interval, indicating that its performance is inferior to that of DL and NISDL Methods I and II.

2.2.2.5 Algorithm Performance

NISDL employs a subtle variation amplification technique (EVM) to enhance skin texture changes. To address inter-individual variability, Cheng et al. [32] proposed SSI, which is correlated with skin saturation. Both NISDL Methods I and II are deep learning–based approaches. The performance of NISDL with SSI is superior to that of DL without SSI. Although both NISDL and DL outperform NIPST, there remains a significant gap between NISDL and DL, with experimental results demonstrating that NISDL performs better than DL. The primary reason lies in the incorporation of SSI in NISDL. Furthermore, NISDL Method II exhibits better performance than NISDL Method I. This improvement is particularly notable when SSI features are extracted in the latter part of the network and concatenated with the image features of ROI. Conversely, when SSI is removed from the network (DL), the performance deteriorates significantly. These findings confirm that SSI enhances skin temperature prediction accuracy in deep learning networks.

2.2.3 Skin Temperature Measurement Based on Remote Photoplethysmography Method

PPG is a non-invasive and cost-effective method for measuring physiological parameters. It utilizes a light source and photodetector to detect changes in blood volume beneath the skin. When tissue is illuminated, subtle variations in light reflection or transmission intensity caused by blood flow are captured by the photodetector, generating the PPG signal (the signal generation principle is illustrated in Figure 2.4). The validity of PPG signals is based on Beer–Lambert law, which states that the amount of light absorbed by blood is proportional to the hemoglobin concentration in both the skin and blood. Consequently, during the cardiac cycle, minute changes in hemoglobin concentration cause fluctuations in the absorbed light within blood vessels, thereby altering skin light intensity values. PPG technology is widely employed in clinical settings for detecting and monitoring various physiological indicators and has also been integrated into consumer devices such as fitness watches and fingertip pulse oximeters.

Researchers have developed a framework that employs ICA and adaptive filtering techniques to enhance PPG signals in videos, which reflect blood volume changes in cutaneous microvascular beds. By analyzing PPG signal amplitude variations, this method correlates skin temperature with other physiological responses (e.g., heart rate) to assess individual thermal comfort. In experiments, thermal sensation data

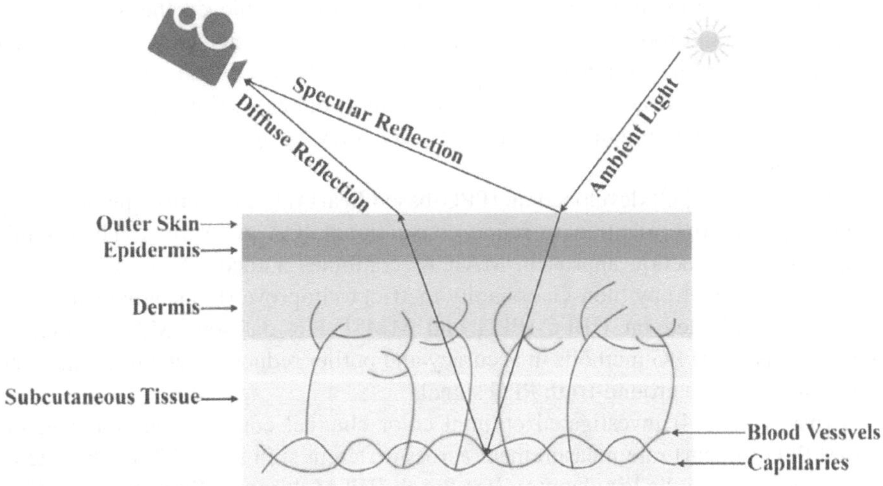

FIGURE 2.4 Schematic diagram of remote PPG signal generation. (Source: Reference [38], Copyright 2024 by Elsevier.)

were collected from 15 participants in 20°C–30°C environments without prolonged acclimatization. Results demonstrated positive correlations among visual indicators, skin temperature, and thermal sensation in ten participants, highlighting the potential of video-based thermal comfort assessment.

Edgar et al. [39] introduced the DETECT-1 study, which developed and validated an algorithm for automated circulatory arrest detection using PPG wristbands. This technology aims to improve out-of-hospital cardiac arrest survival rates by triggering emergency medical systems earlier. The algorithm exhibited high sensitivity and low false-positive rates in detecting circulatory cessation. A comparison between traditional heuristic signal processing and DNN trained on raw signals for continuous atrial fibrillation (AF) monitoring revealed similar AF detection performance. However, the DNN demonstrated superior motion artifact tolerance, processed larger datasets, and through interpretability analysis, showed capabilities in motion artifact suppression, arrhythmia evaluation, and learning inter-beat variability during AF. External validation across diverse populations and PPG sensors confirmed the DNN's stronger suitability for continuous PPG monitoring.

Bachir [40] proposed an innovative diffuse reflectance spectroscopy technique using smartphone flashlights for PPG and vital sign measurement. Their study confirmed that smartphone LED flashes, as alternative broadband light sources, enable accurate, noise-free PPG and SpO_2 measurements. A forward model for diffuse reflectance spectra at varying SpO_2 levels identified the 610–635 nm wavelength range (peak: 624 nm) as optimal for normal oxygen saturation. This underscores smartphones' potential as compact visible-light sources for multispectral PPG recording in fingertip transmission mode.

Bradley et al. [41] investigated near-infrared spectroscopy PPG (NIRS–PPG) for non-invasive intracranial pressure estimation in traumatic brain injury patients.

RF models incorporating long-distance NIRS features achieved the lowest mean absolute error and root mean square error. Meanwhile, Xiong et al. [42] presented GraphPhys, a graph neural network framework for (rPPG) extraction from facial videos. Its three-module architecture (physiological signal extraction, graph denoising, and metric calculation) enhances heart rate accuracy by aggregating signals across facial regions.

Macwan et al. [43] developed an rPPG-based heart rate estimation method using multi-objective optimization to refine conventional ICA algorithms. Their semi-blind source extraction approach, MAICA, combines autocorrelation (periodicity metric) and negentropy (non-Gaussianity metric) to improve rPPG signal extraction accuracy. Evaluated on UBFC-rPPG and MMSE-HR datasets, MAICA outperformed existing rPPG methods in accuracy and outlier reduction, yielding heart rate estimates closer to ground-truth PPG signals.

Ernst et al. [44] investigated optimal color channel combinations for camera-based PPG in heart rate measurement across different skin tones. Through systematic grid search on the Binghamton-Pittsburgh-RPI Multimodal Spontaneous Affect Database (BP4D+), they compared signal-to-noise ratio (SNR) and accuracy (ACC) of 25 color channels across eight regions of interest and 8 color spaces, evaluating linear combinations of RGB channels for heart rate measurement. The study revealed that the hue channel (H) in HSV color space and the chrominance channel (Q) in NTSC color space delivered the highest ACC and SNR in the forehead and glabellar regions. Artemyev et al. [45] developed a novel robust algorithm for accurate heart rate detection via rPPG from standard webcam videos under real-world conditions. The algorithm demonstrates robustness against lighting variations and body movements while being capable of tracking rapid pulse changes. Bhattachrjee et al. [46] introduced SimPPG, a self-supervised learning framework for precise heart rate estimation via rPPG. Leveraging PPG signal similarities across different facial regions through instance discrimination training, this approach enhances noise robustness and effectively handles heart rate prediction in data-limited scenarios. Experimental results show SimPPG outperforms across diverse skin types and datasets, matching state-of-the-art models even with only 75% training data.

Bobbia et al. [47] proposed an unsupervised skin segmentation method for rPPG, which implicitly selects living skin tissue by extracting temporally superpixels (TSPs) with pulsatile characteristics from videos, eliminating complex ROI selection/tracking. By computing pulsatility metrics for each TSP and merging signals, the method generates reliable PPG estimates. Luguern et al. [48] presented wavelet variance maximization (WVM), a contactless respiration rate estimation method via rPPG. This technique improves blood-volume-pulse signal quality by maximizing SNR through continuous wavelet transform and generalized eigenvalue decomposition to determine optimal color channel combinations. Tests with 12 healthy adults showed WVM reduces estimation errors by approximately 20% compared to existing methods. Xi et al. [49] developed a novel rPPG framework combining weighted channel fusion and singular spectrum analysis (SSA) for heart rate estimation under low-light conditions. The method decomposes filtered signals via SSA, refines candidate components using spectral masking against reference heart rates, and then fuses qualified components into final pulse signals. Evaluations on Multi-Illumination HR dataset and PUlse RatE Detection (PURE) dataset demonstrated superior robustness: 35% SNR improvement and 28% enhanced pulse detection stability in low-light environments compared to conventional approaches.

2.3 HUMAN POSTURE DETECTION TECHNOLOGY

2.3.1 Type of Posture Detection Algorithms

In recent years, human pose detection techniques have been an important research direction in the field of computer vision, which aims at recognizing and detecting various parts of the human body and different poses from images or videos. Some of the major human pose detection techniques are listed below.

Traditional methods mainly include edge detection, feature point detection, and template matching. Edge detection is mainly used to detect the edge information in an image by using operators such as Sobel and Canny [50, 51]. Feature point detection is mainly used to recognize key points in an image by using Harris corner point detection, Shi–Tomasi, and other methods [52, 53]. Template matching is mainly done by searching for specific parts in an image using predefined templates [54–57]. With the continuous updating and advancement of artificial intelligence technology, more and more human posture detection techniques have emerged.

3D-based methods include 3D pose estimation and multi-view analysis. 3D pose estimation, also known as 3D human pose estimation, refers to the use of computer vision techniques to determine the pose of the human body in three-dimensional space. This technique identifies and tracks key points of the human body and estimates the location and orientation of these points in 3D space. 3D pose estimation has a wide range of applications in several fields, including augmented reality (AR), VR, motion capture, health and fitness, game development, human–computer interaction, and more. Multi-view analysis of human posture is a technique that combines multiple camera views to understand and infer human posture in 3D space. This approach can provide richer and more accurate information about human posture than a single view and is especially effective when dealing with occlusions, viewpoint limitations, or complex movements. Ye et al. [58] proposed a system that can accurately estimate 3D posture from a single depth image, which combines both posture detection and posture optimization. The input depth map is matched with a set of pre-captured motion examples to generate body configuration estimates, as well as semantic labeling of the input point cloud. The initial estimates are then optimized by directly fitting the body configurations to the observations (e.g., input depth). In addition to the new system architecture, contributions include modifications to the point cloud smoothing technique to handle very noisy input depth maps, view-independent and efficient point cloud alignment and pose search algorithms.

Deep learning–based methods include convolutional neural networks (CNNs), recurrent neural networks (RNNs), graph convolutional networks (GCNs), and generative adversarial networks (GANs). CNNs have been applied in the field of human posture detection with remarkable results. CNNs can effectively extract features from images and be used for the identification and localization of key points on the human body. Tang et al. [59] proposed an intelligent system based on an electronic

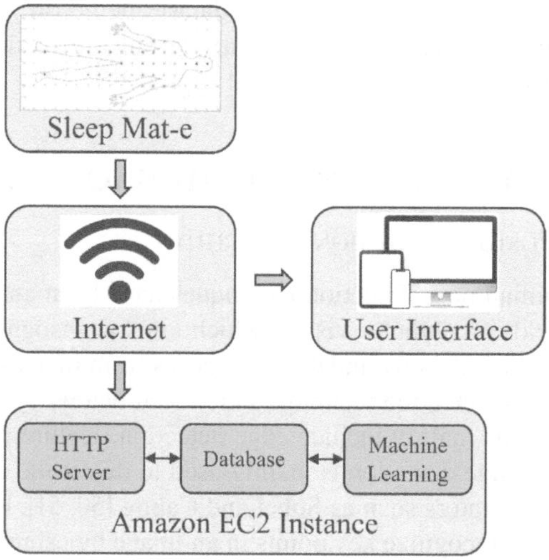

FIGURE 2.5 System architecture.

sleep pad called Sleep Pad-e for monitoring the sleep activity and sleep posture of individuals residing in a home care facility, and its system architecture is shown in Figure 2.5. The system uses a pressure-sensing pad made of piezoresistive material placed on a mattress. The sensors detect the pressure distribution of the body on the pad during sleep and use a CNN to collect and analyze the data to identify different sleep postures. The system recognizes four main positions: face up, face down, right side, and left side. A real-time feedback mechanism is also provided through an accompanying smartphone app for recording postures and sending alerts to the user in case there is a risk of falling out of bed. It also generates a summary of posture and activity for a given duration. Finally, the classification accuracy of the prototype system was evaluated experimentally, and the system achieved a classification accuracy of about 90%. Haghi et al. [60] measured 3D spinal postures by using a non-invasive flexible wearable device to monitor and classify physical activities. To evaluate the applicability of the flexible wearable device in human activity recognition, it developed a comprehensive protocol that included indoor, outdoor, and transitional state activities in three categories: static, dynamic, and transitional. The study used three neural networks, long short-term memory (LSTM) neural network, CNN, and a hybrid model (CNN-LSTM), and comprehensively compared their performances. It was found that LSTM has an overall classification accuracy of up to 98% for all activities. However, the CNN model with accelerometer data provided superior performance ACC for lying (100%), static (standing=82%, sitting=75%), and dynamic (walking=100%, running=100%).

RNNs and their variants (e.g., LSTM network and gated recurrent unit) are mainly used to process time-series data in human posture detection. Zheng et al. [61] extended the application of LSTM to human posture detection and proposed an LSTM-based human posture detection method. Through the time series data collected in real time by accelerometers, gyroscopes, barometers, and orientation sensors that are generally included in current smartphones, a human posture dataset containing 3336 human posture datasets with manually annotated data was produced, and five daily persistent behavioral postures, namely, walking, running, going up the staircase, descending the stairs, and calming, were predicted and classified together with the four emergent behavioral postures, namely, falling down, getting up, sitting down, and jumping. Comparing the LSTM network with the shallow learning algorithms, deep learning fully connected neural networks and CNNs commonly used in this research area, the experimental results show that the proposed method using end-to-end deep learning increases the correctness rate by 4.49% compared to the human posture detection algorithm model based on the produced dataset, which verifies the generalization ability of the network structure and is more suitable for gesture detection.

GCN is a deep learning model designed for graph-structured data, which can capture the complex relationships between individual nodes. In human posture detection, the human body can be modeled as a graph, where joints are nodes in the graph and the connectivity between joints (e.g., skeletal connections) are edges of the graph. Although GCNs have very good potential for human posture detection, as of now, compared to other deep learning methods such as CNNs, GCNs have relatively few applications in this field. With the further development of graph neural network technology, the application of GCN in human pose detection may be more widely explored and applied. Liu et al. [62] proposed a new approach for automatic posture recognition, where skeletal data of human subjects obtained from Kinect sensors are used as the input features for training the AI-driven recognizer. They formulated the potential human pose recognition problem as a classical multi-classification problem after training the GCN and by using 3D skeletal coordinates to recognize human poses. The results show that their proposed GCN has 92.2% classification accuracy in automatic human posture recognition.

GAN is a deep learning model that consists of two parts: generator and discriminator. The application of GAN in human posture detection is relatively new, but it has already shown some unique advantages and potential uses, and with the continuous development and improvement of GAN technology, its application in human posture detection will be more extensive and in-depth. Zhang et al. [53] proposed a human polymorphic motion behavior anomaly recognition method based on GAN. A Gaussian model is used to segment the human multi-gesture motion behavior image, and the image foreground of the segmentation result is the human motion object detection result. The Shi–Tomasi algorithm is chosen to extract the contour feature points from the human motion target detection results, and the extracted contour features are set as hidden spatial random variables and inputted into the GAN. The GAN uses a generator and a discriminator to recognize the multi-pose human motion behaviors and to determine whether the multi-pose human motion behaviors are abnormal or not. The results show that the proposed algorithm is able to accurately identify abnormal multi-pose human motion behaviors with recognition accuracy higher than 99% and average recognition time of less than 200 ms. Its research results provide a reliable basis for research in related fields.

Key point-based methods mainly include multi-joint modeling, which can be used to detect multiple key points of the human body, such as wrists, shoulders, knees, etc.

Safety and automatic control are extremely important when operating manipulators. For large construction manipulators, the main challenge is to accurately recognize the pose of all arm segments. In classical sensing methods, the accuracy of the inclinometer is easily affected by the elastic deformation of the robot arm. This leads to a large error accumulation when sensing the joint angles between arms one by one. In addition, machine vision-based sensing methods are not yet suitable for this outdoor condition. Yao et al. [63] proposed a multi-joint manipulator attitude localization method based on wireless sensor network localization. The pose sensing problem is formulated as a nearest Euclidean distance matrix (NEDM) model. The resulting method is called EDM-based pose localization method, which needs to satisfy the following guiding principles: (1) the pose of each arm segment of a multi-jointed manipulator must be estimated as accurately as possible; (2) the computational speed of the method must be fast; and (3) the designed method should not be susceptible to the influence of obstacles. To further improve the accuracy, its research team explored the inherent structure of the manipulator, i.e., fixed arm length. This is naturally represented as a linear constraint in the NEDM model. A typical multi-jointed manipulator with the mechanical property that all arm segments always lie in the 2D plane is used for the reduced-dimensional operation. Simulation and experimental results show that its proposed method provides an effective solution to the attitude perception problem and achieves better localization performance than existing localization methods.

Time series–based methods mainly include optical flow methods, which can estimate the motion of objects (including human body) by analyzing the pixel motion between video frames. Fujiwara et al. [64] proposed a low-cost fiber-optic force electromyography transducer for non-invasive hand pose recognition. The transducer is attached to the user's forearm via a strap to monitor the proximal posterior radius, medial anterior ulna, and distal posterior radius muscles, and a five-hidden-layer, 20-neuron artificial neural network classifier with a back propagation architecture is used to correlate the captured electromyographic light signals with the performed gesture, followed by a competing layer. Overall results for nine postures and six subjects show a sensitivity of 98.4% and a mean accuracy of 99.7%, which is comparable to EMG methods. Furthermore, compared to current setups, its proposed method allows the recognition of poses characterized by different configurations of finger and wrist displacements, using only three transducers and a simple interrogation scheme, which is more suitable for further applications in human–machine interfaces.

2.3.2 Posture Detection Algorithm Based on OpenPose

The main human skeleton-based method is OpenPose, a popular real-time multiplayer pose estimation system that enables detection of key points of the human skeleton. OpenPose is an advanced real-time multiplayer human pose estimation system that utilizes deep learning techniques to detect and estimate key points and pose information of the human body. Yang et al. [65] proposed a new method based on non-contact measurement of human thermal discomfort. Images of occupant poses associated with thermoregulatory mechanisms are captured by a digital camera, and

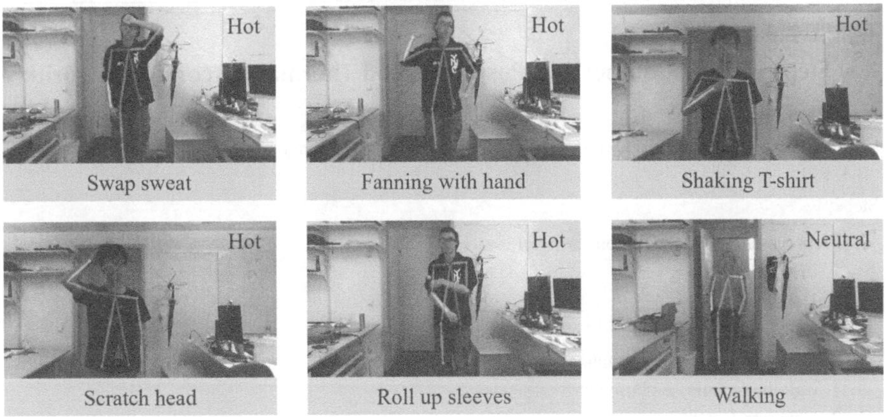

FIGURE 2.6 Recognition results of hot and neutral poses.

the corresponding 2D coordinates are obtained. These poses were converted into skeletal configurations. A new method based on OpenPose was developed, as shown in Figure 2.6, which is an algorithm that recognizes 12 human poses associated with thermal discomfort. The key steps of the algorithm used in this are shown in Table 2.2.

The human body naturally adjusts its position to maximize its thermal comfort. These movements of the bones and joints produce various variations in space. sp_i is the key point of the human skeleton, and since ordinary cameras capture images as 2D images, sp_i is defined as follows:

$$sp_i = [x_i, y_i], \qquad i = 0, 1, \dots \dots, k \tag{2.12}$$

where x_i, y_i denote the image space horizontal and vertical coordinates, i denotes the different key points of the human skeleton and their number, and k is the maximum value of the number of key points. If sp_i can be obtained accurately, corresponding algorithms can be constructed to compute the motion of human body and recognize its pose.

OpenPose is a sequence prediction framework for machine learning, enriched with implicit spatial models that can predict human posture [66–68]. On this basis, the human skeleton can be constructed in digital form. Figure 2.7 shows that there are 19 key points ($k = 18$). The key points of the skeleton are nose, neck, right shoulder, right elbow, right wrist, left shoulder, left elbow, left wrist, right hip, right knee, right ankle, left hip, left knee, left ankle, right eye, left eye, right ear, and left ear. The corresponding numbers range from 0 to 17. Note that $i = 18$ indicates the scene background.

If pooling and convolution are used as implicit layers, a total of 34 implicit layers are used. OpenPose is used to acquire human skeletal key points directly. Based on the skeletal key point information, the human pose capture algorithm and ten sub-algorithms are constructed. An image of a person working in an office is captured by an ordinary camera. After processing by preprocessing and the OpenPose module,

TABLE 2.2
Key Steps of Non-Contact and Posture-Based Thermal Discomfort Algorithm

Algorithm

Output: posture category, range of thermal comfort levels: [−3,3], range of thermal preference levels (−1, 0, 1)

Steps:

1. Preprocess the extracted monitoring video
   1. Frame extraction
   2. Denoise
   3. ROI
2. Search for key point coordinates
   1. Call the OpenPose platform
   2. Generate Jason
   3. Save valuable frames based on confidence level (=0.5)
3. Generate a coordinate matrix based on Jason (18×3)
4. Calculate standard distance (α) and define closest distance threshold (τ=1.5)
5. Postural and thermal comfort assessment
   1. Calculate relative Euclidean distance
   2. Calculate slope
   3. Calculate movement speed
   4. Calculate ten sub-algorithms and call out 12 postures. For example, "walking" and "stomping" belong to the same sub-algorithm, and "hand around the neck" and "huffing and puffing to warm the hands" belong to the same sub-algorithm.
   5. The ten sub-algorithms are as follows: (a) wiping sweat, (b) fanning with hands, (c) shaking T-shirt, (d) scratching head, (e) rolling sleeves, (f) walking, stomping feet, (g) shaking shoulders, (h) crossing arms, (i) crossing legs, (j) wrapping hands around the neck, and warming hands by breathing in and out.
   6. Setting parameters: (a) wiping sweat Lr=1.8, (b) fanning with hands: Lr max=120, Lr min=80, (c) shaking T-shirt: Lr=1.8, Lr max=120, Lr min=80, (d) scratching the head: Lr=1.8, (e) rolling up sleeves: Lr=0.9, (f) walking: Lr=1.8, (g) stomping the feet: the slope of the difference between the "walking" with a slope difference of 30, (h) shaking shoulders: Lr=1.5, (i) crossing arms: Lr=2.0, (j) crossing legs: Lr=1.0, (k) wrapping hands around the neck and huffing to warm hands: Lr=3.
   7. For different postures, the key points for calculating Lr are different.
6. Parameter optimization algorithm

the output is the key point coordinates with confidence level z. If the confidence is less than a threshold ($z<\varepsilon$), the image is discarded. When $z\geq\varepsilon$, the image is fed into the pose recognition module. Finally, the type of pose and its corresponding thermal discomfort are obtained. In this section, the value is 0.5. When the algorithmic platform works, the video is imported from a normal camera to OpenPose. In one second, 24–30 pictures can be collected. The algorithmic platform will discard all images with a confidence level of less than 0.5. Images with a confidence level greater than or equal to 0.5 are stored in a folder. All the key point data of the human skeleton will also be stored in a file. The method gives sub-algorithms for each posture estimation, and to simplify the technical processing, "walking" and "stomping"

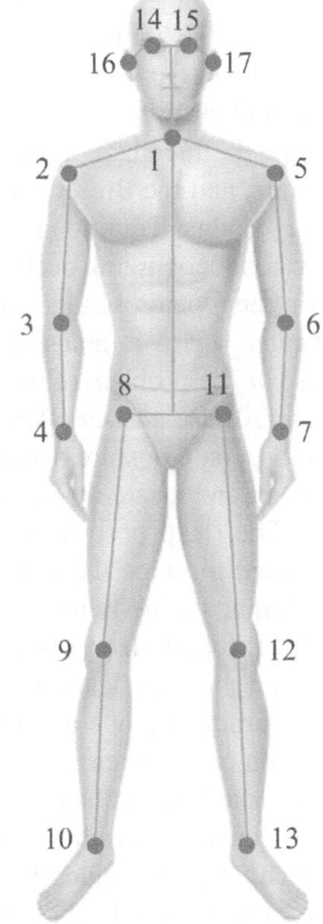

FIGURE 2.7 Human skeleton and key points.

belong to the same algorithm. Similarly, "wrapping the hand around the neck" and "warming the hand with a puff of air" belong to the same sub-algorithm. Therefore, a total of ten sub-algorithms are constructed for different postures in this method.

The distance L between the key points of the skeleton is Euclidean distance, and a standard distance is defined in this section to facilitate the calculation:

$$L_s = |SP_7 - SP_6| \tag{2.13}$$

where SP_7 denotes the left wrist and SP_6 denotes the left elbow; the relative distance between the key points can be calculated according to Equation (2.13).

$$L_r = \frac{L_s}{L} \tag{2.14}$$

Different relative distances L_r are calculated, and different relative distance thresholds are set for different gesture recognition. In addition to this, L_{r_max} and L_{r_min} will be used for recognizing certain poses, which represent the extreme values of L_r. In addition, the algorithm assumes mathematical slopes and utilizes coordinate changes of key points in consecutive image frames for pose estimation. The detailed information of the algorithm is shown in Table 2.2.

Shu et al. [69] used an OpenPose-based single-person pose estimation algorithm and a multi-person pose detection algorithm to process the camera-captured images of people's movements to realize the localization of face key points, human hand key points, and human body joints, and the Bresenham algorithm was used to draw the lines of the human body key points. Traditional human pose recognition algorithms have reduced accuracy when recognizing occluded body parts, and this study improves this problem based on the OpenPose algorithm. The experimental results show that the adopted method can adapt to complex backgrounds with high recognition accuracy and runtime. To solve the problem of recognizing different postures of people during sleep, Yang et al. [70] use an OpenPose-based single-person posture estimation algorithm and a multi-person posture detection algorithm to process the sleep images captured by a camera and locate the key points of human face, human hand, and human body joints. The sleep posture recognition algorithm based on human key point extraction is used to judge the four sleep postures of the human body, namely, left-lying, right-lying, supine-lying, and prone-lying, and the effect of OpenPose joint extraction for the four sleep postures. The experimental results show that its proposed method has good applicability in complex backgrounds. The method has high recognition accuracy and operation rate; its recognition accuracy can reach 92.5%, and the operation rate can reach 9.8 frames/sec. The sleep posture detection system designed by this method can be effectively used for sleep monitoring of diseased people, such as sleep apnea syndrome patients.

Other real-time detection technologies include Google's deep learning models for real-time posture detection: MoveNet model and BlazePose model, a deep learning model for real-time posture detection that quickly and accurately detects key points on the human body. The MoveNet model runs at over 50 frames per second on a wide range of devices and can be used in fitness, medical, and health applications. The MoveNet model consists of two versions with performance tradeoffs: lightning version is faster in time but may be slightly less accurate, and Thunder version is more accurate but slightly slower. Washabaugh et al. [71] used an existing dataset that contained video and synchronized motion capture data of 32 able-bodied participants while walking. Sagittal plane videos were analyzed using pre-training algorithms for four open-source pose estimation methods, OpenPose, TensorFlow MoveNet Lightning, TensorFlow MoveNet Thunder, and DeepLabCut, to extract key points (i.e., landmarks) and to compute hip and knee kinematics and spatiotemporal gait parameters. The absolute error of each markerless pose estimation method is calculated for the traditional marker-based optical motion capture, and the error of the statistical parameter mapping-based pose estimation methods is compared. The results show that the different pose estimation methods differ in their ability to measure kinematics. MoveNet Thunder and OpenPose have the lowest errors in measuring spatiotemporal gait parameters but are not statistically different from each other.

BlazePose, on the other hand, is a lightweight CNN architecture for human pose estimation, designed for real-time inference on mobile devices, capable of running at more than 30 frames per second on a cell phone, and ideally suited for real-time application scenarios, such as fitness tracking and sign language recognition. Shishodia et al. [72] evaluated the difference in accuracy between the BlazePose algorithm and traditional neural networks for human gesture detection. The results show that there is a significant difference in the calculation results ($p < 0.05$), i.e., BlazePose is better than traditional neural networks in improving the accuracy of human posture detection.

With the continuous progress of artificial intelligence technology, the application of computer vision in the field of human posture monitoring is becoming more and more widespread. High-precision image capture and deep learning algorithms are able to monitor and analyze multiple movements of the human body in different states in real time, bringing revolutionary changes to several industries. At the same time, it is also facing multiple challenges in terms of technology, ethics, and privacy. With the continuous progress of technology and the maturity of the market, future posture detection algorithms will develop in a more intelligent and humanized direction.

2.4 PERSONNEL DETECTION, POSITIONING, TRACKING TECHNOLOGY

CNNs have been widely applied in computer vision for tasks such as image classification, object detection, face recognition, image segmentation, video recognition, and image synthesis. For computer vision-based localization and tracking technologies, the key challenge lies in accurately identifying objects from images, making the precision and efficiency of object detection critical considerations in practical applications. CNN-based human detection algorithms can be categorized into two types of object detection architectures: single-stage and two-stage. Single-stage methods directly predict bounding boxes and class probabilities from the entire image, whereas two-stage methods first generate potential bounding boxes using region proposal techniques and then refine each box to predict final coordinates and class probabilities. While two-stage frameworks generally achieve higher detection accuracy, their real-time performance is inferior to single-stage frameworks.

YOLO (You Only Look Once) algorithm [73] processes the entire image, transforming object detection into a regression problem without requiring region proposals or separate calculations for each ROI. This approach significantly improves both detection accuracy and speed. Its core idea is to use a CNN to directly predict bounding boxes and class probabilities from input images [73]. The network architecture consists of four components: input layer, backbone network, neck network, and head output layer.

During the prediction phase of YOLO, the input image is divided into grid cells at three different scales, with each cell assigned three initial anchor boxes. For each anchor box, the algorithm predicts the coordinates (x, y) of the box center relative to the grid cell, along with its width (w), height (h), confidence score, and class probabilities [73]. The confidence score reflects the intersection over union (IOU) between the predicted anchor box and the ground-truth bounding box if the anchor contains an object. Finally, the optimal bounding boxes are selected using non-maximum suppression based on distance-based IOU. During the training phase, initial anchor

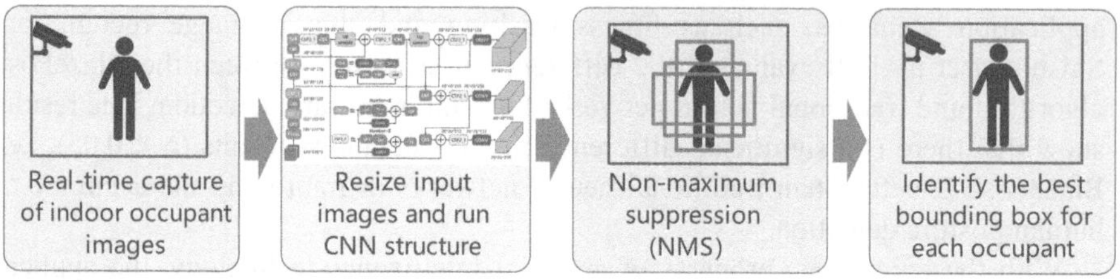

FIGURE 2.8 YOLO algorithm detection process.

box dimensions for the three scales are determined via the k-means clustering algorithm based on ground-truth bounding boxes in the training dataset [73]. The loss function is formulated as a regression problem, optimized through gradient descent to minimize prediction errors.

In practical applications, occupant-centric control strategies for HVAC systems typically require detection systems to meet real-time and long-term operational demands. Numerous studies have employed YOLO algorithms to detect occupancy in buildings, achieving detection accuracy rates between 80% and 97%. For instance, Yang et al. [74] implemented a YOLOv5s-5.0-based human detection method (see Figure 2.8) and evaluated its performance through field experiments. Their results recommend installing cameras near side walls at ceiling height to capture frontal or side-view images of occupants. In office environments, the method achieved a normalized root mean square deviation of 0.0526 during detection.

Choi et al. [75] also employed the YOLOv5 algorithm to detect indoor occupant numbers. Experimental results demonstrated good performance in small offices with fewer than five occupants (normalized root mean square error, NRMSE: 0.0435), while performance degraded in larger offices (NRMSE: 0.0918). Wang et al. [76] proposed a method for occupant localization by dividing indoor spaces into multiple zones using predefined boundary curves. Personnel positions (i.e., assigned zones) were determined based on the coordinate equations of the boundary curves and the bottom-center coordinates of detected bounding boxes. Experimental results showed that YOLO-based detection algorithm achieved a detection accuracy of 92.5%, which is acceptable for practical applications. Mutis et al. [77] introduced a novel vision sensor–based method incorporating human motion analysis for occupant detection. Their experiments revealed that YOLO framework achieved a detection accuracy of 83%, demonstrating feasibility for deployment in building energy efficiency-related applications.

Single Shot MultiBox Detector (SSD) [78] is another single-stage object detection method that utilizes multi-scale feature maps to detect objects, achieving real-time performance and high accuracy. Since the aforementioned methods are trained on very large image datasets, they exhibit robust detection performance.

Common detection methods based on region proposal, such as R-CNN [79], Fast R-CNN [80], and Faster R-CNN [81], deliver high detection accuracy but lack real-time capabilities. The flowchart of the R-CNN algorithm is shown in Figure 2.9. To address limitations in Faster R-CNN's two-stage framework, Li et al. [82] introduced Light-Head R-CNN, a novel two-stage detector. He et al. [83] proposed Mask

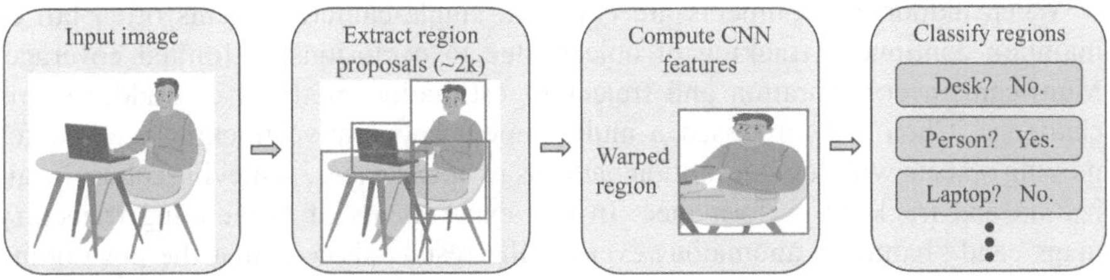

FIGURE 2.9 R-CNN detection process.

R-CNN, a framework for object instance segmentation that effectively detects objects while generating high-quality segmentation masks. Pang et al. [84] developed Libra R-CNN, focusing on balanced learning for improved detection performance. Cai et al. [85] introduced Cascade R-CNN, a multi-stage detection architecture comprising detectors trained with increasing IOU thresholds.

Video image processing technologies can also be integrated with traditional sensors to enhance camera-based detection methods, enabling broader functionalities and energy savings. Erickson et al. [86] combined a wireless camera network with a PIR wireless sensor network, employing a particle filter to fuse occupancy data from both systems with occupancy prediction models, thereby determining the most accurate real-time occupancy for each building zone. Newsham et al. [87] integrated webcam-based motion detection with keyboard and mouse activity monitoring. Their results showed higher detection accuracy compared to commercial ceiling-mounted PIR sensors, achieving over 90% overall precision with very low false negative rates.

In occupant localization and tracking, Razzaq et al. [88] investigated multi-occupant detection and tracking in smart home environments using a novel low-resolution thermal vision sensor. Their approach involved three steps: (1) detecting, tracking, and monitoring occupants by analyzing robust frame differences between consecutive frames; (2) employing CNN layers to process frame sequences from the sensor for activity recognition (e.g., walking, standing, sitting, falling); and (3) fusing detection results from both methods via a time interval–based sliding window to correlate identified activities with individual occupants. The framework achieved 88.46% occupant detection and tracking accuracy, with 90.99% activity classification accuracy. Wang et al. [89] proposed CIOPS-Stereo detection method, which integrates camera calibration and data fusion algorithms using dual cameras for spatial occupant localization. Results demonstrated that CIOPS-Stereo achieved positioning accuracy within 50 cm, outperforming ultra-wideband (UWB) detection methods. Building on this work, Wang et al. [90] further developed skeletal key point data fusion and 3D reconstruction algorithms, enabling occupant posture detection and indoor 3D spatial reconstruction with 92.3% localization accuracy. Dynamic gesture recognition relies heavily on hand trajectory features, necessitating precise hand tracking to capture spatiotemporal information. Yang et al. [91] utilized Microsoft Kinect's human joint recognition and localization capabilities, incorporating Kalman filtering to refine hand positions returned by the Kinect. Experimental results showed that the tracking algorithm produced smoother and more stable hand trajectory curves.

When indoor environments are complex, single-camera systems often fail to maintain continuous tracking of objects due to occlusions or limited coverage. Multi-camera collaboration and trajectory estimation methods can address this challenge. Chen [92] proposed a multi-camera cooperative tracking method for moving objects with regular motion patterns in open spaces, achieving robust localization and tracking performance. In two experiments of "generating trajectory maps" and "handling anomalous events", the results showed that the positioning coordinates mapped by the objects after crossing the multi-camera field of view did not change abruptly, which was consistent with the actual movement of the objects. This validated the efficiency and reliability of the multi-camera collaborative tracking approach. Luo [93] designed and implemented DeepTrack, a model and algorithmic framework for security applications in large indoor spaces, integrating multi-source video data with object recognition and tracking. Experimental results demonstrated that in scenarios with single or non-occluded multiple objects, DeepTrack achieved high localization accuracy and trajectory reconstruction fidelity, with an average positioning precision of 3.81 cm. Under occlusion conditions, the reconstructed trajectories remained coherent, though the average positioning precision decreased to 29.1 cm.

2.5 CONCLUSIONS AND FUTURE DIRECTIONS

This chapter first compares contact, semi-contact, and non-contact measurement methods, followed by a comprehensive review of computer vision-based non-contact occupant information detection technologies (thermally relevant). It introduces algorithmic frameworks and detection performance, covering techniques such as infrared thermography, EVM, rPPG, skeletal key point detection algorithms, YOLO-based occupant detection, R-CNN algorithms, occupant localization, and tracking methods. Specific details include the principles of skin temperature prediction using EVM and human posture recognition based on OpenPose skeletal key point detection. Building on these technologies, Chapters 3 and 4 will elaborate on their applications in sleep state monitoring and environmental regulation, as well as on-demand control of HVAC systems.

Current non-contact measurement technologies for detecting human skin temperature and posture face several challenges: (1) how to improve detection accuracy in complex and dynamically changing indoor environments; (2) how to develop cost-effective solutions for practical applications; (3) privacy concerns associated with image-based occupant detection technologies, and how to enhance occupants' acceptance and trust in computer vision systems; (4) how to adapt predictive models in detection methods to account for differences among diverse user groups, such as the elderly, children, and individuals with special health conditions.

Future research directions could focus on the following aspects: (1) developing detection technologies with higher precision and robustness to dynamically respond to changes in human thermal states and spatial distributions; (2) by effectively integrating different monitoring technologies into a single system, such as infrared thermography, pressure sensors, PIR sensors, and pose detection technology, multimodal monitoring is used to comprehensively assess human thermal state; (3)

ensuring privacy and security of user data as detection technologies advance; (4) building databases for diverse user groups and leveraging machine learning and AI algorithms to perform in-depth data analysis, thereby reducing detection errors caused by group differences.

REFERENCES

1. B. Yang, X. Li, Y. Hou, A. Meier, X. Cheng, J.-H. Choi, F. Wang, H. Wang, A. Wagner, D. Yan, others, Non-invasive (non-contact) measurements of human thermal physiology signals and thermal comfort/discomfort poses-a review, Energy and Buildings 224 (2020) 110261.
2. D. Luong, R. Richman, M. Touchie, Towards window state detection using image processing in residential and office building facades, Building and Environment 207 (2022) 108486.
3. J.-H. Choi, D. Yeom, Study of data-driven thermal sensation prediction model as a function of local body skin temperatures in a built environment, Building and Environment 121 (2017) 130–147.
4. Y. Chen, B. Lu, Y. Chen, X. Feng, Breathable and stretchable temperature sensors inspired by skin, Scientific Reports 5 (2015) 11505.
5. K. Nkurikiyeyezu, G. Lopez, Toward a real-time and physiologically controlled thermal comfort provision in office buildings, in: the 3rd International Workshop on Smart Sensing Systems, IOS Press, 2018: pp. 168–177.
6. C. Jiang, M.K. Masood, Y.C. Soh, H. Li, Indoor occupancy estimation from carbon dioxide concentration, Energy and Buildings 131 (2016) 132–141.
7. S. Wei, J. Calautit, Development of deep learning-based equipment heat load detection for energy demand estimation and investigation of the impact of illumination, International Journal of Energy Research 45 (2021) 7204–7221.
8. K. Weekly, H. Zou, L. Xie, Q.-S. Jia, A.M. Bayen, Indoor occupant positioning system using active RFID deployment and particle filters, in: 2014 IEEE International Conference on Distributed Computing in Sensor Systems, IEEE, 2014: pp. 35–42.
9. H. Zou, H. Jiang, J. Yang, L. Xie, C. Spanos, Non-intrusive occupancy sensing in commercial buildings, Energy and Buildings 154 (2017) 633–643.
10. P.W. Tien, S. Wei, J.K. Calautit, J. Darkwa, C. Wood, A vision-based deep learning approach for the detection and prediction of occupancy heat emissions for demand-driven control solutions, Energy and Buildings 226 (2020) 110386.
11. F. De Oliveira, S. Moreau, C. Gehin, A. Dittmar, Infrared imaging analysis for thermal comfort assessment, in: 2007 29th Annual International Conference of the IEEE Engineering in Medicine and Biology Society, IEEE, 2007: pp. 3373–3376.
12. X. Li, R. Wei, Q. Li, W. Pang, G. Chen, C. Sun, Application of infrared thermal imaging technique in in-situ temperature field measurement of hydrate-bearing sediment under thermal stimulation, Energy 265 (2023) 126321.
13. M. Burzo, M. Abouelenien, V. Pérez-Rosas, C. Wicaksono, Y. Tao, R. Mihalcea, Using infrared thermography and biosensors to detect thermal discomfort in a building's inhabitants, in: ASME International Mechanical Engineering Congress and Exposition, American Society of Mechanical Engineers, 2014: p. V06BT07A015.
14. J. Ranjan, J. Scott, ThermalSense: determining dynamic thermal comfort preferences using thermographic imaging, in: Proceedings of the 2016 ACM International Joint Conference on Pervasive and Ubiquitous Computing, 2016: pp. 1212–1222.
15. A. Ghahramani, G. Castro, B. Becerik-Gerber, X. Yu, Infrared thermography of human face for monitoring thermoregulation performance and estimating personal thermal comfort, Building and Environment 109 (2016) 1–11.

16. D. Li, C.C. Menassa, V.R. Kamat, Non-intrusive interpretation of human thermal comfort through analysis of facial infrared thermography, Energy and Buildings 176 (2018) 246–261.
17. A.C. Cosma, R. Simha, Thermal comfort modeling in transient conditions using real-time local body temperature extraction with a thermographic camera, Building and Environment 143 (2018) 36–47.
18. A.C. Cosma, R. Simha, Machine learning method for real-time non-invasive prediction of individual thermal preference in transient conditions, Building and Environment 148 (2019) 372–383.
19. D. Li, C.C. Menassa, V.R. Kamat, Robust non-intrusive interpretation of occupant thermal comfort in built environments with low-cost networked thermal cameras, Applied Energy 251 (2019) 113336.
20. B. Tejedor, M. Casals, M. Gangolells, M. Macarulla, N. Forcada, Human comfort modelling for elderly people by infrared thermography: Evaluating the thermoregulation system responses in an indoor environment during winter, Building and Environment 186 (2020) 107354.
21. Y. He, H. Zhang, E. Arens, A. Merritt, C. Huizenga, R. Levinson, A. Wang, A. Ghahramani, A. Alvarez-Suarez, Smart detection of indoor occupant thermal state via infrared thermography, computer vision, and machine learning, Building and Environment 228 (2023) 109811.
22. H.-Y. Wu, M. Rubinstein, E. Shih, J. Guttag, F. Durand, W. Freeman, Eulerian video magnification for revealing subtle changes in the world, ACM Transactions on Graphics (TOG) 31 (2012) 1–8.
23. F. Jazizadeh, S. Pradeep, Can computers visually quantify human thermal comfort? Short Paper, in: Proceedings of the 3rd ACM International Conference on Systems for Energy-Efficient Built Environments, 2016: pp. 95–98.
24. W. Jung, F. Jazizadeh, Non-intrusive detection of respiration for smart control of HVAC system, in: Computing in Civil Engineering 2017, 2017: pp. 310–317.
25. K. Alghoul, S. Alharthi, H. Al Osman, A. El Saddik, Heart rate variability extraction from videos signals: ICA vs. EVM comparison, IEEE Access 5 (2017) 4711–4719.
26. F. Jazizadeh, W. Jung, Personalized thermal comfort inference using RGB video images for distributed HVAC control, Applied Energy 220 (2018) 829–841.
27. W. Jung, F. Jazizadeh, Vision-based thermal comfort quantification for HVAC control, Building and Environment 142 (2018) 513–523.
28. X. Cheng, B. Yang, A. Hedman, T. Olofsson, H. Li, L. Van Gool, NIDL: A pilot study of contactless measurement of skin temperature for intelligent building, Energy and Buildings 198 (2019) 340–352.
29. A.R. Smith, Color gamut transform pairs, ACM SIGGRAPH Computer Graphics 12 (1978) 12–19.
30. M.-Z. Poh, D.J. McDuff, R.W. Picard, Non-contact, automated cardiac pulse measurements using video imaging and blind source separation, Optics Express 18 (2010) 10762–10774.
31. R.C. Gonzalez, Digital Image Processing, Pearson Education India, 2009.
32. X. Cheng, B. Yang, K. Tan, E. Isaksson, L. Li, A. Hedman, T. Olofsson, H. Li, A contactless measuring method of skin temperature based on the skin sensitivity index and deep learning, Applied Sciences 9 (2019) 1375.
33. M. Rubinstein, N. Wadhwa, F. Durand, W.T. Freeman, H.Y. Wu, Revealing invisible changes in the world, Science 339 (2013) 519–519.
34. N. Wadhwa, M. Rubinstein, F. Durand, W.T. Freeman, Phase-based video motion processing, ACM Transactions on Graphics (ToG) 32 (2013) 1–10.
35. M. Elgharib, M. Hefeeda, F. Durand, W.T. Freeman, Video magnification in presence of large motions, in: Proceedings of the IEEE Conference on Computer Vision and Pattern Recognition, 2015: pp. 4119–4127.

36. X. Cheng, B. Yang, T. Olofsson, G. Liu, H. Li, A pilot study of online non-invasive measuring technology based on video magnification to determine skin temperature, Building and Environment 121 (2017) 1–10.
37. U. Rubins, A. Miscuks, M. Lange, Simple and convenient remote photoplethysmography system for monitoring regional anesthesia effectiveness, in: EMBEC & NBC 2017: Joint Conference of the European Medical and Biological Engineering Conference (EMBEC) and the Nordic-Baltic Conference on Biomedical Engineering and Medical Physics (NBC), Tampere, Finland, June 2017, Springer, 2018: pp. 378–381.
38. H. Xiao, T. Liu, Y. Sun, Y. Li, S. Zhao, A. Avolio, Remote photoplethysmography for heart rate measurement: A review, Biomedical Signal Processing and Control 88 (2024) 105608.
39. R. Edgar, N.T. Scholte, K. Ebrahimkheil, M.A. Brouwer, R.J. Beukema, M. Mafi-Rad, K. Vernooy, S.-C. Yap, E. Ronner, N. van Mieghem, others, Automated cardiac arrest detection using a photoplethysmography wristband: Algorithm development and validation in patients with induced circulatory arrest in the DETECT-1 study, The Lancet Digital Health 6 (2024) e201–e210.
40. W. Bachir, Diffuse transmittance visible spectroscopy using smartphone flashlight for photoplethysmography and vital signs measurements, Spectrochimica Acta Part A: Molecular and Biomolecular Spectroscopy 303 (2023) 123181.
41. G.R. Bradley, P.A. Kyriacou, Evaluating the effectiveness of non-invasive intracranial pressure monitoring via near-infrared photoplethysmography using classical machine learning methods, Biomedical Signal Processing and Control 96 (2024) 106517.
42. J. Xiong, W. Ou, Z. Liu, J. Gou, W. Xiao, H. Liu, GraphPhys: Facial video-based physiological measurement with graph neural network, Computers and Electrical Engineering 113 (2024) 109022.
43. R. Macwan, Y. Benezeth, A. Mansouri, Heart rate estimation using remote photoplethysmography with multi-objective optimization, Biomedical Signal Processing and Control 49 (2019) 24–33.
44. H. Ernst, M. Scherpf, H. Malberg, M. Schmidt, Optimal color channel combination across skin tones for remote heart rate measurement in camera-based photoplethysmography, Biomedical Signal Processing and Control 68 (2021) 102644.
45. M. Artemyev, M. Churikova, M. Grinenko, O. Perepelkina, Robust algorithm for remote photoplethysmography in realistic conditions, Digital Signal Processing 104 (2020) 102737.
46. S. Bhattachrjee, H. Li, J. Xia, W. Xu, SimPPG: Self-supervised photoplethysmography-based heart-rate estimation via similarity-enhanced instance discrimination, Smart Health 28 (2023) 100396.
47. S. Bobbia, R. Macwan, Y. Benezeth, A. Mansouri, J. Dubois, Unsupervised skin tissue segmentation for remote photoplethysmography, Pattern Recognition Letters 124 (2019) 82–90.
48. D. Luguern, R. Macwan, Y. Benezeth, V. Moser, L.A. Dunbar, F. Braun, A. Lemkaddem, J. Dubois, Wavelet variance maximization: A contactless respiration rate estimation method based on remote photoplethysmography, Biomedical Signal Processing and Control 63 (2021) 102263.
49. L. Xi, X. Wu, W. Chen, J. Wang, C. Zhao, Weighted combination and singular spectrum analysis based remote photoplethysmography pulse extraction in low-light environments, Medical Engineering & Physics 105 (2022) 103822.
50. S. Bei, Z. Xing, L. Taocheng, L. Qin, Sitting posture detection using adaptively fused 3D features, in: 2017 IEEE 2nd Information Technology, Networking, Electronic and Automation Control Conference (ITNEC), IEEE, 2017: pp. 1073–1077.
51. E.T. Wijaya, A.G. Sulaksono, Implementation of neural network and canny edge detection to recognize the crime through surveillance cameras, in: ICASI 2019: Proceedings

of The 2nd International Conference On Advance And Scientific Innovation, ICASI 2019, 18 July, Banda Aceh, Indonesia, European Alliance for Innovation, 2019: p. 324.
52. J. Wang, Z. Huang, W. Zhang, A. Patil, K. Patil, T. Zhu, E.J. Shiroma, M.A. Schepps, T.B. Harris, Wearable sensor based human posture recognition, in: 2016 IEEE International Conference on Big Data (Big Data), IEEE, 2016: pp. 3432–3438.
53. N. Zhang, J. Ren, Q. Xu, H. Wu, M. Wang, Anomaly recognition algorithm for human multipose motion behavior using generative adversarial network, Wireless Communications and Mobile Computing 2022 (2022) 2656001.
54. I.B. Ozer, W.H. Wolf, A.N. Akansu, Relational graph matching for human detection and posture recognition, in: Internet Multimedia Management Systems, SPIE, 2000: pp. 150–161.
55. Y. Han, Reliable template matching for image detection in vision sensor systems, Sensors 21 (2021) 8176.
56. N.D. Thanh, W. Li, P. Ogunbona, A novel template matching method for human detection, in: 2009 16th IEEE International Conference on Image Processing (ICIP), IEEE, 2009: pp. 2549–2552.
57. D.T. Nguyen, W. Li, P. Ogunbona, A part-based template matching method for multiview human detection, in: 2009 24th International Conference Image and Vision Computing New Zealand, IEEE, 2009: pp. 357–362.
58. M. Ye, X. Wang, R. Yang, L. Ren, M. Pollefeys, Accurate 3D pose estimation from a single depth image, in: 2011 International Conference on Computer Vision, IEEE, 2011: pp. 731–738.
59. K. Tang, A. Kumar, M. Nadeem, I. Maaz, CNN-based smart sleep posture recognition system, IoT 2 (2021) 119–139.
60. M. Haghi, A. Ershadi, T.M. Deserno, Recognizing human activity of daily living using a flexible wearable for 3D spine pose tracking, Sensors 23 (2023) 2066.
61. Y. Zheng, F. Li, L. Zhang, S. Liu, Human posture detection method based on long short term memory network, Journal of Computer Applications 38 (2018) 1568.
62. G. Liu, R. Xie, H.-C. Wu, S.-H. Fang, K. Yan, Y. Wu, S.Y. Chang, Automatic human posture recognition using Kinect sensors by advanced graph convolutional network, in: 2022 IEEE International Symposium on Broadband Multimedia Systems and Broadcasting (BMSB), IEEE, 2022: pp. 1–7.
63. Z. Yao, Y. Dai, Q.-N. Li, D. Xie, Z.-H. Liu, A novel posture positioning method for multi-joint manipulators, IEEE Sensors Journal 20 (2020) 14310–14316.
64. E. Fujiwara, C.K. Suzuki, Optical fiber force myography sensor for identification of hand postures, Journal of Sensors 2018 (2018) 8940373.
65. B. Yang, X. Cheng, D. Dai, T. Olofsson, H. Li, A. Meier, Real-time and contactless measurements of thermal discomfort based on human poses for energy efficient control of buildings, Building and Environment 162 (2019) 106284.
66. Y. Yao, Z. Lian, W. Liu, C. Jiang, Y. Liu, H. Lu, Heart rate variation and electroencephalograph–the potential physiological factors for thermal comfort study, Indoor Air 19 (2009).
67. T. Chaudhuri, Y.C. Soh, H. Li, L. Xie, Machine learning based prediction of thermal comfort in buildings of equatorial Singapore, in: 2017 IEEE International Conference on Smart Grid and Smart Cities (ICSGSC), IEEE, 2017: pp. 72–77.
68. C. Dai, H. Zhang, E. Arens, Z. Lian, Machine learning approaches to predict thermal demands using skin temperatures: Steady-state conditions, Building and Environment 114 (2017) 1–10.
69. Z. Shu, P. Wang, W. Zhan, The research and implementation of human posture recognition algorithm via OpenPose, in: 2020 2nd International Conference on Artificial Intelligence and Advanced Manufacture (AIAM), IEEE, 2020: pp. 90–94.
70. M. Yang, J. Li, R. Guo, X. Tang, Implementation and research of human sleep posture recognition based on OpenPose, Physics Experimentation 39 (2019) 45–49.

71. E.P. Washabaugh, T.A. Shanmugam, R. Ranganathan, C. Krishnan, Comparing the accuracy of open-source pose estimation methods for measuring gait kinematics, Gait & Posture 97 (2022) 188–195.
72. A. Shishodia, M. Gunasekaran, An efficient approach for human pose detection using blaze pose algorithm to improve accuracy in comparison with convolutional neural networks, in: AIP Conference Proceedings, AIP Publishing, 2024.
73. J. Redmon, S. Divvala, R. Girshick, A. Farhadi, You only look once: Unified, real-time object detection, in: Proceedings of the IEEE Conference on Computer Vision and Pattern Recognition, 2016: pp. 779–788.
74. B. Yang, Y. Liu, P. Liu, F. Wang, X. Cheng, Z. Lv, A novel occupant-centric stratum ventilation system using computer vision: Occupant detection, thermal comfort, air quality, and energy savings, Building and Environment 237 (2023) 110332.
75. H. Choi, C.Y. Um, K. Kang, H. Kim, T. Kim, Application of vision-based occupancy counting method using deep learning and performance analysis, Energy and Buildings 252 (2021) 111389.
76. J. Wang, J. Huang, Z. Feng, S.-J. Cao, F. Haghighat, Occupant-density-detection based energy efficient ventilation system: Prevention of infection transmission, Energy and Buildings 240 (2021) 110883.
77. I. Mutis, A. Ambekar, V. Joshi, Real-time space occupancy sensing and human motion analysis using deep learning for indoor air quality control, Automation in Construction 116 (2020) 103237.
78. W. Liu, D. Anguelov, D. Erhan, C. Szegedy, S. Reed, C.-Y. Fu, A.C. Berg, SSD: Single shot multibox detector, in: European Conference on Computer Vision, Springer, 2016: pp. 21–37.
79. R. Girshick, J. Donahue, T. Darrell, J. Malik, Rich feature hierarchies for accurate object detection and semantic segmentation, in: Proceedings of the IEEE Conference on Computer Vision and Pattern Recognition, 2014: pp. 580–587.
80. R. Girshick, Fast R-CNN, in: Proceedings of the IEEE International Conference on Computer Vision, 2015: pp. 1440–1448.
81. S. Ren, K. He, R. Girshick, J. Sun, Faster R-CNN: Towards real-time object detection with region proposal networks, Advances in Neural Information Processing Systems 28 (2015).
82. Z. Li, C. Peng, G. Yu, X. Zhang, Y. Deng, J. Sun, Light-head R-CNN: In defense of two-stage object detector, arXiv Preprint arXiv:1711.07264 (2017).
83. K. He, G. Gkioxari, P. Dollár, R. Girshick, Mask R-CNN, in: Proceedings of the IEEE International Conference on Computer Vision, 2017: pp. 2961–2969.
84. J. Pang, K. Chen, J. Shi, H. Feng, W. Ouyang, D. Lin, Libra R-CNN: Towards balanced learning for object detection, in: Proceedings of the IEEE/CVF Conference on Computer Vision and Pattern Recognition, 2019: pp. 821–830.
85. Z. Cai, N. Vasconcelos, Cascade R-CNN: High quality object detection and instance segmentation, IEEE Transactions on Pattern Analysis and Machine Intelligence 43 (2019) 1483–1498.
86. V.L. Erickson, S. Achleitner, A.E. Cerpa, POEM: Power-efficient occupancy-based energy management system, in: Proceedings of the 12th International Conference on Information Processing in Sensor Networks, 2013: pp. 203–216.
87. G.R. Newsham, H. Xue, C. Arsenault, J.J. Valdes, G.J. Burns, E. Scarlett, S.G. Kruithof, W. Shen, Testing the accuracy of low-cost data streams for determining single-person office occupancy and their use for energy reduction of building services, Energy and Buildings 135 (2017) 137–147.
88. M.A. Razzaq, J.M. Quero, I. Cleland, C. Nugent, U. Akhtar, H.S.M. Bilal, U.U. Rehman, S. Lee, uMoDT: An unobtrusive multi-occupant detection and tracking using robust Kalman filter for real-time activity recognition, Multimedia Systems 26 (2020) 553–569.

89. H. Wang, C. Liang, G. Wang, X. Li, Energy-saving potential of fresh air management using camera-based indoor occupancy positioning system in public open space, Applied Energy 356 (2024) 122358.
90. H. Wang, G. Wang, X. Li, An RGB-D camera-based indoor occupancy positioning system for complex and densely populated scenarios, Indoor and Built Environment 32 (2023) 1198–1212.
91. L. Yang, M. Wang, T. Li, Hand tracking based on the Kinect and Kalman filter, in: 2020 IEEE 3rd International Conference on Information Systems and Computer Aided Education (ICISCAE), IEEE, 2020: pp. 708–714.
92. Q. Chen, Research on Multi-camera collaborative target tracking method based on Deep learning, Master's Thesis, East China Normal University, 2019.
93. J. Luo, Indoor human location, tracking and trajectory Reconstruction based on multi-source video information, Master's Thesis, East China Normal University, 2023.

3 Sleep State Monitoring and Environment Regulation

Bo'an Wei, Haichuan Zhao, Yihang Liu, and Bin Yang

3.1 INTRODUCTION

Sleep monitoring technology is a series of methods used to assess and analyze an individual's sleep patterns, sleep quality, and possible sleep disorders. With the development of technology, sleep monitoring devices have gradually moved from professional monitoring in hospitals into homes and become part of a health-conscious lifestyle. Polysomnography (PSG) is the gold standard in the field of sleep monitoring and the most comprehensive method of clinical sleep monitoring, which is usually performed in sleep laboratories in hospitals and other medical institutions. It records brain waves, eye movements, electromyographic activity, heart rate, blood pressure, oxygen saturation, respiration, and snoring by installing a variety of sensors on the patient. Famous manufacturers include COMPUMEDICS®, Philips, etc.

With the gradual increase of social pressure, people pay more and more attention to the quality of sleep, and the demand for sleep monitoring at home is increasing day by day. People prefer to relieve the fatigue of the day through a night of efficient sleep to meet the work and challenges of the next day. As a result, more and more portable sleep monitoring devices are emerging. Portable sleep monitoring devices mainly include smartwatches and smart bracelets, and famous manufacturers include Fitbit, Garmin, ActiGraph, and so on. The emergence of these devices meets the needs of the majority of users for home sleep monitoring, which can be carried out anytime, anywhere, to understand their current sleep quality, timely detection of sleep disorders, and provide more convenience for people's lives. Zhang et al. [1] compared the accuracy of Fitbit Alta, a portable smartwatch, with that of the PSG. The comparison results showed that Fitbit performed well in measuring TST (total sleep time), SE (sleep efficiency), SOL (sleep onset latency), and WASO (wake after sleep onset) but performed poorly in monitoring sleep staging in young adults' sleep.

Whether it is PSG monitoring or portable smartwatches and bracelets, the wearable feeling brought by contact monitoring devices will more or less have some impact on our sleep, so non-contact sleep monitoring methods have emerged. Non-contact sleep monitoring methods mainly include non-contact bio-radar as well as computer vision technology. The non-contact bio-radar technology mainly utilizes electromagnetic waves emitted by radar to detect human vital signs, including

DOI: 10.1201/9781003611455-3

signals such as breathing and heartbeat. The advantage of this technology is that it can monitor the human body's state across objects such as clothing and bedding at a distance, non-contact, and without constraints, which is particularly suitable for sleep monitoring because it does not interfere with the user's normal sleep. Toften et al. [2] evaluated the nocturnal sleep of healthy individuals over 71 nights by using PSG and BioRadar technology, and the time-by-time analysis showed that the BioRadar Somnofy's sensitivity (accuracy in detecting sleep) and specificity (accuracy in detecting arousal) are 0.97 and 0.72, respectively, compared to 0.99 and 0.85 for PSG. BioRadar Somnofy shows high accuracy in sleep staging in healthy individuals and has the potential to assess sleep quality in healthy samples (mainly young adults). The application of computer vision technology in sleep monitoring mainly involves the use of cameras to capture physiological signals and behavioral patterns of individuals and analyze them through algorithms to assess sleep quality. It can mainly be implemented for facial expression and eye movement monitoring, body posture analysis, heart rate and respiration monitoring, deep learning with sleep staging, real-time drowsiness detection system, etc. Zhu et al. [3] monitored respiration through computer vision techniques and achieved a high level of accuracy in their results as compared to the PSG results.

3.2 INFLUENCING FACTORS AND TRADITIONAL MONITORING METHODS OF SLEEP QUALITY

Sleep quality is crucial to human health. According to the World Health Organization, 27% of people around the world have sleep problems, 38.2% of Chinese adults suffer from insomnia, and more than 300 million Chinese people suffer from sleep disorders. Sleep thermal comfort refers to how the human body feels and adapts to the surrounding environment during sleep, and it is closely related to the physiological changes and environmental factors of the human body in sleep state. The main influencing factors of sleep thermal comfort include indoor temperature, relative humidity, carbon dioxide (CO_2) concentration, noise during sleep, indoor light source and light intensity, thermal resistance of clothing and bedding, etc. In this chapter, the influencing factors of sleep thermal comfort and the indoor environments that are most conducive to sleep will be comprehensively analyzed.

The U.S. National Sleep Foundation (NSF) recommends that the optimal temperature for sleeping in the bedroom should be between 15.5°C and 19.4°C. The NSF recommends that the temperature of the bedroom should be between 15.5°C and 19.4°C. It has also been shown that sleep quality is the highest in 26°C environments, followed by 23°C, while 30°C is the worst. Zhang et al. [4] recruited a total of 104 healthy subjects without sleep disorders and investigated the effects of naturally ventilated bedroom environments on subjective perception and sleep quality during the transitional seasons (spring and fall). The results of the study showed that the neutral temperature was estimated to be 23.8°C in the awake state and 26.5°C in the sleep state, and for naturally ventilated bedrooms and mild outdoor climate in the transitional seasons, it was found that slightly warmer bedtime environments may contribute to the quality of sleep of the residents. In summary, indoor temperature has a significant impact on sleep, with either too high or too low temperatures being detrimental to sleep. By setting a reasonable temperature range and a dynamic thermal environment, sleep quality can be improved by decreasing the length of awakening and increasing the length of deep sleep and the total amount of sleep.

The World Health Organization recommends that indoor humidity should be maintained between 40 and 70% throughout the year, while a relative humidity between 45 and 65% is most comfortable. When the indoor humidity is below 40%, it will lead to indoor dryness, which may trigger dryness of the skin, throat, and respiratory tract, which can easily lead to asthma and other respiratory-related diseases. When the air humidity is higher than 70%, the human body will feel stuffy and hot, while when the air humidity is higher than 80%, it may lead to difficulties in heat dissipation and symptoms such as increased body temperature, rapid heartbeat, dizziness, and nausea. Research shows that the humidity of the sleeping environment is most suitable between 40 and 60%. If the environmental humidity is too high or too low, it may have a negative impact on health. In an overly dry environment, the resistance of a person's respiratory system decreases, easily triggering or aggravating respiratory diseases. When the relative humidity is lower than 30%, the air is dry, and the mucous membrane of the respiratory tract in the nose and lungs will lose a lot of water, mucus secretion is reduced, and the movement of the cilia on the mucous membrane is slowed down, making it easy to get infected with diseases.

Compared with indoor temperature and relative humidity, indoor CO_2 concentration is more likely to be ignored by people, but in fact, indoor CO_2 concentration also affects the quality of people's sleep. First of all, CO_2 is a colorless, tasteless, and odorless gas, consisting of one carbon atom and two oxygen atoms, which exists naturally in the Earth's atmosphere. Plants absorb CO_2 and release oxygen through photosynthesis, while animals and humans inhale oxygen and exhale CO_2 through respiration. CO_2 has a variety of uses in industry, including as a refrigerant, fire extinguishing agent, and as a source of air bubbles in the food and beverage industry. In indoor environments, increased CO_2 concentrations may affect comfort and cognitive functioning, and high concentrations may also lead to headaches and drowsiness. Xu et al. [5] conducted a sleep experiment with 16 young subjects (8 males, 8 females) without sleep disorders on a night of three different CO_2 concentrations and monitored their sleep throughout the night with a PSG. The results of a subjective questionnaire showed that the composite score of sleep quality decreased significantly with increasing CO_2 concentration. Objective data showed that SOL (sleep onset latency) was linearly and positively correlated with CO_2 concentration, and SWS (slow-wave sleep) was linearly and negatively correlated with CO_2 concentration, and the results consistently indicated that the higher the CO_2 concentration, the lower the SE and the worse the sleep quality.

Compared with the above indoor environmental parameters, noise is more likely and more intuitive to affect our SE and sleep quality. The effects of noise on sleep are mainly reflected in the following aspects: (1) Noise can cause sleep interruption, causing people to wake up from the deep sleep stage, which in turn affects the continuity of sleep; (2) Continuous background noise or sudden loud noises can reduce the duration of deep sleep (SWS), which is a critical stage for the body's recovery and restoration; (3) Noise disturbances can reduce the efficiency of sleep, i.e., the ratio of the actual time of sleep to the time spent in bed; (4) Excessive noise can make it difficult to fall asleep, especially when a quiet environment is required for falling asleep; (5) Excessive noise can make it difficult to fall asleep, especially more pronounced in people who need a quiet environment to fall asleep; and (6) Prolonged noise exposure is more likely to cause anxiety and stress, which further affects the quality of sleep. Bevan et al. [6] compared the effects of different noise environments on sleep in hospitals and at home through a combination of subjective and objective methods and showed that the median sound level measured in hospitals averaged 48.6 dBA, which exceeds the World Health Organization's recommendation of 30 dBA, which is also higher than the sound level measured at home. Due to the higher sound levels in hospitals, this also leads to more sleep deprivation and a severe reduction in the quality of sleep compared to sleep at home. Therefore, a reasonable and effective reduction in noise levels in hospitals and wards may improve the quality of sleep for both children and parents, reducing stress in more ways than one.

It has been found through research that not all sounds during sleep negatively affect sleep quality, for example, white noise. White noise is a special type of sound signal that has the same or uniform energy output at all frequencies that can be heard by the human ear. In other words, the sound spectrum of white noise is flat, similar to the characteristic of white light having the same energy at all visible frequencies, hence the name "white noise". White noise does not contain any particular pitch or rhythm, which makes it an ideal background sound to mask other louder or unpleasant sounds. Because of its even distribution of frequencies, white noise can be effective in masking or blurring out other sounds such as traffic noise, human voices, or other environmental noises. In a work or study environment, white noise can be used to minimize distractions and help people concentrate. And for some people, white noise can help improve sleep quality, especially in noisy environments. It can reduce the number of nighttime awakenings by providing a steady background sound. Taranto-Montemurro et al. [7] monitored the clinical polysomnograms of eight healthy subjects under two different conditions, one with normal ambient noise and the other with louder, continuously filtered white noise. The monitoring results showed a strong tendency for subjects to exhibit shorter delays to sleep when exposed to filtered white noise. Total sleep duration did not change between the two (448 ± 62 vs. 451 ± 90 mins, respectively, $p>0.5$), as well as SE (84 ± 7 vs. $83 \pm 13\%$ time in bed, $p>0.5$) and sleep architecture ($p>0.5$). In the experimental model of transient insomnia, sleep latency was reduced by approximately 40% when subjects slept in the background of filtered white noise compared to normal ambient noise. These studies amply demonstrate that filtered white noise can be used to minimize sleep onset insomnia.

Just as important as noise is the influence of light sources and lighting. While appropriate lighting can have a positive impact on sleep quality, too much unnecessary light during sleep can have the following effects on the body and sleep quality: (1) Suppression of melatonin secretion. Melatonin is a hormone secreted by the pineal gland, which plays a key role in regulating the sleep–wake cycle. Darkness promotes melatonin secretion, while light, especially blue light, inhibits melatonin production, thereby interfering with sleep. (2) Affects the biological clock. The human biological clock, or circadian rhythm, is strongly influenced by light. Bright light at night can delay the biological clock, making it difficult to fall asleep, while light in the morning can help reset the biological clock and improve daytime alertness. (3) Altering sleep structure. Exposure to light sources at night, especially short-wavelength light (e.g., blue light), may reduce the duration of deep sleep (SWS) and rapid eye movement (REM) sleep, affecting the quality of sleep. (4) Affects mood and behavior. Prolonged exposure to artificial light sources at night may lead to sleep disturbances and may be associated with mood problems (e.g., depression and anxiety). (5) Prolonged sleep latency. Sleep latency, the time it takes from going to bed to falling asleep, may be lengthened by excessive indoor light. (6) Increased nighttime awakenings. When waking up at night, indoor light sources may make falling asleep again more difficult. Birgit et al. [8] found that melatonin secretion was not significantly different under the effect of the two light sources by using two different intensities of blue light illumination (280 Lux blue light-enriched and 240 Lux blue light-suppressed) on 29 elderly people over 65 years of age and that sleep quality improved in the female subjects after receiving blue light-enriched illumination, and the improvement in sleep quality was not significant. The results showed that there was no significant difference in melatonin secretion between the two light sources, and the sleep quality of female subjects improved after receiving blue light-enriched illumination, and the improvement effect was more significant than that of male subjects, which also indicates that the appropriate use of light sources within a reasonable range can have a certain effect on the promotion of the quality of sleep, but it should be reasonably controlled to avoid the opposite.

The thermal resistance of clothing and bedding also has a significant impact on the quality of sleep, so their selection is particularly important. Thermal resistance is the ability of a material to prevent heat transfer under certain conditions. During sleep, there is a heat exchange between the human body and the surrounding environment, and the thermal resistance of clothing and bedding determines the efficiency of this exchange. First of all, clothing thermal resistance is an important physical parameter in thermal comfort research, which directly affects the human body's requirements for a thermal environment. In the Chinese region, due to the lack of methods for calculating the thermal resistance of garments, researchers measured the thermal resistance of common garments and explored calculation methods suitable for use in field investigations. It was found that different building types would have different requirements for occupant attire, which may lead to changes in occupant demands on the environment as well. Second, the total thermal resistance of bedding, as an important factor in thermal environment affecting sleep, is inversely proportional to thermal neutral temperature of sleeping state. When the total thermal resistance of bedding is small, both hot and cold stimuli will disturb sleep. By adjusting the comforter and pajamas, the quality of sleep can be kept stable in a wide range of low temperatures. In addition, the thermal resistance of bedding, as an important variable in sleep thermal comfort equation, will be affected by subjective factors such as quilt structure, material, air permeability, and quilt combination layer sequence, as well as by objective factors such as wind speed, air humidity, and air layer gaps in the sleep environment. By analyzing the effects of quilt cover layer sequences, quilt thickness, and the thickness of the air layer present between quilts on satisfying sleep insulation performance, researchers proposed a computational model and established a mathematical model for calculating the thermal resistance of quilts [9]. Zhang et al.

[10] categorized 36 subjects according to different degrees of warm and cold preferences into cold preference, neutral preference, and warm preference. It was found that subjects with warm preference had higher skin temperature, body–bed interface temperature, bed microenvironmental temperature, and bedding system insulation for good sleep quality, but lower heart rate than subjects with cold preference. Gender differences were evident in subjects with the same preference, with females having higher temperatures than males in all three temperatures (skin temperature, body–bed interface temperature, and bed microenvironmental temperature). These results also suggest that we can improve the sleep quality of subjects with different preferences by rationally altering the thermal resistance of their sleep garments as well as their bed microenvironments to have a more efficient sleep.

The above content mainly introduces a variety of influencing factors of sleep quality, from which it can be seen that if the indoor temperature and humidity, CO_2 concentration and ventilation, noise, and clothing and bedding thermal resistance are reasonably adjusted and controlled, the sleep quality of whole night may be altered and enhanced, to efficiently complete the next day's work and tasks with more abundant energy. In the following, this chapter will provide a systematic and complete detailed introduction to the existing sleep quality assessment and monitoring methods.

Sleep quality assessment and testing methods can be categorized into subjective and objective methods. Subjective methods are mainly through the use of subjective sleep quality questionnaires, of which the most classic and widely used is the Pittsburgh Sleep Quality Index (PSQI), which is a widely used self-assessment scale that comprehensively measures the quality of sleep of the participant in the past month. The PSQI scale consists of 24 items involving 19 self-assessment items and 5 other-assessment items. The PSQI scale consists of 24 items, involving 19 self-rated and 5 other-rated items. Among the self-rated items, 18 are used for scoring. These items comprise seven factors, including sleep quality, time to sleep, sleep duration, SE, sleep disorders, hypnotic medication use, and daytime dysfunction. Each factor is scored on a scale of 0–3. The cumulative score for each factor is the total PSQI score, which ranges from 0 to 21, with higher scores indicating poorer sleep quality. Researchers generally consider a PSQI score of less than or equal to 5 to be good quality sleep. In many experimental studies investigating the effects of the indoor environment on sleep quality, experimentalists often use PSQI questionnaires to conduct preliminary screening of subjects to select the appropriate group of people to act as experimental subjects. Xu et al. [5], in an experiment investigating the effects of indoor CO_2 concentration on sleep quality, selected a total of 16 subjects (8 males and 8 females) to conduct the experiment through the use of PSQI questionnaires for screening experiment. Szymczak et al. [11] used the PSQI questionnaire to evaluate the subjective sleep of 32 subjects for one week in the area of average altitude of 4524 m and the area of sea level. The results showed that the mean score of PSQI in the area of high altitude was 7.6±4.8, while the mean score of the sleep quality in the area of sea level was 2.9±1.7. They found that the high altitude through the use of PSQI investigation significantly reduced subjects' sleep quality. A more appropriate questionnaire to find out the sleep quality of the subjects for the previous night's sleep would be Groningen Sleep Quality Scale (GSQS), which is a quick and simple scale to assess the quality of the subject's sleep during the previous night. The questionnaire consists of a series of yes/no questions and takes only 1 minute to complete, making it ideal for scenarios such as diary studies where frequent assessments are required. The GSQS has a score range of 0–14, with a score of 6 or more indicating poor sleep quality in the previous night. Jafarian et al. [12] assessed the subjective sleep performance of 100 subjects during an altitude gain of up to 3500 m using the GSQS questionnaire. Overall, 46% of the subjects would experience varying degrees of sleep disturbance (GSQS score ≥6). In addition to this, there are other subjective sleep quality assessment scales such as Athens Insomnia Scale (AIS) [13], Epworth Sleepiness Scale (ESS) [14], and Self-Rating Scale of Sleep Status (SRSS) [15].

Objective sleep quality assessment and testing methods are mainly based on the use of instruments and equipment to monitor sleep and derive objective data. Among them, PSG is the international "gold standard" for sleep staging and diagnosis of sleep disorders, and it can record and analyze several sleep physiological indexes. Liao et al. [16] investigated the effects of a 20-minute warm footbath of 40°C before bedtime on the sleep quality of elderly people with good and poor sleep quality through the use of PSG. Åkerstedt et al. [17] selected eight subjects sleeping according to an irregular schedule of 6 hour sleep and 1 hour naps designed to provide normal bedtime (1/3 of the total number) but with varying quality of sleep, while continuously monitoring with PSG. The results of the study found that objective measures of sleep continuity were closely related to perceived sleep quality, which essentially implies sleep continuity. Continuous monitoring of sleep quality throughout the night using PSG not only provides a general understanding of the subjects' sleep-related indicators but also allows continuous monitoring of their respiration, blood oxygenation, and periodic leg movements to determine whether the subjects suffer from obstructive sleep apnea (OSA) [18], center sleep apnea (CSA) [19], mixed sleep apnea (MSA) [20] and other sleep-related disorders, so that reasonable and effective measures can be taken to assist in the treatment, and the occurrence of sleep-related disorders can be reduced while improving SE.

In recent years, with the increasing prevalence of sleep problems, commercial companies have introduced sleekly designed, low-cost and user-friendly sleep monitoring devices, including wearable and contactless devices. Wearable devices, such as smartwatches and bracelets, collect biological data through built-in sensors, while non-contact devices, such as sleep mattresses and sleep bands, monitor sleep through pressure sensors or radar technology. In addition to the emergence of a large number of portable wearable devices, mobile sleep quality monitoring applications are also widely popular, and some cell phones have a built-in sleep monitoring function, which analyzes the user's physiological data during sleep through algorithms, such as heart rate, blood oxygen, and respiratory rate, etc. Flernández et al. [21] found that wearing an Actiwatch on 31 students who had ingested a non-alcoholic beer resulted in a reduction in sleep latency and an improvement in subjective PSQI scores.

3.3 SLEEP QUALITY PREDICTION BASED ON COMPUTER VISION

The current mainstream methods for sleep quality assessment are contact-based, such as PSG. However, these invasive approaches tend to exacerbate sleep disorders. PSG experiments are typically conducted in specialized sleep laboratories, with a rather complex procedure. Participants are required to wear multiple electrodes, wires, and respiratory leads on their bodies, which imposes a significant burden and discomfort [22]. Moreover, the output data must be interpreted by trained professionals. Given these limitations, the use of PSG in daily life is impractical. Other measurement methods also have varying degrees of impact on the accuracy of the experiment and may cause a sense of foreign objects. Therefore, there is an urgent need to explore effective non-contact methods for sleep monitoring, such as computer vision-based monitoring techniques. Many scholars, both domestically and internationally, have already begun to apply non-contact methods to sleep monitoring. Non-contact measurement methods should meet the following criteria: (1) They should not interfere with the natural sleep state of the individual; (2) They should enable real-time collection of physiological parameters; (3) They should provide accurate and reliable measurements. Currently, the two most widely used non-contact measurement methods are infrared sensor-based detection methods (e.g., infrared thermography) and camera-based detection methods (e.g., skeletal keypoint detection technology).

Sleep breathing information is highly correlated with sleep quality. Liao et al. [23] proposed a thermography-based method for detecting and classifying sleep apnea. By capturing the temporal information of chest and abdominal movements, they constructed a two-dimensional complex feature space for central and OSA under different physiological mechanisms. Based on the statistical characteristics of this space, they introduced two features: respiratory effort intensity and respiratory effort asynchrony feature. These features enabled classification and detection of two types of sleep apnea. Akbarian et al. [24] employed five distinct computer vision-based algorithms to process infrared video data, tracking and analyzing body movements to differentiate between OSA and central sleep apnea. Wang et al. [25], based on the characteristics of actual respiratory signals, proposed a novel and efficient respiratory simulation model to generate rich, high-quality training data. They applied a gated recurrent unit (GRU) neural network with bidirectional and attention mechanisms (BI-AT-GRU) to classify six clinically significant respiratory patterns. The performance of the BI-AT-GRU was tested using data measured by a depth camera (as shown in Figure 3.1). Gaetano Scebba et al. [26] proposed a novel non-contact system for monitoring respiratory activity by integrating near-infrared and far-infrared cameras. Multiple respiratory information streams were extracted from independent video channels, and two data fusion models were subsequently applied to determine whether the subject was breathing (apnea detection) and to measure the respiratory rate. Compared to wakefulness, respiratory efforts during sleep are more

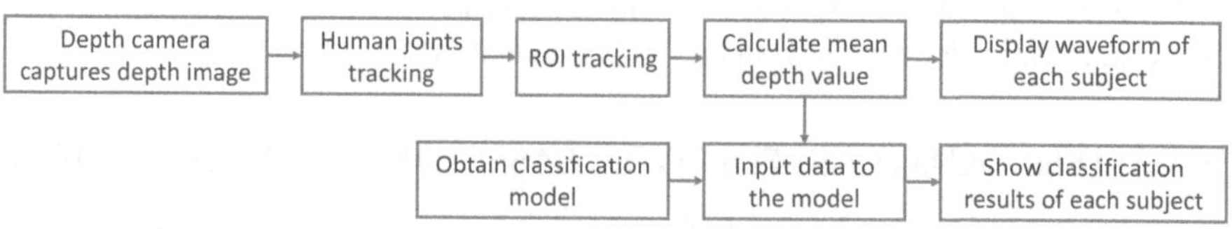

FIGURE 3.1 Flowchart of an online system for real-time identification and classification of multi-subject respiratory pattern.

periodic and consistent. Additionally, these efforts exhibit higher sound intensity, making the patterns more distinct relative to the background noise level.

Dafna et al. [27, 28] proposed a novel sleep quality analysis system based on sound signal processing and audio signal detection (Figure 3.2 illustrates the audio signal of a subject), which was subsequently validated. The results indicated that the detection accuracy for sleep/wake states was 82.1% (on a per-epoch basis), the

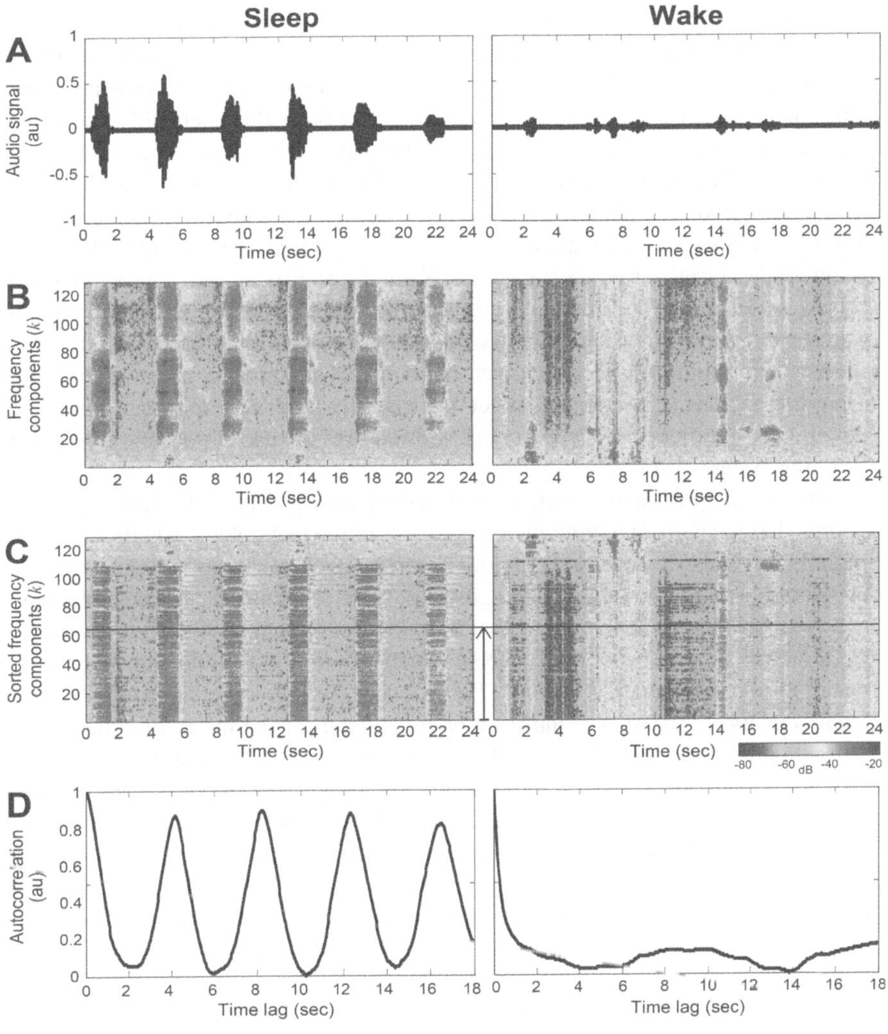

FIGURE 3.2 An example of 24-second audio signal intervals from a subject: (A) Audio signal following noise suppression. (B) The corresponding spectrogram (frequency components) of the audio signal in (A). (C) The sorted frequency components according to periodicity measure. (D) The enhanced autocorrelation of the interval calculated from the lower half of the sorted frequency components (C), visualized by the vertical solid line. (Source: Reference [28], CC BY 4.0, https://creativecommons.org/licenses/by/4.0/.)

detection error for sleep latency was 3.9%, the estimation error for TST was 11.4%, and the estimation error for SE was also 11.4%. Wu et al. [29] introduced a new method for assessing sleep quality based on sound events recorded in bedroom. Using subjective sleep quality as the training label and integrating multiple machine learning methods, including kernelized self-organizing map, hierarchical clustering, and hidden Markov model, they developed a model representing sleep patterns under specific quality levels. Xue et al. [30] investigated a sleep stage detection method aimed at categorizing sleep states into four distinct stages: wakefulness, rapid eye movement sleep, light sleep, and deep sleep. In addition to extracting time-domain and frequency-domain features from respiratory sounds, they also incorporated non-linear features associated with snoring sounds to better characterize snoring-related signals in respiratory audio. To effectively integrate these three sets of features, they proposed a feature fusion technique combining generalized canonical correlation analysis with ReliefF algorithm for discriminative feature selection. The final sleep stage detection was implemented using well-established classifiers, including decision trees, support vector machines, k-nearest neighbors, and ensemble classifiers. An internal dataset was concurrently established, comprising sleep audio data from a sleep laboratory. Experimental results demonstrated that the proposed method outperforms existing related approaches and holds promise for large-scale non-contact sleep monitoring applications. Non-contact monitoring methods based on respiratory signals or sleep apnea detection can only monitor respiratory conditions. More precise prediction of sleep quality still requires the use of portable EEG–EOG recording systems [31].

Human body posture is closely related to sleep quality. In recent years, significant progress has been made in human posture recognition technology, whether targeting single individuals or multiple individuals. In the area of single-individual posture recognition, Chen et al. [32, 33] proposed posture recognition methods and structure-aware networks. Toshev et al. [34] introduced a human pose estimation method based on deep neural networks (DNNs), which was the first application of DNNs in human pose estimation. Chen et al. [35] proposed a novel structure-aware convolutional network. In the area of multi-individual posture recognition, Pishchulin et al. [36] first proposed DeepCut algorithm. In the same year, Insafutdinov et al. [37] introduced DeeperCut as an improvement over DeepCut. Cao et al. [38] proposed OpenPose algorithm, which has propelled the development of multi-individual posture recognition. Güler et al. [39] presented a dense human pose estimation algorithm. Li et al. [40] proposed a novel and effective method to address the problem of pose estimation in crowded scenes. In recent years, numerous additional pose estimation methods have emerged, offering significant insights for the study of sleep thermal comfort.

Among studies employing non-contact, vision-based methods to monitor physiological parameters, the use of thermography to extract local skin temperature is the most widely applied approach. Many studies have analyzed the relationship between skin temperature and thermal sensation by extracting local skin temperature from infrared images [41]. Pavlin et al. [42] found that the temperature extracted from the forehead region using an infrared thermographic camera is highly correlated with thermal comfort. Li et al. [43, 44] developed a comprehensive system framework for predicting human age, gender, and thermal comfort based on facial data collected using infrared thermographic sensors. They discussed the differences between RGB images and infrared thermograms in predicting age and gender. Cosma et al. [45] used infrared cameras and Kinect sensors and found that changes in facial temperature are correlated with thermal comfort. Faridah et al. [46] reviewed the feasibility of predicting thermal sensation votes using infrared thermography. By measuring the body and facial skin temperatures of Indonesian subjects under heating and cooling conditions, they investigated the relationship with thermal sensation. The study results indicated that predicting thermal sensation using skin temperatures from cheeks, chin, nose, and forehead can achieve an accuracy rate of 69%.

He et al. [47] proposed a method for predicting human thermal state by integrating infrared thermography, computer vision, and machine learning (as illustrated in Figure 3.3). This approach measures the distribution of skin temperature on the face and hands, leveraging the temperature differences both within and between these regions to mitigate the inevitable calibration drift in thermal infrared cameras. The study was conducted in both an outdoor carport and an indoor environmental chamber, where thermal infrared cameras and computer vision techniques were used to collect a total of 48,422 sets of skin temperature data from cheeks, nose, and hands, as well as 715 subjective responses regarding thermal sensation. To predict human thermal state, a random forest classification model was constructed using either absolute skin temperatures (the highest and central temperatures of cheeks and hands, as well as the central temperature of the nose) or intra- and inter-segment temperature differences of the cheeks, hands, and nose. These measurements can accurately predict human thermal state with an accuracy of 92–96%. Baek et al. [41] utilized the temperature distribution and relationships of face and body to predict individual thermal comfort. This method classifies personal thermal sensation votes using preprocessed infrared images, which are fed into a deep convolutional neural network (DCNN) to display temperature distribution, achieving an accuracy and F1 score of 96%. The DCNN supports analysis that considers the location and numerical information of temperature features appearing in the image, using features extracted from the DCNN to interpret human body temperature, clothing surface temperature, and body temperature distribution. The study also explored the body regions with the greatest impact on human thermal comfort classification using explainable artificial intelligence (AI) methods. This approach overcomes the uncertainty of extracting local body temperature statistics without considering the surrounding environment.

In summary, the results of the aforementioned studies have demonstrated the effectiveness and feasibility of thermography as a means of predicting human thermal comfort [41] and its potential application in monitoring the thermal state of individuals in sleep environments. Sleep quality prediction methods based on computer vision technology typically encompass the following four aspects:

1. *Data Acquisition*: Collecting multi-dimensional data related to human posture, facial expressions, respiratory signals, and indoor environment using cameras and sensors. These data form the foundation for subsequent analysis.
2. *Feature Extraction*: Extracting useful feature information from the collected data, such as sleep posture, facial expressions, respiratory rate, and

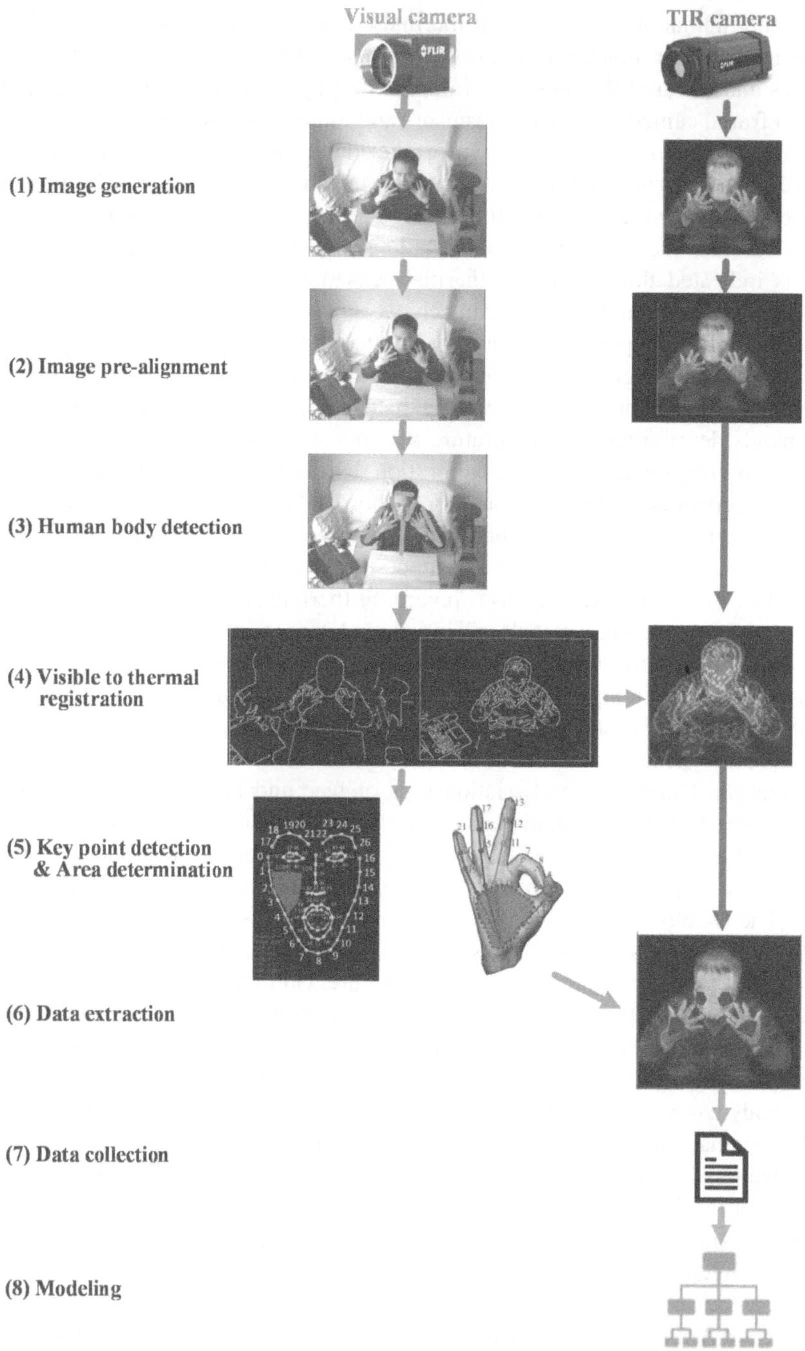

FIGURE 3.3 The thermal state prediction process integrates camera imaging, body detection, image registration, data extraction, and modeling. (Source: Reference [47], Copyright 2023 by Elsevier.)

skin temperature. These features aid computers in more accurately assessing sleep comfort.

3. *Model Development*: Establishing predictive models using machine learning and deep learning algorithms based on the extracted features.
4. *Prediction and Evaluation*: Utilizing the developed models to predict new data and evaluating their performance in real-world scenarios. By continuously adjusting and optimizing the models, prediction accuracy and reliability can be enhanced, ultimately leading to the validation of model performance.

3.4 ON-DEMAND REGULATION OF SLEEP THERMAL ENVIRONMENT

Non-contact measurement methods can monitor human physiological parameters and postures in real time and accurately and are used to predict human thermal comfort. Based on this information, Heating, Ventilation and Air Conditioning (HVAC) systems can be automatically adjusted to meet human sleep thermal comfort requirements. As discussed in Section 3.1, thermal environment is a key factor affecting sleep quality, with both excessively high and low ambient temperatures disrupting normal sleep. Therefore, to ensure sleep quality, it is essential to create a comfortable thermal environment [48]. Different sleep stages require different ambient temperatures, and human thermal sensation and thermal comfort vary with environmental changes [49]. During sleep, individuals are unable to actively control devices or provide timely feedback. Therefore, creating a comfortable sleep environment poses a significant challenge. Regulating the air conditioning system based on real-time monitoring of sleeper's thermal state to ensure thermal comfort, preventing excessive cold or hot, has become a crucial approach to improving sleep quality. Many people do not maintain the same posture for extended periods during sleep and unconsciously change positions to relieve fatigue or discomfort. Thus, it is necessary to monitor sleep posture and other physiological parameters during sleep, with the hope of obtaining signals that can be used to adjust the air conditioning system [50].

Liu et al. [51, 52] employed a deep learning method based on convolutional neural networks (CNNs) to detect six of the most common sleep postures. Zhang et al. [53] proposed the application of passive infrared (PIR) sensing technology in sleep posture recognition and concurrently utilized an action database [54] that defined ten distinct sleep postures. These ten sleep postures encompass various representative dimensions, involving alternating changes in full-body sleep positions. Mohammadi et al. [55] applied transfer learning methods to implement supervised machine learning strategies on non-contact vision-based infrared camera data, successfully quantifying the sleep postures of subjects covered by blankets (as shown in Figure 3.4). This represents the first successful application of this machine learning method to detect four predefined postures and the empty bed state in 8–10-hour nocturnal sleep events, representing real-world home sleep scenarios. The method was evaluated based on manually scored sleep postures and postures estimated using clinical PSG measurement techniques. Among 12 healthy subjects, the study found that

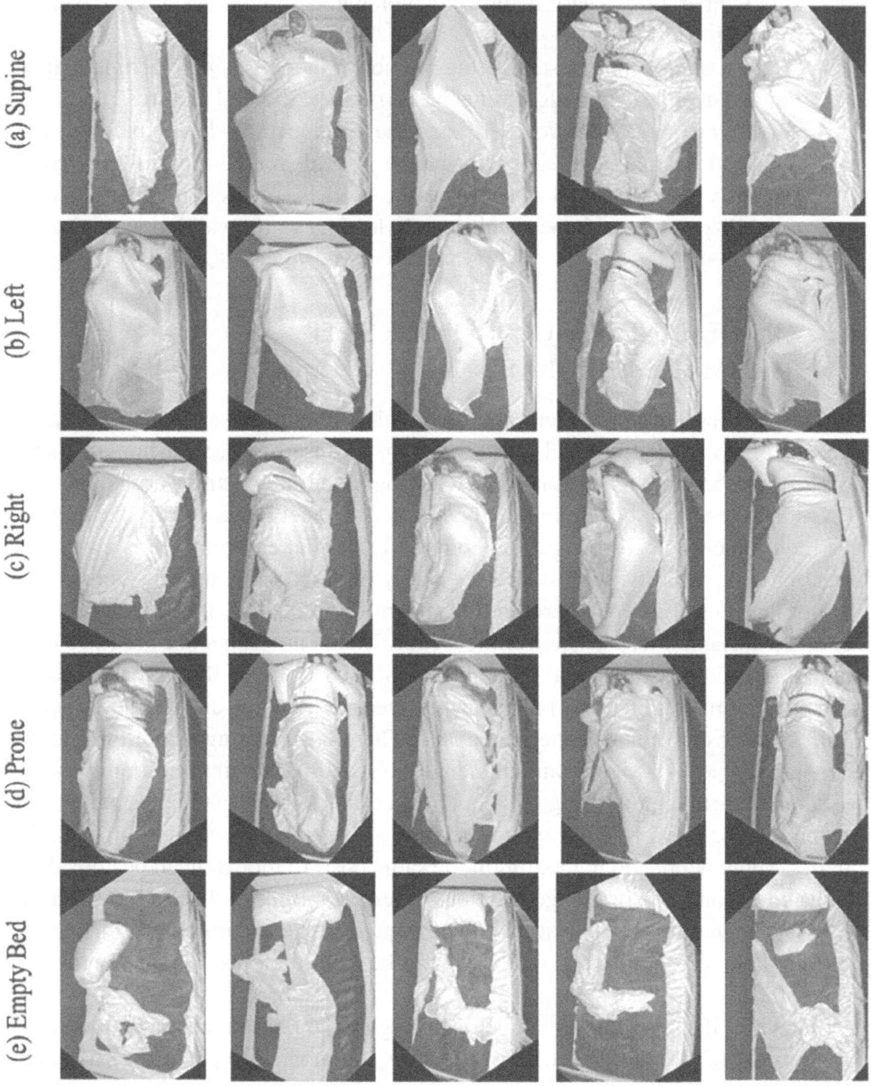

FIGURE 3.4 Examples of sleep postures for different subjects: (a) supine, (b) left, (c) right, (d) prone, (e) empty bed. (Source: Reference [55], CC BY 4.0, https://creativecommons.org/licenses/by/4.0/.)

ResNet-152 pre-trained network achieved the best performance compared to standard CNN networks and other pre-trained networks. Compared to the clinical standard of posture estimation using PSG position sensors, this method outperformed other video-based sleep posture estimation methods and yielded higher performance.

It is evident that the combination of infrared video capture and deep learning AI can successfully quantify sleep postures and transitions between postures under real nocturnal conditions. Compared with the currently accepted clinical methods, this non-contact approach offers superior posture estimation. However, most sleep posture recognition studies focus on common sleep positions and are unable to achieve real-time identification of multiple postures during deep sleep. This limitation necessitates the integration of more precise recognition algorithms with diverse data collected during sleep for further research. Additionally, the majority of studies on sleep posture recognition are based on traditional machine learning models. Although many studies have achieved the goal of recognition to some extent, there is still a need for objective indicators to optimize the design of sleep posture recognition systems in different application scenarios [50].

The development of novel microwave radar technology has facilitated sleep posture recognition. Additionally, it offers advantages such as anti-occlusion, non-contact operation, and privacy protection. Zhou et al. [56] utilized millimeter-wave radar to acquire radar signals under sleep postures. By parallelly combining multiple single-channel two-dimensional radar features, they enhanced the system's representation capability and improved its generalization ability. Based on the principle that the effective radar cross-section is affected by different sleep postures, Islam et al. [57] employed a non-contact, concealed microwave Doppler radar to measure cardiopulmonary motion patterns for sleep posture assessment (Figure 3.5). Although radar technology for sleep recognition appears effective, concerns remain regarding the potential harm of microwaves to the human body, an issue that requires further investigation.

Non-contact methods for skin temperature measurement using infrared technology have been widely applied across various research fields, including thermal comfort. Wu et al. [58] employed a machine learning-based model for predicting thermal sensation to regulate air conditioning temperature, as illustrated in Figure 3.6. Initially, a comprehensive study was conducted on the development process of individual thermal comfort models, from theory to practice. Subsequently, an intelligent air conditioning system was developed, capable of predicting thermal sensation based on infrared sensors that measure occupants' skin temperature and implementing automatic control. A thermal sensation prediction model was established using a random forest algorithm, with skin and ambient temperatures as input parameters, achieving a high prediction accuracy of 0.84. Additionally, addressing the measurement errors of low-cost infrared sensors, they conducted simulation experiments based on human models and proposed a calibration method using multiple linear regression. Finally, experimental results demonstrated that the air

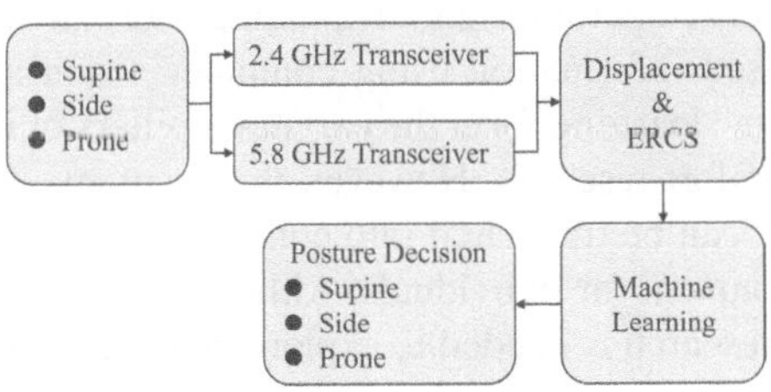

FIGURE 3.5 The detection method of sleep postures using radar transceivers and machine learning.

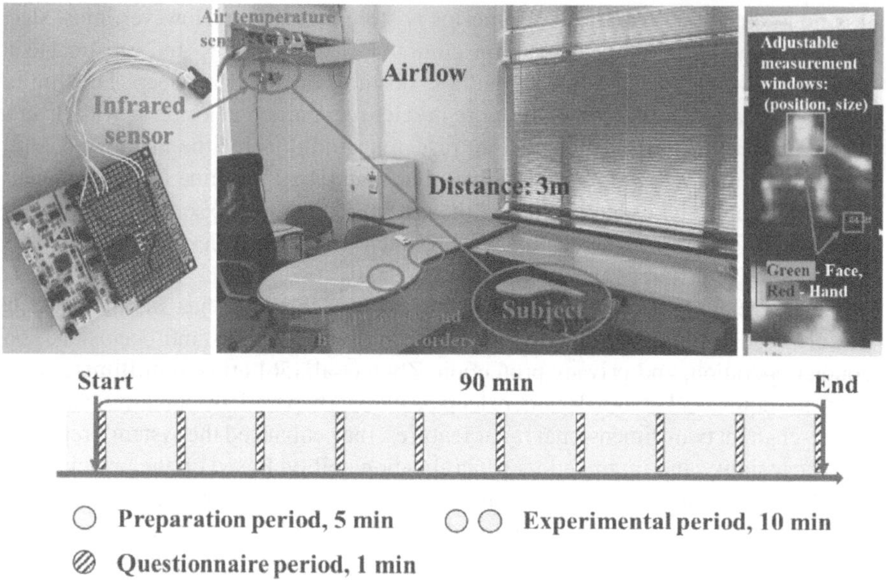

FIGURE 3.6 Experimental scenarios and scheduling for the performance testing of air conditioning prototypes. (Source: Reference [58], Copyright 2023 by Elsevier.)

conditioning system could effectively implement automatic control based on predicted thermal sensation, preventing extreme temperatures and ensuring occupant thermal comfort.

Metzmacher et al. [59] focused on evaluating control algorithms and machine learning-based heuristic algorithms used to train individual thermal comfort models to tailor the system to user preferences. They proposed a personalized thermal regulation test system that provides a server-based modular platform for testing actuators and control strategies, and facilitating monitoring and data collection, as shown in Figure 3.7. They also compared ON/OFF control methods, proportional-integral-derivative (PID) control, and model predictive control (MPC); briefly introduced the modular system architecture; and presented experimental results for the controllers. The results demonstrated that the integrated modular system could consolidate sensor data, actuator information, and thermal imagery for assessing thermal comfort. Ghahramani et al. [60] developed a non-invasive thermal/visual camera system for measuring facial skin temperature and evaluating thermal sensation. Methods for extracting image-based information using computer vision and modeling thermal comfort using machine learning form the two main pillars of infrared technology in personal thermal comfort research. However, the accuracy of prediction is crucial. Accurate predictions can be translated into control commands for devices to create a comfortable environment for individuals. Although prediction accuracy continues to improve, further research is needed to explore its integration with precise control [58]. Infrared technology is non-invasive and can minimize interference with sleep; thus, the aforementioned methods can be further extended for predicting sleep thermal sensation.

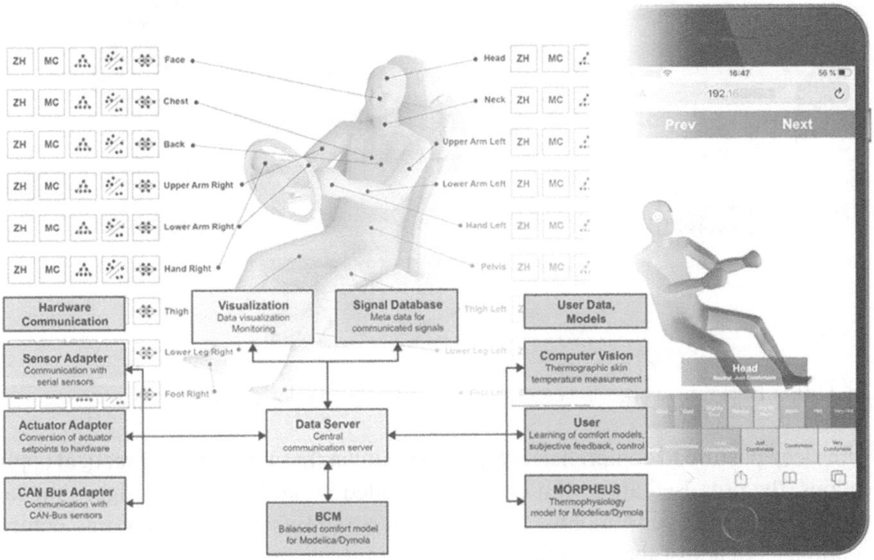

FIGURE 3.7 An overview of the design of personalized thermal regulation test system, model allocation, and user feedback. (Source: Reference [59], Copyright 2022 by Elsevier.)

Building on the aforementioned studies, sleep thermal environment regulation based on sleep states primarily encompasses the following two aspects. The first approach involves the measurement of skin temperature, which plays a crucial role in thermoregulation and sleep [61]. Compared to proximal skin temperature, distal skin temperature is particularly important for sleep regulation [62]. Although body parts may be covered during sleep, infrared technology can be used to monitor facial skin temperature and the microenvironment around bed. The second approach involves sleep posture [50], using computer vision technology to monitor changes in body posture during sleep, which may reflect sleep thermal comfort to some extent [63].

Su et al. [63] proposed a novel method for estimating sleep comfort based on a partial mixture model, which utilizes remote infrared cameras and temperature sensors to collect information on human sleep posture and real-time temperature. Based on extensive experimental analysis, they also defined categories of sleep thermal comfort and provided detailed descriptions of sleep comfort features. A sleep thermal comfort prediction model was established based on human posture and ambient temperature. Experimental results indicated that this method performs well in online sleep comfort analysis. Cheng et al. [64] employed a skeletal keypoint detection algorithm to obtain human keypoint information and constructed a human posture recognition model using keypoint coordinates, as illustrated in Figure 3.8. They categorized sleep postures into two types: stretched and curled. Based on video stream data, they developed a human posture estimation algorithm to estimate sleep-related postures and evaluate thermal comfort during sleep. Additionally, an estimation model for quilt coverage was proposed, integrating the identified quilt-covered area

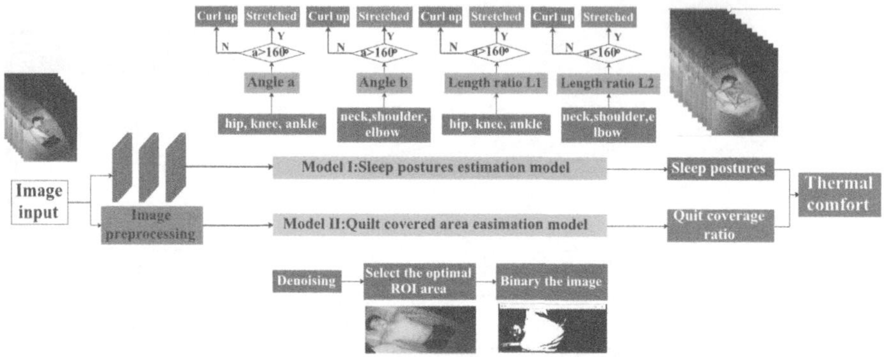

FIGURE 3.8 Framework for sleep thermal comfort prediction algorithm.

with sleep posture to comprehensively assess human thermal state. In practical applications, the proposed algorithm can be embedded in hardware. This algorithm can detect the thermal comfort state of sleepers in real time and translate it into control commands to adjust the air conditioning system. If the user feels cold, the terminal can send a heating command to the air conditioner via an infrared transmitter, and similarly, it can send a cooling command. The aforementioned methods can regulate the indoor environment on demand based on the predicted human sleep thermal state to ensure sleep thermal comfort. Currently, there are few studies related to on-demand regulation of sleep thermal environments, and future researches should focus on the application effects of existing thermal comfort prediction methods in regulating sleep environments.

The advent of Internet of Things (IoT) has enhanced the accuracy of sleep posture recognition, providing feedback on sleep posture information to clinicians or sleep researchers to prevent sleep disorders and improve sleep quality. Tang et al. [65] designed an intelligent recognition system that employs CNNs to analyze collected data and identify various sleep postures. This system not only monitors sleep posture but also offers early warnings during sleep. It transmits users' posture data to the cloud for processing via the internet and integrates IoT technology to develop an Android application, providing users with statistical data on their sleep postures. This IoT-based sleep posture recognition system facilitates clinical research and can be applied to telemedicine monitoring [66]. The technology can also be integrated with smart home systems to control air conditioning devices. While IoT has broadened the application scope of sleep posture recognition systems, the high cost remains a challenge for their widespread use in daily life [50].

Non-contact measurement methods may raise concerns regarding personal privacy and data security, particularly in sleep environments, where potential security risks can cause alarm and anxiety. Although infrared sensors are considered to pose fewer privacy issues compared to RGB cameras [41], personal information is sensitive and must be taken into account. Previous studies have indicated that privacy concerns can be alleviated by discarding images after parameter optimization or by encrypting and storing data in a secure environment [41, 47]. From the perspective of IoT, in addition to technological solutions, enhancing user awareness of potential risks and enacting relevant legislation are also important measures to foster trust and advance the application of emerging technologies [67].

Current non-contact sleep monitoring research primarily focuses on regulating the environment based on thermal comfort of monitored individual. However, the control target is overall sleep environment. How to integrate the different sleep thermal states of multiple occupants to control the air conditioning system has not yet been discussed. Some studies have mentioned adjusting the indoor temperature based on the proportion of occupants with non-neutral thermal states [68, 69]. Future research should focus on balancing different thermal comfort needs in shared spaces and energy conservation.

The future development prospects of non-contact sleep monitoring technology are broad. It offers a more natural, accurate, and convenient method for sleep monitoring. Through continuous technological innovation and interdisciplinary collaboration, we can anticipate that shortly, everyone will have access to high-quality, personalized sleep management and improvement services. This will not only significantly enhance people's quality of life but also bring breakthroughs and developments to the field of sleep science research.

3.5 CONCLUSIONS AND FUTURE DIRECTIONS

With the development of the economy and the progress of science and technology, as well as the increasing improvement of the standard of living of the residents, people are more and more concerned about the quality of their sleep and physical health. More and more portable monitoring devices and wearable monitoring devices can be used to continuously monitor the quality of sleep and can continuously monitor several physiological indicators of the body, so that people do not need to go to the hospital to directly understand their sleep situation and physical health. In recent years, with the rapid development of computer vision technology, more and more computer vision technologies have been applied to the field of sleep monitoring and have achieved a lot of novel research results, which greatly reduces the negative impact of foreign body sensation brought by wearable devices on sleep.

The application of computer vision technology in sleep monitoring also faces some problems. First, existing research relies more on public databases or small sample data and lacks validation on a large number of clinical samples, resulting in limited model generalization capabilities. Second, although deep learning models can realize end-to-end mapping, the extracted features are abstract and lack a clinical basis, which is not easy to accept by professional doctors, and PSG is still "gold standard" for clinical sleep research. Third, most sleep event recognition algorithms can only recognize a single category of events, and in-depth research is needed for the recognition of multiple categories of events and event localization methods. Fourth, whether the privacy of subjects and patients can be guaranteed. Finally, there is almost no research on the application of regulating the corresponding equipment according to the predicted sleep thermal comfort and sleep quality to improve sleep quality. The problems listed above are very worthy of continued research and in-depth consideration.

With the development of technology, a large number of researchers have developed air conditioning equipment that shows great potential and innovation in regulating indoor thermal environment. It mainly contains the following advantages. Smart air conditioners monitor indoor and outdoor temperature changes in real time through built-in sensors and intelligent control systems, automatically adjusting indoor temperature to achieve a more comfortable living environment and reduce energy consumption. At the same time, smart air conditioners can link with smart home systems to link air purifiers, curtains and windows to create a healthy and peaceful sleep environment. In the future, the combination of computer vision-based sleep monitoring methods and smart home systems will be the key to improving sleep quality. Based on machine learning algorithms and AI technology, smart air conditioners can adjust the most comfortable sleeping environment temperature as well as other equipment parameters according to human sleep status. In summary, intelligent air conditioning equipment is developing in the direction of being more intelligent and personalized to meet the needs of users for a healthy and comfortable sleep environment. With the continuous progress of technology, more innovative solutions will further enhance sleep quality.

REFERENCES

1. H. Zhang, Y. Yan, M. Kang, X. Fan, P. Wargocki, L. Lan, The comparison between Fitbit Alta HR and polysomnography in monitoring sleep of young adults, in: International Conference on Building Energy and Environment, Springer, 2022: pp. 2099–2102.
2. S. Toften, S. Pallesen, M. Hrozanova, F. Moen, J. Grønli, Validation of sleep stage classification using non-contact radar technology and machine learning (Somnofy®), Sleep Medicine 75 (2020) 54–61.
3. K. Zhu, M. Li, S. Akbarian, M. Hafezi, B. Taati, Vision-based contactless respiratory rate monitoring during sleep, Sleep Medicine 64 (2019) S428.
4. X. Zhang, G. Luo, J. Xie, J. Liu, Associations of bedroom air temperature and CO_2 concentration with subjective perceptions and sleep quality during transition seasons, Indoor Air 31 (2021) 1004–1017.
5. X. Xu, Z. Lian, J. Shen, T. Cao, J. Zhu, X. Lin, K. Qing, W. Zhang, T. Zhang, Experimental study on sleep quality affected by carbon dioxide concentration, Indoor Air 31 (2021) 440–453.
6. R. Bevan, S. Grantham-Hill, R. Bowen, E. Clayton, H. Grice, H.C. Venditti, A. Stickland, C.M. Hill, Sleep quality and noise: Comparisons between hospital and home settings, Archives of Disease in Childhood 104 (2019) 147–151.
7. L. Taranto-Montemurro, L. Messineo, S. Sands, A. Azarbarzin, M. Marques, A. Wellman, 0394 effect of background noise on sleep quality, Sleep 40 (2017) A146–A147.
8. B. Sander, J. Markvart, L. Kessel, A. Argyraki, K. Johnsen, Can sleep quality and wellbeing be improved by changing the indoor lighting in the homes of healthy, elderly citizens? Chronobiology International 32 (2015) 1049–1060.
9. M. Huang, X. Hao, K. Zhang, Q. Xing, K. Yang, Thermal resistance of different quilt combinations during sleep, Journal of Hunan University of Science and Technology (Natural Science Edition) 35 (2020) 32–37. (In Chinese)
10. H. Zhang, L. Shen, Thermal preference effects on physiological responses and bedding system insulation selection during good quality sleep in the transitional season, Building and Environment 253 (2024) 111302.

11. R.K. Szymczak, E.J. Sitek, J.W. Sławek, A. Basiński, M. Siemiński, D. Wieczorek, Subjective sleep quality alterations at high altitude, Wilderness & Environmental Medicine 20 (2009) 305–310.
12. S. Jafarian, F. Gorouhi, A. Taghva, J. Lotfi, High-altitude sleep disturbance: Results of the Groningen Sleep Quality Questionnaire survey, Sleep Medicine 9 (2008) 446–449.
13. W. Dong, A. Cao, H. Gu, G. Deng, H. Yu, Investigation on sleep status of soldiers of a frontier troops stationed in Plateau area, Journal of PLA Hospital Administration 20 (2013) 272–274. (In Chinese)
14. H. Guan, J. De, Y. Chen, L. Zhaxi, D. Awang, L. Yang, P. Tian, A case-control study on sleep quality and related factors in Tibetan patients with hypertension living in plateau, Plateau Scientific Research 3 (2019) 75–82. (In Chinese)
15. Q. Li, K. Ding, Y. Yu, T. Qin, Analysis of sleep status and influence path of soldiers at high altitude, PLA Journal of Preventive Medicine 35 (2017) 251–254. (In Chinese)
16. W.-C. Liao, L. Wang, C.-P. Kuo, C. Lo, M.-J. Chiu, H. Ting, Effect of a warm footbath before bedtime on body temperature and sleep in older adults with good and poor sleep: An experimental crossover trial, International Journal of Nursing Studies 50 (2013) 1607–1616.
17. T. Åkerstedt, K. Hume, D. Minors, J. Waterhouse, The meaning of good sleep: A longitudinal study of polysomnography and subjective sleep quality, Journal of Sleep Research 3 (1994) 152–158.
18. H. Gotsopoulos, J.J. Kelly, P.A. Cistulli, Oral appliance therapy reduces blood pressure in obstructive sleep apnea: A randomized, controlled trial, Sleep 27 (2004) 934–941.
19. S. Sharma, V. Malik, C. Vasudev, A. Banga, A. Mohan, K. Handa, S. Mukhopadhyay, Prediction of obstructive sleep apnea in patients presenting to a tertiary care center, Sleep and Breathing 10 (2006) 147–154.
20. M. Yamauchi, S. Tamaki, M. Yoshikawa, Y. Ohnishi, H. Nakano, F.J. Jacono, K.A. Loparo, K.P. Strohl, H. Kimura, Differences in breathing patterning during wakefulness in patients with mixed apnea-dominant vs obstructive-dominant sleep apnea, Chest 140 (2011) 54–61.
21. L.F. Flernández, R.B. Santos, C.G. de Isla, A. Beatríz, R. Moratinos, C.B. Ibars, J.C. Juánez, Analysis of improved sleep quality and anxiety in college students under stress by consuming non-alcoholic beer, Revista Española de Nutrición Comunitaria 18 (2012) 218–223.
22. J.-H. Byun, K.T. Kim, H. Moon, G.K. Motamedi, Y.W. Cho, The first night effect during polysomnography, and patients' estimates of sleep quality, Psychiatry Research 274 (2019) 27–29.
23. C. Liao, Z. Huang, F. Qin, Y. Wang, T. Wang, Y. Tong, Non-contact sleep apnea detection and classification using thermal imaging, Journal of University of Chinese Academy of Sciences 42 (2025) 538–546.
24. S. Akbarian, N. Montazeri Ghahjaverestan, A. Yadollahi, B. Taati, Distinguishing obstructive versus central apneas in infrared video of sleep using deep learning: Validation study, Journal of Medical Internet Research 22 (2020) e17252.
25. Y. Wang, M. Hu, Y. Zhou, Q. Li, N. Yao, G. Zhai, X.-P. Zhang, X. Yang, Unobtrusive and automatic classification of multiple people's abnormal respiratory patterns in real time using deep neural network and depth camera, IEEE Internet of Things Journal 7 (2020) 8559–8571.
26. G. Scebba, G. Da Poian, W. Karlen, Multispectral video fusion for non-contact monitoring of respiratory rate and apnea, IEEE Transactions on Biomedical Engineering 68 (2020) 350–359.
27. E. Dafna, A. Tarasiuk, Y. Zigel, Sleep-quality assessment from full night audio recordings of sleep apnea patients, in: 2012 Annual International Conference of the IEEE Engineering in Medicine and Biology Society, IEEE, 2012: pp. 3660–3663.

28. E. Dafna, A. Tarasiuk, Y. Zigel, Sleep-wake evaluation from whole-night non-contact audio recordings of breathing sounds, PLoS One 10 (2015) e0117382.
29. H. Wu, T. Kato, M. Numao, K. Fukui, Statistical sleep pattern modelling for sleep quality assessment based on sound events, Health Information Science and Systems 5 (2017) 1–11.
30. B. Xue, B. Deng, H. Hong, Z. Wang, X. Zhu, D.D. Feng, Non-contact sleep stage detection using canonical correlation analysis of respiratory sound, IEEE Journal of Biomedical and Health Informatics 24 (2019) 614–625.
31. X. Wang, D. Matsushita, Non-contact determination of sleep/wake state in residential environments by neural network learning of microwave radar and electroencephalogram–electrooculogram measurements, Building and Environment 233 (2023) 110095.
32. X. Chen, A.L. Yuille, Articulated pose estimation by a graphical model with image dependent pairwise relations, Advances in Neural Information Processing Systems 27 (2014).
33. Y. Chen, C. Shen, X.-S. Wei, L. Liu, J. Yang, Adversarial posenet: A structure-aware convolutional network for human pose estimation, in: Proceedings of the IEEE International Conference on Computer Vision, 2017: pp. 1212–1221.
34. A. Toshev, C. Szegedy, Deeppose: Human pose estimation via deep neural networks, in: Proceedings of the IEEE Conference on Computer Vision and Pattern Recognition, 2014: pp. 1653–1660.
35. Y. Chen, Z. Wang, Y. Peng, Z. Zhang, G. Yu, J. Sun, Cascaded pyramid network for multi-person pose estimation, in: Proceedings of the IEEE Conference on Computer Vision and Pattern Recognition, 2018: pp. 7103–7112.
36. L. Pishchulin, E. Insafutdinov, S. Tang, B. Andres, M. Andriluka, P.V. Gehler, B. Schiele, Deepcut: Joint subset partition and labeling for multi person pose estimation, in: Proceedings of the IEEE Conference on Computer Vision and Pattern Recognition, 2016: pp. 4929–4937.
37. E. Insafutdinov, L. Pishchulin, B. Andres, M. Andriluka, B. Schiele, Deepercut: A deeper, stronger, and faster multi-person pose estimation model, in: Computer Vision–ECCV 2016: 14th European Conference, Amsterdam, The Netherlands, October 11–14, 2016, Proceedings, Part VI 14, Springer, 2016: pp. 34–50.
38. Z. Cao, G. Hidalgo, T. Simon, S.-E. Wei, Y. Sheikh, OpenPose: Realtime multi-person 2D pose estimation using part affinity fields, IEEE Transactions on Pattern Analysis and Machine Intelligence 43 (2021) 172–186. https://doi.org/10.1109/TPAMI.2019.2929257.
39. R.A. Güler, N. Neverova, I. Kokkinos, Densepose: Dense human pose estimation in the wild, in: Proceedings of the IEEE Conference on Computer Vision and Pattern Recognition, 2018: pp. 7297–7306.
40. J. Li, C. Wang, H. Zhu, Y. Mao, H.-S. Fang, C. Lu, Crowdpose: Efficient crowded scenes pose estimation and a new benchmark, in: Proceedings of the IEEE/CVF Conference on Computer Vision and Pattern Recognition, 2019: pp. 10863–10872.
41. J. Baek, D.Y. Park, H. Park, D.M. Le, S. Chang, Vision-based personal thermal comfort prediction based on half-body thermal distribution, Building and Environment 228 (2023) 109877.
42. B. Pavlin, G. Pernigotto, F. Cappelletti, P. Bison, R. Vidoni, A. Gasparella, Real-time monitoring of occupants' thermal comfort through infrared imaging: A preliminary study, Buildings 7 (2017) 10.
43. P. Li, P. Dai, D. Cao, B. Liu, Y. Lu, Non-intrusive comfort sensing: Detecting age and gender from infrared images for personal thermal comfort, Building and Environment 219 (2022) 109256.
44. P. Li, T.M. Froese, G. Brager, Post-occupancy evaluation: State-of-the-art analysis and state-of-the-practice review, Building and Environment 133 (2018) 187–202.

45. A.C. Cosma, R. Simha, Thermal comfort modeling in transient conditions using real-time local body temperature extraction with a thermographic camera, Building and Environment 143 (2018) 36–47.
46. F. Faridah, M.M. Waruwu, T. Wijayanto, R. Budiarto, R.C. Pratama, S.E. Prayogi, N.M. Nadiya, R.J. Yanti, Feasibility study to detect occupant thermal sensation using a low-cost thermal camera for indoor environments in Indonesia, Building Services Engineering Research and Technology 42 (2021) 389–404.
47. Y. He, H. Zhang, E. Arens, A. Merritt, C. Huizenga, R. Levinson, A. Wang, A. Ghahramani, A. Alvarez-Suarez, Smart detection of indoor occupant thermal state via infrared thermography, computer vision, and machine learning, Building and Environment 228 (2023) 109811.
48. K. Okamoto-Mizuno, K. Mizuno, Effects of thermal environment on sleep and circadian rhythm, Journal of Physiological Anthropology 31 (2012) 1–9.
49. M.H. Kryger, T. Roth, W.C. Dement, Principles and Practice of Sleep Medicine, Elsevier/Saunders, 2005.
50. X. Li, Y. Gong, X. Jin, P. Shang, Sleep posture recognition based on machine learning: A systematic review, Pervasive and Mobile Computing 90 (2023) 101752.
51. Z. Liu, Y. Le, J. Sun, others, A study on sleep position recognition of body pressure image based on KPCA and SVM, in: 2019 IEEE International Conference on Power, Intelligent Computing and Systems (ICPICS), IEEE, 2019: pp. 189–193.
52. Z. Liu, S. Mingliang, K. Lu, others, A method to recognize sleeping position using an CNN model based on human body pressure image, in: 2019 IEEE International Conference on Power, Intelligent Computing and Systems (ICPICS), IEEE, 2019: pp. 219–224.
53. S. Zhang, G. Wang, X. Guo, Sleep-pose recognition based on pyroelectric infrared sensing technology, in: 2018 13th World Congress on Intelligent Control and Automation (WCICA), IEEE, 2018: pp. 495–498.
54. N. Kakuta, S. Yokoyama, M. Nakamura, K. Mabuchi, Estimation of radiative heat transfer using a geometric human model, IEEE Transactions on Biomedical Engineering 48 (2001) 324–331.
55. S.M. Mohammadi, S. Enshaeifar, A. Hilton, D.-J. Dijk, K. Wells, Transfer learning for clinical sleep pose detection using a single 2D IR camera, IEEE Transactions on Neural Systems and Rehabilitation Engineering 29 (2020) 290–299.
56. T. Zhou, Z. Xia, X. Wang, F. Xu, Human sleep posture recognition based on millimeter-wave radar, in: 2021 Signal Processing Symposium (SPSympo), IEEE, 2021: pp. 316–321.
57. S.M.M. Islam, V.M. Lubecke, Sleep posture recognition with a dual-frequency microwave Doppler radar and machine learning classifiers, IEEE Sensors Letters 6 (2022) 1–4.
58. Y. Wu, B. Cao, M. Hu, G. Lv, J. Meng, H. Zhang, Development of personal comfort model and its use in the control of air conditioner, Energy and Buildings 285 (2023) 112900.
59. H. Metzmacher, M. Syndicus, A. Warthmann, C. van Treeck, Exploratory comparison of control algorithms and machine learning as regulators for a personalized climatization system, Energy and Buildings 255 (2022) 111653.
60. A. Ghahramani, Q. Xu, S. Min, A. Wang, H. Zhang, Y. He, A. Merritt, R. Levinson, Infrared-fused vision-based thermoregulation performance estimation for personal thermal comfort-driven HVAC system controls, Buildings 12 (2022) 1241.
61. E.J. VanSomeren, More than a marker: Interaction between the circadian regulation of temperature and sleep, age-related changes, and treatment possibilities, Chronobiology International 17 (2000) 313–354.
62. L. Lan, K. Tsuzuki, Y. Liu, Z. Lian, Thermal environment and sleep quality: A review, Energy and Buildings 149 (2017) 101–113.
63. L. Su, M. Xu, Sleep comfort analysis using a part-based mixture model with nighttime infrared video, Neurocomputing 259 (2017) 66–75.

64. X. Cheng, F. Hu, B. Yang, F. Wang, T. Olofsson, Contactless sleep posture measurements for demand-controlled sleep thermal comfort: A pilot study, Indoor Air 32 (2022) e13175.
65. K. Tang, A. Kumar, M. Nadeem, I. Maaz, CNN-based smart sleep posture recognition system, IoT 2 (2021) 119–139.
66. A. Channa, M. Yousuf, N. Popescu, Machine learning algorithms for posture identification of obstructive sleep apnea patients using IoT solutions, in: 2020 International Conference on E-Health and Bioengineering (EHB), IEEE, 2020: pp. 1–6.
67. Y. He, N. Li, J. Lu, N. Li, Q. Deng, C. Tan, J. Yan, Meeting thermal needs of occupants in shared space with an adjustable thermostat and local heating in winter: An experimental study, Energy and Buildings 236 (2021) 110776.
68. Y. He, N. Li, H. Zhang, Y. Han, J. Lu, L. Zhou, Air-conditioning use behaviors when elevated air movement is available, Energy and Buildings 225 (2020) 110370.
69. Y. He, N. Li, N. Li, J. Li, J. Yan, C. Tan, Control behaviors and thermal comfort in a shared room with desk fans and adjustable thermostat, Building and Environment 136 (2018) 213–226.

4 On-Demand Regulation for Indoor Thermal Environment

Yihang Liu, Min Qi, Xinrou Han, and Bin Yang

4.1 INTRODUCTION

By detecting indoor occupant information and indoor/outdoor environmental parameters, it is possible to more precisely quantify occupants' demands for indoor environments, thereby creating high-quality indoor environmental conditions while reducing building system operational energy consumption. Consequently, researchers have proposed occupant-centric control (OCC) strategies to reduce system energy consumption, improve human thermal comfort, and ensure air quality [1]. OCC strategies are referenced in IEA EBC Annex 79, ASHRAE Standard 62.1, and ASHRAE Handbook [2–4]. Implementing OCC strategies requires integrating indoor occupant information, as both passive presence (e.g., occupancy) and active behaviors (e.g., window opening, human–device interactions) influence the start/stop times and operational parameters of heating, ventilation, and air conditioning (HVAC) systems. Therefore, OCC strategies are generally categorized into two types: identifying occupied zones based on occupant presence, quantity, and location; predicting occupant thermal states by analyzing indoor human behaviors (e.g., window operation, human–device interactions) to comprehensively regulate building system operations [5].

In demand-driven regulation based on passive occupant presence, control strategies primarily address three scenarios: (1) Large-volume spaces (e.g., transportation hubs): Characterized by high passenger flow, varying dwell times, and diverse functional zones, existing HVAC systems can be adjusted hourly based on real-time occupancy. (2) Public buildings (e.g., offices): Featuring regular working hours, low occupant density, and distinct zoning, system parameters can be optimized using occupant distribution data (e.g., numbers, clothing insulation). (3) Local microenvironments: Individual thermal comfort devices serve single users, offering fast response times. Compared to the previous two buildings, this scenario is more suitable for on-demand adjustment, which can achieve real-time adjustment and real-time response.

For demand-driven regulation based on active occupant behaviors, thermal adaptive actions (e.g., drinking frequency, clothing adjustments, window operation) can predict thermal comfort and guide HVAC system regulation. Researchers have compiled extensive occupant behavior databases to inform rational HVAC operation strategies. Active feedback of thermal sensation and preference also falls under active behaviors [6], enabling the development of mathematical models linking environmental/physiological parameters to thermal sensation. These models support

DOI: 10.1201/9781003611455-4

delivering near-neutral thermal conditions in occupied zones. Building on occupant detection and parameter prediction, various automated control methods are explored to optimize dynamic system performance. This chapter will provide a comprehensive introduction to research on full ventilation, personalized ventilation (PV), personal comfort systems (PCS), and natural ventilation in demand-driven regulation.

4.2 INDOOR HUMAN BEHAVIOR CHARACTERISTICS

4.2.1 Occupancy Prediction Models

Zhou et al. [7] analyzed residential occupancy patterns using clustering analysis, function fitting, and correlation analysis, comparing data from 2008 and 2017 to reveal temporal variations in usage patterns and explore influencing factors of behavioral changes. Their findings hold significant implications for predicting future trends and analyzing energy consumption in residential buildings. Dobbs et al. [8] employed automatically trained Markov chains for occupancy prediction, simplifying the algorithm via approximate dynamic programming to forecast future occupancy based on current observations. The stochastic occupancy model was integrated into a model predictive control (MPC) framework to simulate the impact of occupancy on thermal comfort and energy use. Li et al. [9] provided a dataset collected from four residential occupancy sensors and proposed a novel non-homogeneous Markov occupancy prediction model. Results showed that the Markov model achieved an average prediction accuracy of 5% and a maximum error of 11% for 15-minute-ahead occupancy forecasts, outperforming other methods. HVAC systems can adjust operational parameters based on occupancy patterns derived from such prediction models.

4.2.2 Occupant Count

Dong et al. [10] deployed a sophisticated sensor network and trained a hidden Markov model using experimental data. During testing, the average accuracy for occupancy detection reached 73%. Building on these results, they further proposed a hidden Markov model based on a Gaussian mixture model [11], achieving an average occupancy detection accuracy of 83%. Under this framework, the air conditioning system achieved 18.5% energy savings while maintaining indoor thermal comfort. Dong et al. [12] enhanced and implemented advanced machine learning (ML) methods, including adaptive Gaussian processes, hidden Markov models, episode discovery, and semi-Markov models. They designed and implemented a nonlinear MPC strategy based on dynamic programming. Results demonstrated 30.1% energy reduction during heating seasons and 17.8% reduction during cooling seasons compared to conventional fixed-temperature operation condition. By adjusting HVAC system parameters based on occupant distribution data, energy efficiency and comfort requirements can be simultaneously met.

4.2.3 Window-Opening Behavior

Interactions between occupants and windows significantly impact building energy consumption, indoor thermal comfort, and air quality. Haldi et al. [13] developed three distinct modeling approaches (logistic probability distributions, Markov chains, and continuous-time random processes) to predict window-opening/closing behaviors about occupancy patterns and indoor/outdoor parameters, based on seven years of data. Shi et al. [14] investigated window states in eight naturally ventilated residential apartments in Beijing and Nanjing, analyzing correlations between window-opening probability and outdoor air temperature, relative humidity, wind speed, and PM 2.5 concentration. Using univariate and multivariate linear logistic regression, they established stochastic models for occupant–window interactions in monitored households across both cities. Results identified outdoor temperature as the most critical explanatory variable. Yun et al. [15] studied summer window-opening behaviors in private and shared offices, developing stochastic models that link such behaviors to indoor air temperature, time, and prior window states. Their results revealed statistically significant relationships between window-opening actions and indoor temperature during summer. By synthesizing patterns in window operation behaviors, adjusting HVAC parameters and window openings based on detected environmental data can reduce energy waste and enhance indoor comfort.

4.2.4 Thermostat Adjustment Behavior

The interaction between occupants and thermostats is primarily driven by thermal comfort requirements [16], with the research framework for adaptive behaviors illustrated in Figure 4.1. Consequently, understanding and simulating occupant–thermostat interactions are crucial for the design, evaluation, and control of demand response strategies. Vellei et al. [17] proposed a thermostat adjustment behavior model based on user interaction data from approximately 9,000 Canadian thermostats, incorporating dynamic thermal discomfort prediction derived from a two-node thermophysiological model and a dynamic thermal perception model. Fabi et al. [18]

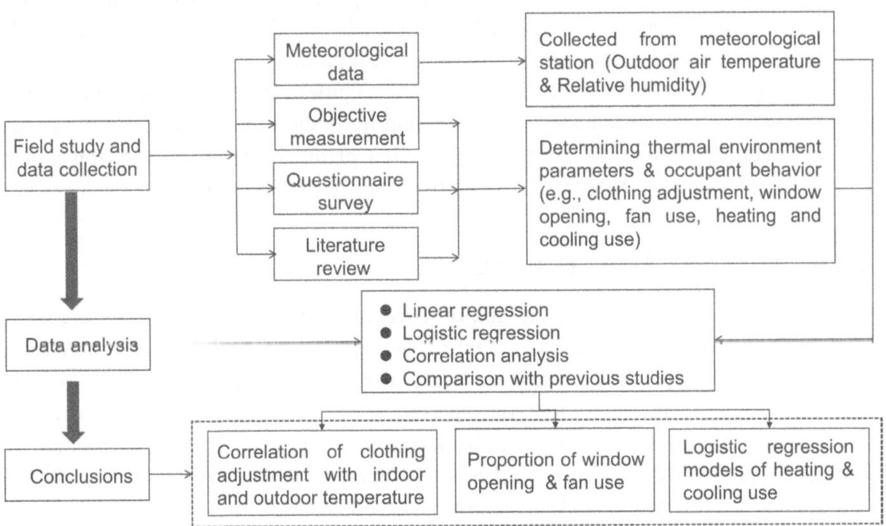

FIGURE 4.1 Research framework of thermal adaptive behavior.

utilized logistic regression methods to infer the probability of thermostat radiator valve setpoint adjustments based on heating setpoint behaviors in 13 Danish households, developing 3 distinct occupant–heating control interaction models and exploring their impacts on indoor environmental quality and energy consumption. He et al. [19] collected 873 questionnaires on occupant behaviors in air-conditioned office buildings in the Pearl River Delta region to investigate relationships between indoor environmental quality and adaptive behaviors. An ordinal logistic regression method was employed to establish a probabilistic prediction model for cooling temperature setpoints. Twenty typical occupant behavior patterns in office buildings were categorized based on control modes of air conditioning, windows, blinds, and lighting systems. Ren et al. [20] conducted surveys and monitoring in 30 households across 8 cities to develop conditional probability models for air conditioner usage, linking it to environmental and event triggers. Results demonstrated that integrating mathematical models of typical usage patterns could effectively describe air conditioner behaviors in residential buildings under environmental and event-driven conditions. Adjusting the operational parameters of ventilation and air conditioning systems based on thermostat adjustment behavior patterns can better satisfy occupants' thermal preferences.

4.2.5 Shading and Lighting Adjustment Behavior

Correia da Silva et al. [21] continuously monitored occupant behaviors related to manual control of lighting and shading devices in eight offices under normal operation for over two months. The study evaluated the consistency between existing behavioral models and actual occupant behaviors, providing insights to refine occupant behavior modeling in building simulation software. Ding et al. [22] investigated light environment preferences and usage patterns of lighting and shading systems among occupants in 12 private offices through questionnaires. They employed Markov models and logistic survival models to quantify predictive models for coupled control behaviors. Inkarojrit [23] developed a predictive control logic model for window shading by integrating environmental parameters, thermal comfort, and visual comfort. Newsham et al. [24] proposed a camera-based lighting and shading control system that regulates lighting and shading devices through daylight harvesting and occupancy sensing. As these behaviors primarily relate to lighting adjustments, they will not be further elaborated in subsequent chapters.

4.2.6 Occupant Active Feedback

Collecting occupants' active feedback on thermal comfort data enables the assessment of their satisfaction with current temperature setpoints and thermal preferences. Numerous studies predict occupants' thermal sensations based on environmental and physiological parameters to better meet their needs, though energy-saving analyses remain limited. While active feedback does not directly correlate with occupant information detection, it indirectly links to occupant distribution patterns, offering guidance to meet thermal preferences in occupied zones.

The aforementioned occupant behavior characteristics are all relevant to the operational adjustments of ventilation and air conditioning systems. These can be integrated with control strategies to balance comfort requirements and energy consumption reduction. Sections 4.2 and 4.3 will detail how to regulate ventilation and air conditioning system operations based on occupant behaviors and introduce existing advanced control methodologies.

4.3 ON-DEMAND REGULATION OF HVAC SYSTEMS BASED ON PERSONNEL INFORMATION

Buildings are designed to create a healthy, comfortable, and safe indoor environment distinct from the external natural environment, meeting occupants' needs. As people increasingly spend more time indoors, the relationship between occupants and indoor building environments has grown more intertwined. HVAC systems regulate temperature, humidity, and air quality to provide thermal comfort and ensure good indoor air quality. However, existing HVAC systems face operational shortcomings, and performance improvements must address three key aspects.

1. System Operation Energy Consumption

   HVAC systems account for approximately 50% of building energy consumption [25]. Traditional mixing ventilation (MV) ensures thermal comfort across entire spaces but wastes energy by conditioning unoccupied zones [26]. Stratum ventilation (SV) [27] and underfloor air distribution (UFAD) [28] systems deliver conditioned air directly to occupied zones, optimizing thermal comfort and air quality at occupant height. Studies highlight the energy-saving potential of non-uniform airflow distribution [29]. However, occupants often occupy localized areas temporarily or for extended periods, making target airflow zones dynamic and complex [30]. Recent pandemics have further heightened demands for ventilation to reduce indoor pollutant concentrations, increasing energy consumption as systems supply more fresh air [31]. Integrating HVAC control strategies with occupancy information to prioritize occupied zones could reduce energy waste while maintaining comfort and air quality.
2. Occupant Behavior

   HVAC control strategies based on occupant data fall into two categories. Active control behaviors: Occupants interact with building devices (e.g., adjusting thermostats or window openings) to meet individual thermal preferences. However, conflicting preferences in multi-occupant spaces require strategies that satisfy the majority. System parameter adjustments based on these behaviors can enhance thermal comfort. Passive occupancy-based control: Sensors and data analytics predict occupancy to regulate system parameters, ensuring localized comfort environment. However, delays between system adjustments and environmental responses cause parameter fluctuations and lagging adjustments, compromising occupant comfort [32].
3. Occupant Detection Technologies

   Indoor environments are dynamic and complex due to occupant presence. HVAC control strategies and energy use are closely tied to occupant data, including device interactions, locations, and occupancy rates [33].

While Wi-Fi sensors, smart meters, and environmental sensors are widely used in HVAC systems, their real-time detection performance in high-density public spaces remains limited due to structural layouts, sensor placement, and usage habits [34]. These shortcomings hinder real-time HVAC adjustments based on occupancy. Computer vision (CV) technologies have emerged as a solution to address these challenges.

In response, extensive research applies CV technology to demand-driven HVAC regulation. Real-time monitoring of dynamic occupancy enables demand-based control strategies, improving comfort and reducing energy consumption [35]. Recent studies, particularly in CV-integrated demand-controlled ventilation (DCV), demonstrate significant potential. DCV method adjusts ventilation rate based on indicators like humidity, occupancy, and CO_2 levels [36–38] to control indoor pollutants. Such strategies align with OCC frameworks, which dynamically optimize HVAC parameters to balance comfort and energy efficiency [6]. OCC targets both full ventilation systems and PV, alongside natural ventilation and PCS.

4.3.1 Full Ventilation and Personalized Ventilation

The traditional ventilation system primarily delivers conditioned fresh air into the entire room based on a fixed fresh air volume or fixed fresh air ratio [39], which does not match indoor demand. Although the existing ventilation methods can dilute or remove polluted indoor air, the demand for fresh air and the concentration of indoor pollutants are dynamically changing. This can easily lead to insufficient fresh air, poor air quality, or over-ventilation and also increases unnecessary energy consumption.

DCV, first proposed by Kusuda [40], suggests the use of indoor pollutant concentration (typically CO_2) as an indicator to predict fresh air volume, aiming to improve indoor air quality and reduce energy consumption. This method has been applied in many buildings. Hesaraki et al. [41] installed CO_2 sensors in the return air ducts of residential rooms to regulate fresh air supply, achieving a 43% reduction in fan and heating energy consumption. Wachenfeldt et al. [42] compared the performance of a ventilation system based on constant air volume, a ventilation system based on CO_2 concentration in return air, and a ventilation system based on infrared radio frequency technology across 157 Norwegian classrooms. Their findings revealed that during teaching periods, the latter two systems reduced energy consumption by 38% and 51%, respectively, compared to the constant air volume system. However, when occupancy varies across different zones within a room, single CO_2 sensor detection proves insufficient to support efficient ventilation control. Subsequent studies improved CO_2 concentration-based ventilation systems by deploying multiple CO_2 sensors in large spaces to estimate occupancy levels, thereby optimizing energy-efficient operation of HVAC systems [43]. While these sensors can adjust fresh air volume more accurately based on real-time occupancy and reduce energy consumption, they exhibit limitations in capturing dynamic variations in building-wide occupancy distribution. Additionally, research indicates that CO_2 concentration estimation accuracy is compromised by time delays, high costs, and indirect detection methods [44], resulting in constrained practical applicability.

Accurate detection of actual occupancy information enables precise control of ventilation systems. For example, robust occupancy detection methods can provide HVAC systems with energy-saving potential of up to 20% [45]. Diverse methods exist for detecting indoor occupancy, with common occupant counting approaches including image recognition, CO_2 concentration-based prediction methods, smartphone positioning, Bluetooth, and Wi-Fi sensors. CV technology has been implemented in various building environments to achieve environmental sensing, perception, and occupant information detection, such as occupant count, location, and behavioral postures, as shown in Chapter 2. To address occupancy calculation in single-zone scenarios, Yang and Pantazaras et al. [46] evaluated four distinct methods: (1) a face detection system using Pan–Tilt–Zoom cameras; (2) an overhead video-based occupancy counting system; (3) a physical model based on CO_2 concentration; (4) a statistical model based on CO_2 concentration. Results demonstrated that the face detection method achieved the most stable performance and highest accuracy, with an R^2 value of 0.972, followed by the CO_2 concentration-based statistical model at R^2 0.938. Nevertheless, CO_2 concentration-based models remain particularly promising as a relatively low-cost and non-intrusive approach.

In large-volume spaces, such as shopping malls, exhibition halls, and transportation hubs, this type of building has the characteristics of large passenger flow, different regional functions, and dynamic changes in residence time and location of occupants. However, the existing air conditioning system is difficult to adjust in real time according to the actual number and location changes of occupants [47]. Therefore, achieving efficient ventilation control based on non-uniform occupant distribution necessitates real-time monitoring of fluctuations in pedestrian flow, occupancy locations, and zonal occupant density. Meanwhile, given the rapid dispersion of airborne pollutants in enclosed environments, rational adjustment of fresh air volume according to varying ventilation scenarios and demand can simultaneously fulfill indoor air quality requirements and mitigate pollutant transmission [48]. Wang et al. [49] proposed a CV-assisted occupant-centric environmental control framework, specifically an infection-resistive ventilation strategy driven by occupant-related data, as depicted in Figure 4.2. This strategy integrates facial detection technology with a bounding box-based trajectory recognition model to quantify occupant numbers, localizes facial positions in images, and determines entry/exit statuses. The results showed that the proposed strategy can dynamically maintain the infection risk near the target value of 1%. Compared with the ventilation strategy based on static Wells–Riley model, the proposed strategy can save 63.07% of ventilation energy consumption or 34.20% of HVAC system energy consumption. This study provided a methodology to balance ventilation energy efficiency with infection risk control, offering critical references for occupant-centric dynamic environmental management and building occupant health. Dynamically adjusting video extraction frequency according to the characteristics of entrances and exits and the change of human flow in different scenarios can effectively improve the accuracy of occupant information detection and the energy efficiency of ventilation systems, such as offices, conference rooms, hospitals, railway stations, and libraries. To further minimize cross-infection risks, research has expanded to behavioral monitoring, including mask compliance [50] and social distancing adherence. Kaviya et al.

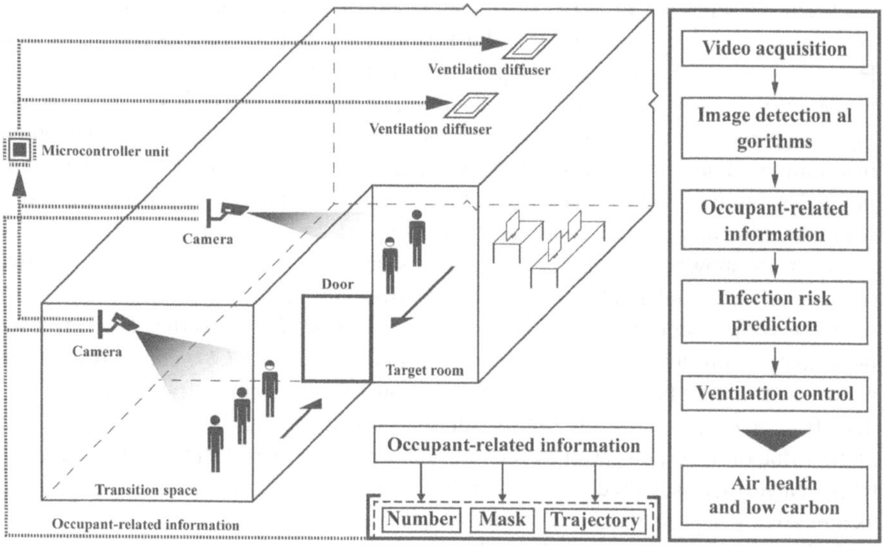

FIGURE 4.2 Framework for occupant centric control of indoor environments. (Source: Reference [49], Copyright 2023 by Elsevier.)

[51] developed a deep learning (DL)-based predictive model coupled with real-time risk analytics, employing digital cameras to detect interpersonal distance and mask-wearing behavior. This system identifies high-risk transmission zones, enabling quantitative evaluation of COVID-19 transmission risks under variable social distancing conditions.

Ventilation demands in large-volume spaces exhibit significant complexity, where the time lag between environmental feedback and corresponding HVAC system adjustments often leads to mismatches between dynamic indoor load variations and thermal environment control, thereby resulting in energy waste [52]. Meng et al. [53] proposed a real-time occupant load estimation method combined with a predictive HVAC control strategy. This approach applies DL to detect real-time occupant distribution in activity centers while establishing a dynamic occupant load estimation model. By predicting indoor temperature trends caused by load variations, the system adjusts the cooling capacity of HVAC system to achieve energy consumption reduction.

In shared small-to-medium-sized spaces such as classrooms, meeting rooms, and offices, while foot traffic variability remains relatively low, static factors like uneven occupant distribution across zones and dynamic factors like temporal variations in occupant density persist [54]. Integrating occupant occupancy and activity data with HVAC control parameters can enable precise on-demand ventilation, avoiding energy waste from conditioning unoccupied zones or over-conditioning occupied zones. Wang and Feng et al. [55] combined video surveillance with CO_2 concentration monitoring to detect room occupancy, as shown in Figure 4.3. These two detection methods operate under distinct conditions and cross-validate to enhance accuracy. Simulations and experiments validated the energy-saving potential of

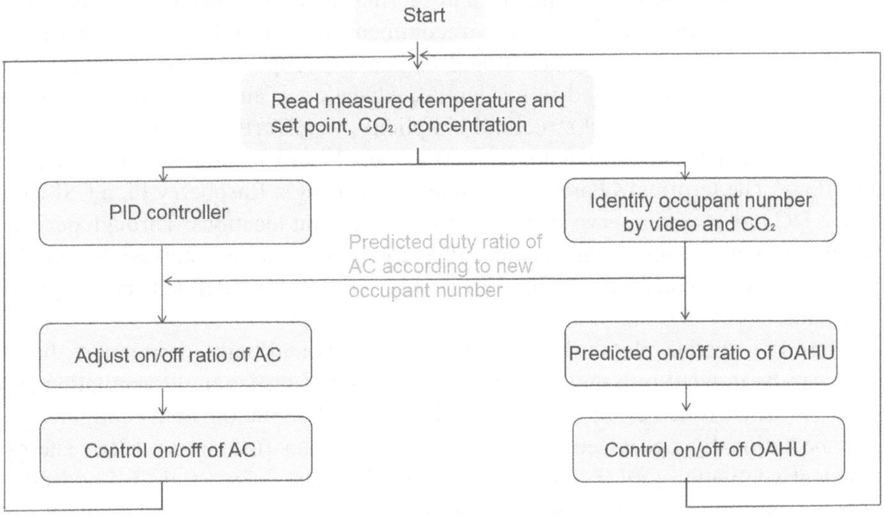

FIGURE 4.3 Predictive control flow chart based on the number of occupants.

occupancy-based predictive control, revealing that the hybrid video-CO_2 method achieves 91% detection accuracy, while the proposed predictive control reduces energy consumption by up to 6.7% for air conditioning and 85.2% for outdoor air handling units. Wang and Li et al. [56] developed an occupant position-oriented ventilation system that directly supplies fresh air to occupied zones, thereby optimizing energy efficiency. To determine the indoor occupant positions, a three-dimensional (3D) camera-based indoor occupancy localization system was developed. This stereo camera provides 3D positional information for targets. Field experiments collected classroom occupancy counts and dwell times, with hierarchical clustering analysis identifying distinct typical occupancy scenarios. Results demonstrated that total classroom occupancy duration constituted only 50.6% of operational hours. When regulating ventilation based on occupancy data, fresh air volume decreased by 60.2%, and total system energy consumption dropped by 67.7%.

Kong et al. [57] employed a side-by-side field experimental approach to quantify the performance of occupant-centric HVAC control strategies in commercial buildings. Results demonstrated that OCC strategies maintained satisfactory thermal comfort and perceived air quality, with user satisfaction exceeding 80%. Although daily energy-saving rates varied depending on occupancy sensor accuracy and outdoor environmental conditions, the weekly average energy conservation ranged between 17% and 24%. Building on these findings, researchers have focused on advanced air distribution methods. Yang and Liu et al. [58] integrated CV-based occupant distribution detection with a demand-driven regulation strategy and a SV system. The study utilized YOLOv5s-5.0 algorithm to dynamically detect occupant counts in office zones and adjust system operation parameters, enabling real-time responses to fresh air demand, indoor cooling load, and dynamic variations in fresh air-cooling load. Experimental results validated the energy-saving potential of this approach while ensuring air quality and thermal comfort. Furthermore, based on occupancy detection outcomes, it is recommended to install cameras at ceiling height near sidewalls to capture frontal or lateral occupant images. Split-type air conditioners, widely adopted in residential, educational, and office settings due to their cost-effectiveness and structural simplicity, were further investigated. Zhang and Gao et al. [59] proposed a novel smart air supply terminal tailored to occupants' positions. The terminal's hardware system, comprising a Raspberry Pi, a CSI camera, a DC motor, and a servo motor, identifies occupant locations. Through periodic automatic adjustments based on detected positions, simulation results confirmed that the smart terminal achieves optimal air supply across five predefined vertical planes in rooms.

Based on spatial relationships between occupants and built environment, buildings can be divided into multiple localized zones and utilize multi-ventilation systems. Whether zoned partitioning or multi-system collaboration, both configurations can be rationally optimized through CV-enabled data fusion to achieve energy-efficient ventilation [60]. For multi-zone HVAC systems, Cui et al. [61] addressed the current limitation of occupancy-based zonal demand ventilation by proposing a CV-assisted zonal DCV method. Image data from networked cameras deployed across zones is processed to identify total occupancy and zonal occupant distribution. The system's outdoor air control module and zonal regulation module jointly respond to ventilation demands at both global and sub-zone levels. This intelligent adjustment of outdoor air volume in target zones effectively reduces energy consumption. Lan et al. [62] employed simulation software to predict predicted mean vote (PMV) based on positional and zonal distribution data, integrating fuzzy logic control algorithms to modulate operational parameters of terminal air outlets in each zone. Results confirmed that this method could enhance thermal comfort. Zhou et al. [63] developed a demand-driven temperature control method tailored to uneven load distribution characteristics. By deploying variable air volume (VAV) systems to meet temperature requirements in occupied zones, this approach reduces supply airflow rates by approximately 20%–30% and cuts energy consumption by over 10% compared to conventional return air temperature-based control strategy.

The demand-based adjustment methods in the above studies are primarily based on the number and location of occupants. However, the actual activities of occupants within the building environment are quite flexible, with variations in their physical activity levels. This also leads to differences in heat gains and ventilation requirements. While CV algorithms can accurately identify human activities through visual data, detection performance is constrained by architectural obstructions and clothing occlusions, resulting in incomplete occupant images [64]. Tien et al. [65] implemented a vision-based DL framework to detect common occupant activities in office buildings, such as standing, sitting, walking, and napping, thereby constructing real-time occupant heat gain profiles. These heat gain profiles are fed into building energy management systems, with simulation results demonstrating improved accuracy in estimating heat gain values and CO_2 concentrations, enabling dynamic ventilation adjustments aligned with occupant activity levels. However, single-modality occupant data detection remains insufficient to support fully adaptive HVAC regulation [66]. Although vision-based models can gather partial occupant information, the diversity of required parameters often necessitates the simultaneous deployment of multiple CV models. To address this, Jung et al. [67] developed a vision-based multi-label detection framework for capturing both behavioral and apparel attributes of individual occupants. By tracking occupant-specific bounding boxes across video frames, the model annotates detected human activity and clothing types within each box. This method facilitates comprehensive occupant data acquisition across real-world scenarios, effectively reducing energy overuse and enhancing indoor environmental quality in diverse spaces.

For full ventilation systems, the primary objective is to fulfill air quality requirements for all occupants across the entire building space. However, such systems often fail to balance individual thermal preferences. Consequently, PV systems effectively complement conventional HVAC systems by better accommodating occupants' heterogeneous thermal comfort demands. PV systems deliver conditioned clean air directly to occupants' breathing zones [68], creating localized satisfactory micro-environments while achieving systemic energy efficiency [69]. The flow interaction between personalized air and thermal boundary layer may play an important role in inhaled air quality and energy efficiency performance of PV system. Nevertheless, interaction between PV-supplied personalized airflow and free convective flow around human body [70] may induce an uncomfortable draft sensation [71]. Xu et al. [72] employed schlieren imaging, Laser Doppler Anemometry, and tracer gas measurements to identify the airflow patterns in breathing zone and the effect of the interactions on ventilation effectiveness. Their study revealed that airflow interaction alters flow distribution in breathing zone and influences inhaled air quality, with these effects dependent on the positioning and direction of the non-uniform invading flow from the nozzle. Alsaad et al. [73] investigated the interaction between human microenvironment and personalized airflow using schlieren imaging to visualize PV-induced variations in near-body air velocity and turbulent intensity, as shown in Figure 4.4. The schlieren analysis demonstrated that PV-generated low-turbulence steady flow successfully penetrate the convective boundary layer surrounding the body. Thermal imaging visualization confirmed that PV-supplied cool air breaches

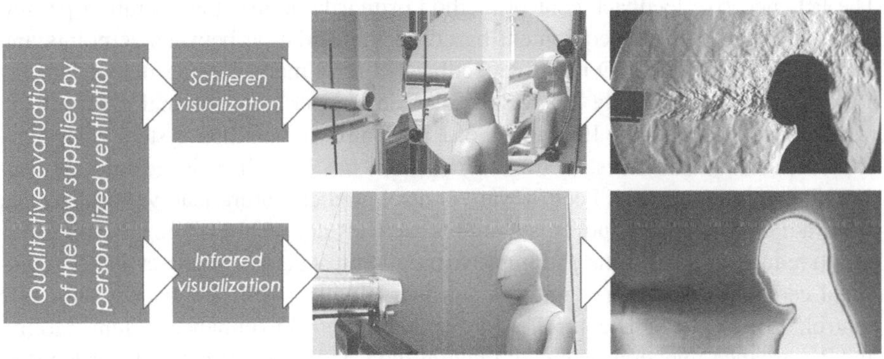

FIGURE 4.4 Evaluation of human interaction with personalized airflow in real-time video using schlieren imaging and thermography. (Source: Reference [73], Copyright 2020 by Elsevier.)

corona-shaped thermal boundary layer, reducing facial temperature and enhancing localized thermal perception.

PV devices hold application potential in office environments, yet existing PV devices predominantly target occupants in static seated postures [74]. In reality, office workers exhibit postural movements within a limited range during seated activities, such as upper torso rotation, sit-to-stand transitions, and head tilting. To address this, Gao and Li et al. [75] developed an imaging-based personalized directional airflow device that adapts to seated occupant movements. A camera recorded dynamically changing postures of subjects working at computers over 8-hour periods, enabling determination of target airflow zones (boundaries and dimensions) and optimization of nozzle angles, dimensions, and airflow parameters. Computational fluid dynamics simulations and experimental results demonstrated that this device successfully delivers air to target zones with enhanced delivery effectiveness. Compared to conventional PV systems, it reduces energy consumption by 88.2% while increasing supply air accessibility by 48% under equivalent energy inputs. Thus, for occupant-centric PV systems, adjusting airflow direction, velocity, and temperature can align with individual preferences [76]. CV technology can further enhance PV performance by detecting interaction between airflow and human body, monitoring occupant positions and physiological parameters, thereby holistically evaluating individual thermal comfort and executing adaptive controls to satisfy localized thermal demands for all occupants.

4.3.2 Advanced Control Methods of Air Conditioning and Ventilation Systems

In the regulation process of air conditioning systems, the primary control methods involved include traditional negative feedback control based on proportional-integral-derivative (PID), feedforward control, optimization control, fuzzy logic control, MPC, and reinforcement learning (RL) control. Advanced control methods help enhance the adaptability and rapid response capabilities of ventilation and air conditioning systems for demand-driven regulation based on occupant information. The PID negative feedback control method primarily adjusts the operating parameters of air conditioning equipment based on the deviation between setpoints and measured values. The PID negative feedback control method using indoor temperature setpoints is currently the widely adopted approach for thermal environment regulation. To address the limitations of PID control in handling system delays and unpredictable disturbances, feedforward control has been introduced into air conditioning control systems. Feedforward control predicts future load variations (e.g., solar radiation, occupant movement) and proactively adjusts the air conditioning system to reduce response time and improve regulation accuracy. Wang et al. [77] combined cerebellar model articulation controller (CMAC) neural networks with PID control, utilizing the PID controller for feedback control to enhance system stability and disturbance rejection, while employing the CMAC neural network for feedforward control to improve response speed and control precision. Simulation results demonstrated that this system exhibits high accuracy, real-time performance, strong anti-interference capability, and robustness. Meng et al. [53] introduced an occupant load control coefficient into air conditioning regulation strategies, adjusting the system's cooling capacity to reduce energy consumption while providing compensation for control system adjustments, thereby optimizing cooling capacity supply method. To account for individual differences among occupants, researchers have begun exploring temperature regulation methods based on thermal sensation prediction, primarily using physiological indicators such as mean skin temperature [78], local skin temperature [79], and heart rate variability [80] to predict thermal comfort.

To achieve specific optimization objectives, researchers have incorporated optimization algorithms into air conditioning control strategies. Ning et al. [81] proposed a neural network-based optimal operation strategy to identify the optimal setpoints for chilled water supply temperature, discharge air temperature, and fan static pressure. This approach minimizes the energy consumption of chillers and fans while maintaining indoor environmental comfort. Simulation results indicated that compared to traditional nighttime reset operation, this strategy achieved energy savings of approximately 10% under full-load conditions and 19% under partial-load conditions. Li et al. [82] developed a multi-objective optimization method, which integrates NSGA-II and TOPSIS techniques to determine the total equivalent ventilation rate for MV system combined with portable air purifiers under two scenarios. A modified Wells–Riley equation was adopted to predict the transmission risk of respiratory infectious diseases. Field experiments demonstrated that optimal solutions for the combined system could be derived based on varying indoor air quality requirements, seasons, and outdoor conditions.

In air conditioning control systems, fuzzy control can handle factors that are difficult to quantify precisely, such as indoor temperature, humidity, and human comfort. By establishing a series of fuzzy rules, it enables intelligent regulation of air conditioning equipment. Subramaniam et al. [83] designed and developed a novel fuzzy logic controller that achieves a nonlinear mapping between PMV, temperature, CO_2 concentration, and control parameters (recirculation flaps, blower's speed, and refrigerant mass flow rate) in vehicle HVAC systems. The results showed that the proposed controller could regulate PMV and CO_2 concentration within all predefined ideal thresholds while increasing energy consumption only marginally. Cui et al. [84] introduced a new fuzzy controller optimized via genetic algorithms, which refined the fuzzy rules of traditional fuzzy control to design an adaptive air conditioning fuzzy controller. The results demonstrated the effectiveness of this fuzzy controller in real-time control of indoor thermal environments, making it suitable for nonlinear indoor environmental systems. Lan et al. [62] used simulation software to predict the PMV of current occupants based on their locations and zonal distribution (Figure 4.5). By integrating fuzzy logic control algorithms to adjust the operating parameters of terminal air outlets in different zones, the system was shown to enhance thermal comfort for occupants.

MPC methods are based on the dynamic model of a system and calculate the optimal control sequence over a future period through online optimization. In air conditioning control systems, MPC comprehensively considers multiple objectives such as indoor temperature, humidity, and energy consumption, employing optimization algorithms to derive the optimal control strategy for overall system performance improvement. Jiang et al. [85] proposed an MPC-based optimization algorithm to

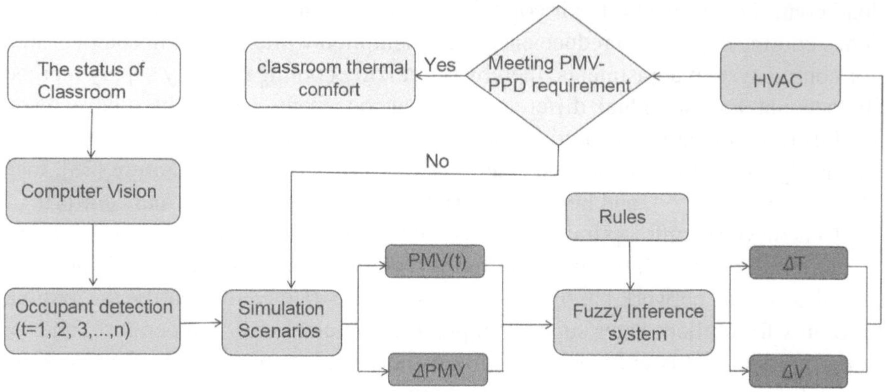

FIGURE 4.5 The framework of fuzzy logic control.

identify optimal control setpoints for VAV air conditioning systems. This strategy integrates model predictive methods with a neural network-based occupancy prediction model. Results demonstrated superior control responsiveness and higher energy-saving potential compared to nighttime reset control strategy, particularly under partial-load conditions. Yang et al. [86] introduced an adaptive robust MPC (ARMPC) method, which incorporates a model adaptation function during MPC controller operation. This function utilizes real-time building operation data to estimate uncertain parameters in the building model online. The approach integrates an additive uncertainty model representing disturbance uncertainties with the building model to achieve robust optimization. Compared to thermostats, when internal load uncertainty increased from 0% to 60%, energy savings achieved by ARMPC decreased from ≈20% to ≈15%. Non-adaptive MPC controllers maintained ≈20% energy savings with rising uncertainty but sacrificed thermal comfort. ARMPC's adaptive modeling and robust optimization avoided constraint violations caused by model inaccuracies and measurement disturbances.

RL enables control systems to interact with their environment, continuously optimizing control strategies based on feedback such as indoor temperature changes and energy consumption. In air conditioning systems, RL can learn user preferences and habits to automatically adjust parameters for personalized comfort. Additionally, RL possesses self-optimization and adaptive capabilities, improving control performance over time. Li et al. [87] established a TRNSYS–MATLAB co-simulation platform for a thermal storage air conditioning system and developed a time-of-use electricity pricing-based demand response strategy. This strategy integrates environmental conditions, thermal comfort, and energy consumption. The proposed RL algorithm adjusts thermostat setpoints during demand response phases. Results showed that the RL-based strategy achieved precise temperature setpoint control, reducing operating costs by 9.17% compared to non-thermal-storage systems with fixed setpoints. Yu et al. [88] reformulated the HVAC energy cost minimization problem for multi-zone commercial buildings as a Markov game, considering stochastic occupancy, thermal comfort, and indoor air quality. An HVAC control algorithm based on multi-agent deep RL (MADRL) was proposed to solve this game. The algorithm requires no prior knowledge of uncertain parameters and operates independently of building thermal dynamics models. Simulation results demonstrated its effectiveness, robustness, and scalability in optimizing energy costs while maintaining comfort and air quality.

4.3.3 Personal Comfort Systems

Full ventilation systems are designed to ensure collective occupant thermal comfort and health within built environments, rather than targeting individual needs. In contrast, PV systems specifically address individual airflow and fresh air requirements. Variations in occupants' perceptions of airflow and thermal demands arise primarily from differences in lifestyle habits and activity levels. Typically, occupants experience comfort within a range determined by ambient environmental conditions and their physiological or health status. For example, when office workers are required to perform activities in a local location for a long time, PV devices installed on the desk can deliver clean and regulated air to a locally occupied area, significantly improving the inhaled air quality of breathing area and providing local thermal comfort in conjunction with individual thermal preferences. However, PV systems face challenges in installation complexity and spatial layout constraints. As an alternative, PCS can integrate with HVAC systems to accommodate individual thermal preferences. This integration proves particularly valuable during extreme seasons (winter/summer), when occupants' perceptions of HVAC-maintained holistic indoor environments may diverge. PCS enhances personal comfort by selectively heating or cooling localized thermal environments based on occupants' body-segment-specific thermal sensations [89].

In recent years, PCS has received most of the researchers' attention, with studies validating their effectiveness in enhancing indoor thermal comfort and reducing energy consumption when appropriately implemented [89]. PCS can be categorized through multiple criteria, such as operational principles, user feedback mechanisms, and chronological development. However, a robust classification should prioritize the regulatory capabilities of PCS, including comfort optimization, energy efficiency, and economic viability. Rawal et al. [90] detailed the various types of PCS devices (heating, cooling, ventilation, heating with ventilation, and cooling with ventilation) while synthesizing localized environmental outcomes and subjective occupant feedback for each category. Their work further quantified the energy-saving potential and economic feasibility of various PCS devices. Reflecting advancements in personal environmental conditioning technologies, Yang et al. [91] chronologically reviewed the task ambient conditioning (TAC) system, personal environmental control system (PECS), PCS, and personal thermal management (PTM) system. The PECS is subdivided into PV and PCS.

Song et al. [92] discussed two PCS subtypes: personal cooling devices and personal heating devices. Based on heat transfer modes, personal cooling devices and personal heating devices were further classified into five categories: conductive heat transfer mechanisms, radiative heat transfer mechanisms, natural convective air-based (NCAB) mechanisms, and multi-mode convective devices (MCDs) combining two or more mechanisms. As can be summarized, current PCS mostly consist of devices designed to create localized personal microenvironments. Based on their usage state, they mainly include wearable, attachable, and standalone types. Wearable types primarily refer to clothing like wearable vests and protective suits incorporating cooling substances (phase change materials) [93, 94], mostly used to meet personnel cooling demands during summer. Attachable types mainly include devices with embedded electric heating pads [95] or cooling pads [96], along with portable desktop devices [97], panels [98], and nozzles [99] that need to be attached to personnel or furniture like desks and chairs. Standalone types mainly include hand and foot warmers [100, 101], warm air blowers [102], and desk fans [103]. These devices are independent of personnel but capable of stimulating local body parts. PCS devices are diverse, user-friendly, and can be easily deployed in different indoor environments.

While the use of PCS can reduce HVAC system energy consumption for holistic thermal conditioning, the effective operation of PCS depends on occupants' perceptions of localized thermal environments and ambient environmental parameters. Currently, most PCS rely on manual adjustments by occupants to regulate localized conditions. Occupants can manually adjust the temperature, humidity, velocity, and other parameters of PCS through control panel, remote control, smartphone app or other devices to meet thermal comfort needs. However, manual control is constrained by occupants' activity patterns and their limited understanding of the relationship between thermal discomfort causes and environmental parameters. Therefore, the precise automatic control of PCS can be combined with human comfort and the local thermal environment of occupants to better play the role of PCS and achieve the best benefits [104].

In recent years, the rapid development of CV, ML, and big data collection technologies has provided rich technical support for the realization of PCS automatic control. Previous studies mainly focused on personalized thermal comfort modeling in the traditional HVAC background and lacked research on the combination of personalized thermal comfort modeling and PCS automatic control. In Figure 4.6, Wang et al. [105] used personal conditioning devices (PCD) to provide local cooling to users according to their personal thermal preferences. Wrist temperature, heart

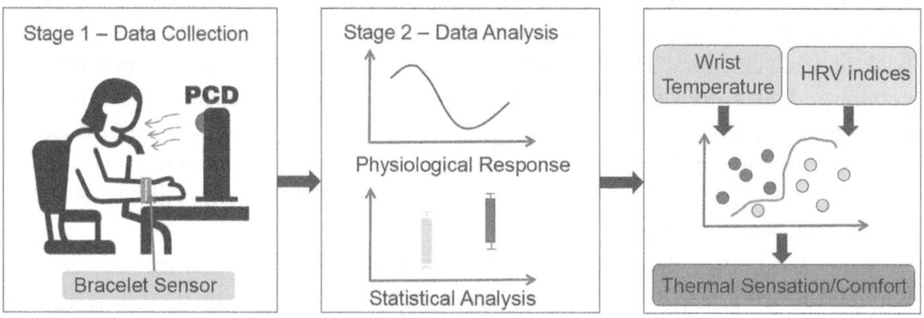

FIGURE 4.6 Real-time thermal sensation and comfort prediction using personal conditioning devices.

rate variability, and thermal sensation/comfort questionnaires were collected to predict thermal comfort of PCD users using ML methods. Studies have shown that the thermal stimulation of PCD is uneven and localized compared to the thermal stimulation of centralized HVAC systems.

The automatic control of PCS usually adopts the combination of visual sensors (such as cameras) and contact sensors to detect indoor environmental parameters and human information and automatically adjust system parameters. Compared to manual occupant adjustments, this approach enables intelligent real-time environmental perception and parameter optimization without requiring user intervention. But at the same time, the realization of automatic control needs to train the historical data of manually controlled PCS with different user feedback in the previous period, to provide a more accurate and comfortable indoor environment. Wu et al. [106] developed an automated PCS (APCS) by integrating PCS with personal comfort models for localized microenvironment regulation, as depicted in Figure 4.7. The dual-spectrum thermal imager is used to identify forehead, eyes, nose, and cheek areas in infrared and RGB images and is transmitted to the central control computer. By synthesizing PCS operational states, skin temperature, and environmental parameters as inputs, the system predicts individual thermal sensations and automatically adjusts PCS settings accordingly. The results of climate chamber experiments show that PCS operates similarly under automatic and manual control conditions. Therefore, the APCS developed is consistent with the results of human active control, and good automatic control can be achieved.

The effectiveness of localized environmental regulation varies across different PCS and even among different usage scenarios of the same PCS. For instance, desktop fans and foot heaters exhibit distinct regulatory outcomes, leading to divergent real-time occupant feedback. Consequently, selecting appropriate sensing technologies

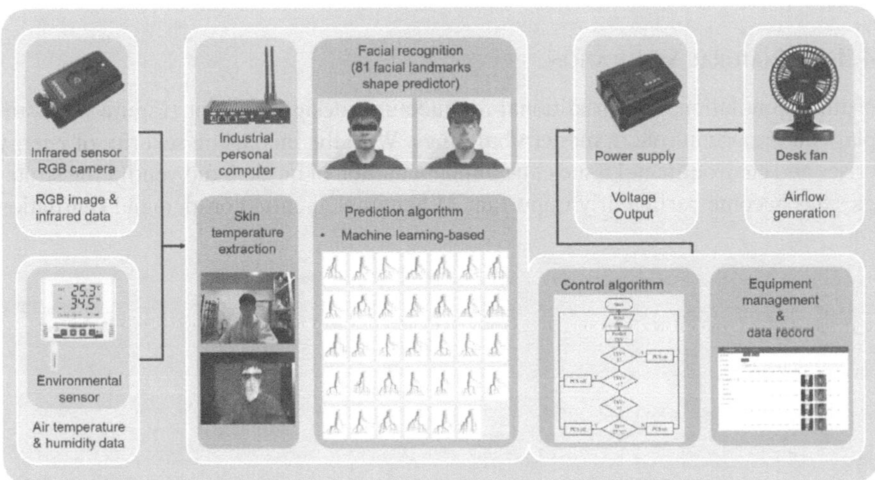

FIGURE 4.7 Automatic adjustment architecture of personal comfort system. (Source: Reference [106], Copyright 2023 by Elsevier.)

is critical. Aryal et al. [107] systematically compared sensing and ML approaches to develop individual comfort models under three PCS operational modes: baseline (no fan/heater), active fan, and active heater. The sensor devices arranged indoors include environmental sensors, wrist skin temperature sensors, and facial thermal imaging. The results showed that the prediction accuracy of environmental sensor data, such as ambient temperature, humidity, wind speed, and radiation temperature, was relatively high compared to wearable devices monitoring wrist skin temperature or thermal imaging monitoring skin temperature in different areas of the face. Furthermore, constructing separate personalized thermal comfort models for scenarios with and without PCS devices is essential to enhance occupant satisfaction. Wu et al. [108] investigated variations in feature parameter selection across PCS types and usage patterns. Three ML algorithms were used to develop personal thermal comfort prediction models for three scenarios based on infrared temperature: fan off, fan blowing toward the face, and fan blowing toward the hands. The study revealed that incorporating parameters representing specific contextual information or independently modeling data categories for distinct individuals and PCS usage modes significantly improves model predictive performance.

Shetty et al. [109] utilized a wireless sensor and actuator network (WSAN) to collect data on desktop fan usage preferences, occupant status, and indoor/outdoor environmental conditions in shared office spaces. This WSAN framework enables the deployment of PCS in built environments, allowing manual, automated, or hybrid activation of PCS based on collected data, thereby enabling intelligent and rapid responses to localized environmental conditioning through occupant thermal comfort feedback [110]. To predict individual preferences for desktop fan usage, fan operational states were modeled as a classification problem, while fan speed was modeled as a regression problem. The prediction results based on decision trees (DTs), random forests (RFs), and boosted trees demonstrated that the RF algorithm achieved the best performance with an average accuracy of 97.73% in predicting fan states.

4.3.4 Natural Ventilation

Natural ventilation, as a traditional architectural design concept (Figure 4.8), still plays an important role in modern buildings. With the increasing severity of energy crises and environmental issues, the implementation of on-demand ventilation strategies has become particularly important. When implementing on-demand ventilation

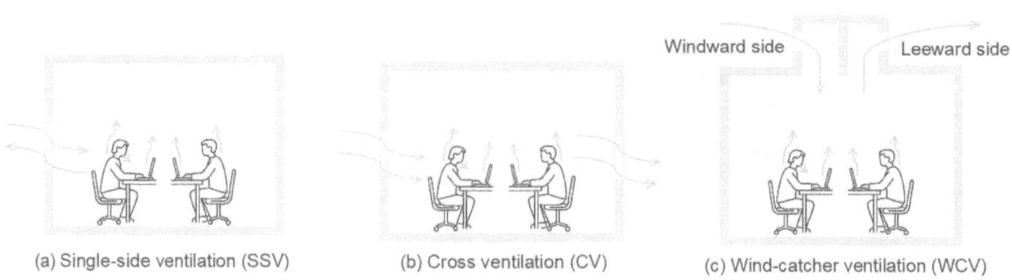

FIGURE 4.8 Three common types of natural ventilation.

strategies, the rational use of natural ventilation resources is an efficient and environmentally friendly method. It promotes air movement and achieves ventilation by utilizing the natural temperature difference and wind pressure difference between indoors and outdoors. This strategy can not only significantly reduce energy consumption and dependence on air conditioning systems but also effectively lower the operating costs of buildings, achieving dual benefits of economy and environmental protection. At the same time, it can ensure thermal comfort and air quality of indoor environment. Natural ventilation can effectively dilute indoor pollutants, such as volatile organic compounds (VOCs), carbon dioxide (CO_2), and other substances that may affect indoor air quality, thereby creating a healthier and more comfortable indoor environment for occupants. By carefully designing the building layout and structure, such as reasonably placing windows, vents, and shading devices, the potential of natural ventilation can be maximized [111].

The window opening grade plays a decisive role in the efficiency of natural ventilation. It not only affects the rate of air exchange between indoors and outdoors but also directly relates to the energy efficiency of building and the comfort of indoor environment. In winter, the outdoor temperature is generally lower than the indoor temperature. If the windows are opened too wide at this time, the heat inside the room will be quickly lost to the outside through heat transfer and convection. This phenomenon is referred to as "heat loss", which increases the burden on the heating system and leads to higher energy consumption to maintain the desired indoor temperature. In extremely cold climates, heat loss can be particularly severe, causing a rapid drop in indoor temperature. This not only affects the thermal comfort of occupants but also increases heating costs and leads to energy waste. Therefore, controlling window opening grade reasonably is crucial for maintaining indoor thermal comfort and reducing energy consumption in winter.

In summer, the outdoor temperature is usually higher than the indoor temperature. However, by implementing effective "natural ventilation" strategies, such as opening windows for ventilation at night or when the outdoor temperature is relatively low, the cool external air can be effectively utilized to lower the indoor temperature. This method reduces the dependence on mechanical ventilation systems, thereby decreasing electricity consumption and associated operating costs. The use of natural ventilation to regulate indoor temperature in this way is known as "passive cooling". Passive cooling not only provides "free cooling" but also significantly improves indoor air quality by reducing the accumulation of indoor pollutants. Good indoor air quality has a positive impact on the health and comfort of occupants.

Therefore, monitoring the window opening grade (Figure 4.9) is crucial for implementing effective natural ventilation strategies [112]. By real-time monitoring of the window opening status, the rate of indoor–outdoor air exchange can be more precisely controlled to optimize the effectiveness of natural ventilation.

Window opening behavior represents the most direct and prevalent stochastic method for utilizing natural ventilation. Early research on window opening behavior primarily focused on analyzing influencing factors, including outdoor temperature, outdoor relative humidity, indoor temperature, indoor relative humidity, CO_2 concentration, PM2.5 concentration, wind speed/direction, and noise levels [113]. Notably, numerous studies have demonstrated that outdoor environmental factors

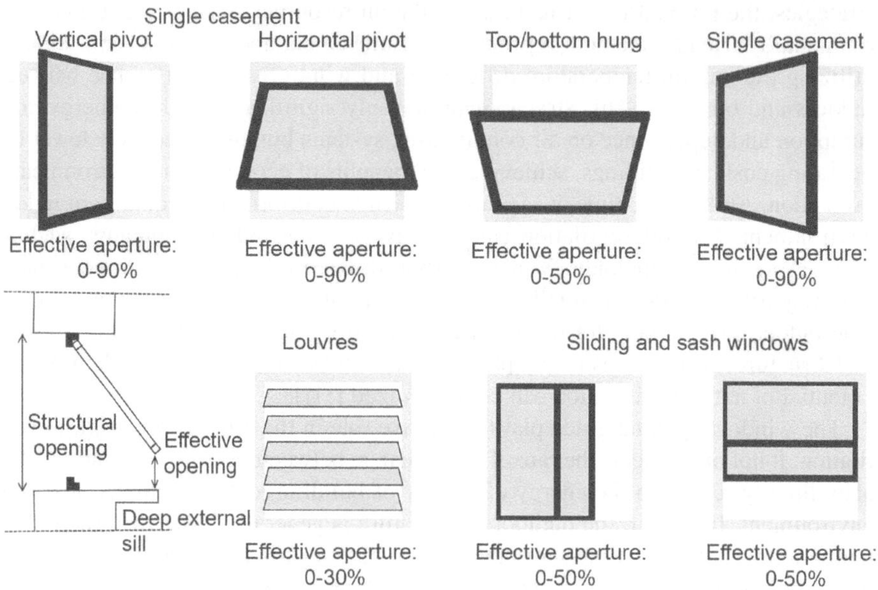

FIGURE 4.9 Main types of windows and the range of effective window area.

exert more significant impacts on window opening behavior [13, 114–117]. Fabi et al. [118] categorized influencing factors into five groups (physical environment, contextual, psychological, physiological, and social) and identified three predominant determinants affecting window opening behavior: outdoor temperature, indoor temperature, and CO_2 concentration.

As early as 1990, Fritsch et al. [119] first developed a mathematical model using a Markov chains model to predict window opening angles. They established four different Markov chains to link outdoor temperature with occupant behavior (opening windows at different angles). The changes in the impact of environmental parameters on window opening behavior indicate that occupants' window opening habits vary across different regions and seasons. When transferring the window state model to buildings in other seasons and regions, the reliability of the model should be carefully examined. Multivariate logistic regression analysis provides a powerful tool that can consider multiple influencing factors simultaneously to predict window state [114, 120]. Compared with traditional univariate logistic regression, the multivariate model can capture the interactions and synergistic effects among different factors, thereby improving the accuracy and reliability of predictions. In the study by Shi et al. [120], they used a multivariate logistic regression model to predict window states in different seasons, including cooling season, transitional season, and heating season. The results showed that the prediction accuracy for each season exceeded 70%, indicating good predictive performance of the model. Especially in the transitional season, the model's accuracy reached as high as 98%. However, in practice, considering more influencing factors does not necessarily lead to better predictions [121]. Haldi et al. [122] found that in naturally ventilated buildings, using both indoor and outdoor temperatures as predictive variables in the logistic regression model may lead to some counterintuitive conclusions. For example, the model may erroneously predict that as the external temperature rises, the frequency of people opening windows decreases. This prediction does not match the actual situation because in naturally ventilated buildings, the indoor temperature often rises with the increase in outdoor temperature, and people are more likely to open windows to achieve a cooling effect. Due to the high correlation between indoor and outdoor temperatures, these two factors may interfere with each other in the model, making it difficult for the model to accurately capture their independent impacts on window opening behavior. Therefore, they suggested that in free-running building environments, indoor and outdoor temperatures should not be used simultaneously as variables to predict window opening behavior. They believed that other factors that can independently influence window opening behavior should be sought, or only one of the temperature variables should be selected for analysis to avoid prediction distortion caused by strong correlation between variables. Thus, the selection of predictive variables should be determined based on the characteristics of the building and the season. Liu et al. [113] proposed a hybrid model combining logistic regression and probability distribution models for analyzing the window opening state of buildings. The results showed that outdoor temperature, relative humidity, and PM2.5 concentration significantly affected the window opening state. The outdoor temperature had the highest prediction accuracy for the logistic regression model, reaching 86.7%. The probability density functions of window opening states varied under different outdoor temperatures. When the outdoor temperature was between 20°C and 25°C, the mean and variance of the probability density function were the largest. However, this method may be affected by the reflection of window glass and objects in the room, leading to errors in the image recognition process.

Window status monitoring technology based on ordinary cameras can be broadly divided into two types: traditional methods and innovative methods [123]. Each of these methods has its characteristics and is suitable for different scenarios and requirements. Traditional methods mainly rely on image processing techniques to identify the status of windows by analyzing the distribution and intensity of pixels in the images captured by cameras. The research work of Bourikas et al. [124] provides an innovative solution for monitoring the status of windows in naturally ventilated offices. They developed a low-cost camera-based system that can predict the open and closed states of windows with an accuracy rate as high as 90%–97% by analyzing the distribution and intensity of pixels in the images. This achievement has been verified in tests conducted in both winter and summer, demonstrating that the system can maintain high performance across different seasons. However, the system also faces some challenges when deployed in practice. Due to the fixed position of the camera, the area of the exterior wall that can be monitored is limited, which may restrict the amount of sample data and thus affect the comprehensiveness and accuracy of the monitoring.

Zheng et al. [125] focused their research on the calculation of the percentage of window opening. They proposed a method that can identify the open state of most windows with an error of about 5% based on the intensity distribution of window pixels. This method performs well in monitoring sliding windows but is currently only applicable to windows that are regularly distributed on the facade. This means that for irregularly distributed or special types of windows, the method may require further optimization and adjustment. Luong et al. [126] developed a method for monitoring the condition of windows on building facades based on CV technology. This method uses image segmentation techniques to automatically segment individual window images and can monitor window status with an accuracy rate of 89%. However, the limitation of this method is that it is mainly applicable to occluded windows and requires manual adjustment of the window status threshold, which limits the automation and scalability of the method. Donges et al. [127] developed and validated a camera-based system for detailed monitoring of the opening angle of windows, the position of shading devices, and the operating status of lighting equipment in buildings. This system can handle multiple windows and blinds simultaneously and allows for continuous monitoring and deployment. The device can also provide detailed estimates or modeling of natural ventilation rates and supports the development of occupant behavior models.

In practice, traditional window monitoring methods that rely on pixel intensity distribution have certain limitations in real applications. Since they mainly identify window status by analyzing the brightness of pixels in the image, they are very sensitive to environmental illumination and weather conditions. For example, under strong light or in shadows, the brightness of pixels may change, leading to misjudgments. Similarly, in adverse weather conditions such as rainy, foggy, or snowy days, the image quality may be affected, further reducing the accuracy of monitoring.

To overcome these challenges, DL algorithms have gradually been employed for monitoring window status. DL algorithms, especially convolutional neural networks (CNNs), are capable of automatically extracting features by learning from a large amount of image data, without the need for manually designed feature extraction rules. This means that even under varying illumination and weather conditions, DL algorithms can maintain high accuracy and robustness. In addition, DL algorithms can adaptively learn and continuously improve performance through ongoing training and optimization.

Markovic et al. [128] proposed a window opening model based on DL methods to improve the shortcomings of existing models and enhance the generalization ability and robustness to individual occupants. Tien et al. [129] developed and validated a DL method that can automatically identify the status of individual windows in real time, with an accuracy rate of 97.29%, demonstrating the significant advantages of DL in automatically recognizing window status. Building on the aforementioned research, Tien et al. [130] developed and implemented a DL-based real-time monitoring system aimed at improving the accuracy and efficiency of monitoring the open and closed status of windows inside buildings. They selected the faster region-based CNN (faster R-CNN) as the core algorithm and used a large-scale image dataset for model training. In a series of experiments, the system achieved an accuracy rate of 92.2% in detecting the open and closed status of windows. In addition, the study also employed DL methods to detect human activities. This method can help remind occupants or managers to avoid unnecessary heat loss caused by opening windows. Moreover, the method can also predict indoor carbon dioxide concentrations and suggest appropriate natural ventilation strategies for occupants.

Sun et al. [131] proposed an innovative method for monitoring window status in cold regions. They developed an automated real-time monitoring system by combining CNN and image processing techniques. Through experiments, they found that in the transitional season, buildings in cold regions tend to use larger window opening angles, and the probability of opening windows facing southeast is higher than other directions. This finding is of great significance for optimizing natural ventilation strategies in cold regions. In addition, the method has strong scalability and can be combined with building energy consumption and other factors for analysis and application in various application scenarios. Wang et al. [132] further explored the application of artificial intelligence (AI) in building environment monitoring. They developed a system based on AI-driven cameras and deep vision algorithms that can monitor occupant behavior and the open/closed status of windows in real time. By using a shallow artificial neural network (ANN) to build a predictive model, the system can predict indoor PMV, heating energy consumption, and carbon dioxide concentration based on different occupant behavior patterns and window open/closed conditions. The application of this technology allows the window opening controller to adjust the window opening degree according to real-time data, thereby effectively maintaining indoor air quality and reducing unnecessary heat loss and waste. Compared with traditional control strategies, the window opening controller developed by Wang et al. shows excellent performance in energy saving. According to research data, heating energy consumption can be reduced by 0.6%–29.0%, and indoor thermal comfort can be improved by 0%–58.8%. This achievement not only demonstrates the potential of AI technology in the field of building energy saving but also provides new ideas and solutions for future intelligent buildings. It is worth mentioning that the research work of Tien et al. and Wang et al., who applied DL models to window status monitoring, not only improved the accuracy of monitoring but also provided new solutions for building energy saving and indoor environment quality optimization. These studies show that the application prospects of DL technology in the field of construction are broad, and it can provide strong technical support for the realization of more intelligent and automated building management systems.

Compared with traditional optical cameras, window monitoring methods based on thermal infrared (TIR) cameras have a series of significant advantages. First, TIR cameras can capture the infrared radiation emitted by objects, which means that they are not limited by visible light and can observe the status of windows even in low-light or completely dark environments. This characteristic makes TIR cameras particularly suitable for monitoring windows at night, as the poor light conditions make it difficult for optical cameras to obtain clear images. Second, TIR cameras can penetrate certain types of glass, allowing them to detect whether windows are open even if they are covered by curtains or blinds. This is particularly important for applications that need to monitor the status of windows for energy management and security surveillance. In addition, the thermal imaging capability of TIR cameras also enables them to detect temperature differences around windows, which helps to identify whether windows are well-sealed and whether there are issues such as air leakage or heat loss. This is of great significance for improving the energy efficiency of buildings and the comfort of occupants. Chen et al. [133] have made significant progress in the field of window status monitoring technology. They adopted an innovative remote sensing method using TIR cameras to identify indoor temperature. This technology achieves precise quantification of indoor temperature by analyzing the infrared images of fully open windows, especially in heating, overheating, and cooling states. The research results show that in heating and overheating states, the absolute deviation between the predicted indoor temperature and the actual temperature can be controlled within 0.5°C, which is an excellent level of accuracy in the field of indoor temperature monitoring. However, in the cooling state, the measurement deviation is relatively large, which may be due to the large temperature difference between indoors and outdoors during cooling, increasing the difficulty of parsing infrared images.

In summary, window status monitoring technology has evolved from relying on traditional image processing techniques to adopting advanced DL algorithms. Although traditional methods are effective under certain conditions, they are susceptible to light and weather conditions, limiting their scope of application and accuracy. Innovative methods represented by DL algorithms, on the other hand, provide more accurate and robust monitoring capabilities by automatically extracting image features, maintaining high accuracy even under changing environmental conditions. As AI and DL technologies continue to advance, window status monitoring technology will become more mature and diversified, thus better meeting the needs of different building types and usage requirements. These technologies will be integrated with Internet of Things devices, wireless communication technologies, and advanced data analysis methods to provide key support for building energy-saving, indoor environment control, and intelligent building management. This will drive the construction industry toward greater efficiency and environmental friendliness. Personalized and customized solutions will learn and adapt to the behavior patterns and preferences of occupants, automatically adjusting window opening grade and optimizing indoor environment. Intelligent building management systems will achieve more refined energy management and environmental control, reduce energy waste, and improve indoor air quality and comfort.

Moreover, data-driven approaches will help building managers and occupants make more informed decisions. The importance of interdisciplinary cooperation, the updating of policy regulations, and the involvement and feedback of users will also be emphasized. Education and training will adapt to technological development to ensure that users can fully utilize these advanced technologies. In summary, the future development of window status monitoring technology will create a more comfortable, healthy, and energy-efficient living and working space for occupants, while making an important contribution to the sustainable development of construction industry.

4.4 DATA-DRIVEN THERMAL COMFORT PREDICTION METHOD

HVAC systems play a vital role in providing satisfactory indoor thermal comfort and air quality for occupants, but they also cause a lot of energy consumption [134], and there is the problem of personalization differences caused by factors such as human habit preference. Especially for shared buildings, such as offices, classrooms, and dormitories. At present, HVAC systems use a fixed temperature setpoint to meet the thermal comfort needs of most occupants. According to a survey of more than 90,000 respondents in about 900 buildings, roughly two-fifths of building occupants felt the thermal comfort interfered with their ability to complete their work, and the dissatisfaction rate of temperature reached 39% [135]. When building an indoor environment changes dynamically, the one-size-fits-all control approach of HVAC systems has limitations in meeting the preferences of occupants. To better meet the thermal comfort of most occupants, the thermal sensation prediction method was used to predict the thermal comfort needs of personnel in part of the studies mentioned in Section 4.2. This chapter will introduce the main thermal sensation prediction methods.

Among the traditional thermal comfort evaluation models, PMV model proposed by Fanger [136] and adaptive thermal comfort model proposed by De Dear and Brager [137] have been widely used and identified by Building Environmental Performance Assessment Criteria [138, 139] as the main methods to evaluate the thermal comfort created by HVAC systems. This type of model has been used by many researchers and has proved to be reliable under uniformly controlled stable conditions. Traditional thermal comfort assessment models are mainly based on basic physical principles, which are suitable for standard environments and theoretical derivation, but they have shortcomings in dealing with individual differences, dynamic environmental changes, and complex thermal sensation factors. With the development of CV technology, non-invasive measurement methods can more accurately collect personnel information under non-uniform environmental conditions to improve the accuracy of thermal comfort assessment. The researchers assessed thermal comfort using a mobile robot quipped with a thermal camera and an RGB-D camera, as well as thermal and humidity sensors and developed a method of estimating the clothing insulation using a depth-thermal camera system. The results of the comparison between PMV predicted value and user questionnaire show that the method has a certain degree of accuracy [140]. Although the evaluation accuracy has been improved to a certain extent by this method, many parameters that affect actual thermal perception have not been fully considered for the dynamic and non-uniform distribution of built environment.

To overcome these problems, AI technologies such as CV and ML are being used to capture personal characteristics data as well as data about the dynamic built environment. The problem of traditional thermal comfort evaluation methods can be solved by a data-driven method to obtain a more accurate thermal comfort prediction model. Compared with the traditional thermal comfort evaluation methods, the data-driven thermal comfort prediction model can use various monitoring data to analyze multiple input parameter combinations and find the parameter type with the best prediction effect [141]. In addition, while there are differences in occupant thermal preferences, the data-driven model can achieve predictions at both individual level and population average. The data-driven approach involves cleaning and preprocessing the collected data and discovering patterns and relationships between the data by training the model algorithmically. Therefore, data collection is the basis of model building. In addition to the four environmental parameters (air velocity, air temperature, average radiation temperature, and relative humidity) and individual factors (clothing resistance and metabolic rate) involved in the traditional PMV–PPD

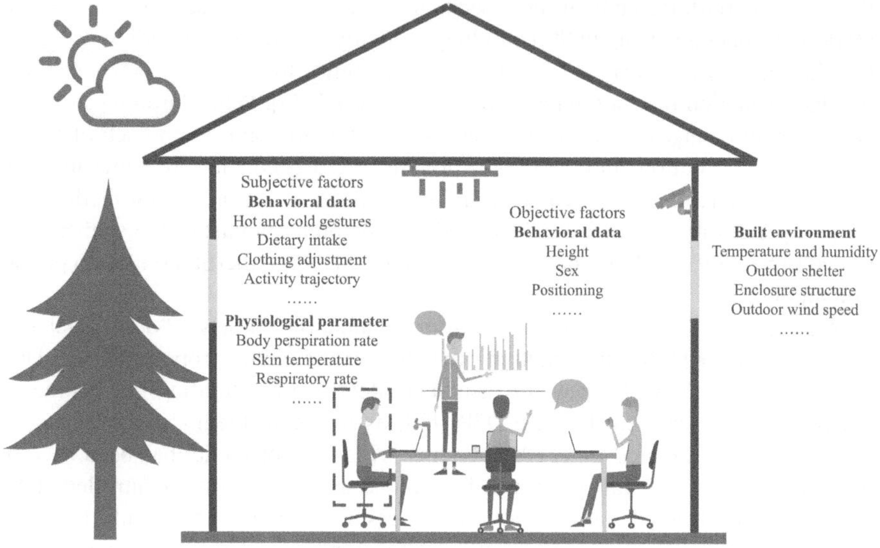

FIGURE 4.10 Factors affecting the prediction accuracy of occupants' thermal comfort.

model, personal parameters and environmental parameters are also key to establish a thermal comfort prediction model. At present, most researchers have explored various types of influence parameters, including personal parameters and environmental parameters, to generate a data-driven thermal comfort prediction model [142]. As shown in Figure 4.10, factors affecting prediction accuracy can be divided into subjective factors and objective factors.

The subjective factors can be divided into human physiological parameters and human behavior data. Both types of parameters are closely related to occupants. Physiological information includes body temperature, skin temperature in different parts, heart rate, respiratory rate, sweat production, electrodermal activity, and blood pressure. These parameters are not controlled by human; it is physiological feedback information generated unconsciously by occupants in an architectural environment [143]. Therefore, many researchers have proposed to develop individual thermal comfort prediction models using heart rate, electroencephalogram (EEG), or the most representative local skin temperature [144–146]. Human behavior is self-adjustment made by occupants according to the temperature change of building environment to maintain a relatively stable comfort level. This kind of behavior change is a manifestation of the active adjustment of human body, including changes in posture and activity intensity [140], aiming to make the body adapt to different environmental conditions and ensure its thermal comfort. Objective factors already exist objectively in building environment and will not change due to occupant willingness, mainly including built environment (envelope structure parameters, ambient temperature and humidity, air velocity, etc.) and personnel information (gender, age, height, personnel distribution, etc.) [147]. These data are the basis for the development and optimization of thermal comfort prediction models.

Adjusting HVAC system operation according to the predicted thermal comfort index can not only achieve building energy saving but also take into account thermal comfort of occupants. Existing studies have used ML models or DL models to predict thermal comfort, such as support vector machine (SVM), decision tree (DT), k-nearest neighbor (KNN), RF, backpropagation (BP), CNN, ANN, and other algorithms. Feng et al. [148] reviewed 25 studies published in the last ten years and summarized the data-driven modeling process of human thermal comfort models in the built environment into three steps: experimental design, data collection, and model establishment. In model establishment, information about thermal sensation, thermal preference, or thermal acceptability as reported by a person is often considered a categorical value. Ren et al. [149] used three ML classification algorithms (BP, RF, and CNN) to predict human thermal comfort. The data source for model training was an experimental database containing physiological parameters of ten male and ten female volunteers under three temperature conditions (a total of 3,600 sets of data). The results show that RF model has a significant advantage with an overall accuracy of 89.3%, which is 8% higher than the overall accuracy achieved by CNN and 14.1% higher than BP. In addition, comparing the effects of different parameter combinations on model accuracy, it was found that the prediction accuracy could be improved to 95.6% by combining the datasets of all physiological parameters, such as environmental parameters, electrocardiogram (ECG), and EEG.

Katić et al. [150] explored two different approaches based on output data (thermal sensation and thermal comfort votes) and five different input variables composed of multiple influencing variables (air temperature, relative humidity, time, radiation temperature, PCS, average skin temperature, hand skin temperature). This model was developed by SVM with four different kernel functions (Linear, Quadratic, Cubic, and Gaussian) and ensemble algorithms (boosted trees, Bagged trees, and RUS boosted trees). As shown in Figure 4.11, the results showed that the best median accuracy of 0.84 for the human thermal comfort prediction model was achieved by using the RUS Boosted tree. The model with input combinations consisting of PCS control behavior, average skin temperature,

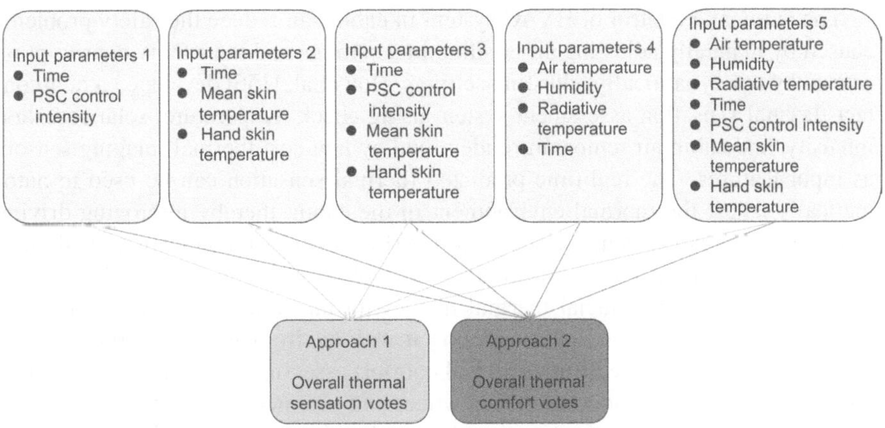

FIGURE 4.11 Input and output parameters of human overall comfort prediction model.

and hand skin temperature had the best prediction accuracy. However, the limitations of this study are the small number of participants and insufficient data support.

When training the prediction models, some ML and DL algorithms need through hyperparameters to maximize the model performance [151]. Luo et al. [152] applied nine ML algorithms and three data sampling methods to predict 3-point and 7-point thermal sensation vote (TSV) in ASHRAE Comfort Database II. The results show that the prediction accuracy of RF for 3-point TSV and 7-point TSV is 66.3% and 61.1%, respectively, which is the best among the tested algorithms. This study also addressed other common considerations in ML comfort model establishment, such as tuning hyperparameters, splitting of training and testing data, and encoding methods. The results found that tuning parameters and selecting input features are important for ML models. Yang et al. [153] conducted a comparative analysis of four ML approaches, including SVM, DT, an ensemble algorithm, and KNN. By incorporating both physiological parameters related to human skin temperature and environmental variables as model inputs, the researchers developed a TSV model that achieved a remarkable prediction accuracy of 95.8%. Furthermore, the hyperparameters of ML algorithms were tuned using a Bayesian optimization technique, specifically designed to enhance model performance.

In the field of combining CV with human thermal sensation prediction, Yang et al. [154] proposed a novel algorithm based on OpenPose for real-time and contactless measurements of human thermal comfort status, and the technical details are shown in Section 2.3.2. By combining with the thermal discomfort perception obtained from the questionnaire survey of hundreds of subjects, 12 thermal discomfort-related poses were defined. Li et al. [155] trained individual thermal comfort prediction model by collecting facial image data of cold, comfort, and warm categories, respectively, in a controlled laboratory environment through thermal imaging technology. In this study, a spatial attention mechanism module (Attention-ResNet50) was incorporated to enhance the prediction accuracy. Thermal comfort tests and predictions were also conducted for different genders. The results show that the prediction accuracy of female thermal comfort model is higher which is reflected in that female facial features are more sensitive to ambient temperature changes.

The automatic control of HVAC system in cabin can reduce the safety problems caused by manually adjusting air conditioning settings and the energy consumption caused by not dynamically adjusting settings. Hou et al. [156] developed a non-contact thermal sensation assessment system using cheek temperature, solar radiation intensity, and cabin air temperature identified by low-cost thermal imaging sensors as input features. The real-time predicted thermal sensation can be used to automatically adjust the thermal environment in the cabin, thereby improving driving safety and energy efficiency. Zhou et al. [157] introduced a non-invasive and dual-phase prediction model. First, the model first employs a DL neural network to extract pertinent features from facial thermal images for an initial prediction. Then, this prediction model is refined by integrating it with environmental data through ML algorithms, facilitating real-time thermal comfort assessment. The results show that the characteristic data collected during winter and summer seasons enables the construction of a comprehensive dataset to build a highly accurate predictive model for on-demand control of the cabin HVAC system.

To sum up, traditional indoor environment control methods cannot meet the actual thermal comfort needs. The data-driven thermal comfort prediction model can predict the actual thermal comfort state of individuals or groups by analyzing and learning various subjective and objective parameters. Adjusting the operating parameters of HVAC systems according to the thermal comfort predicted by the model can create a more comfortable indoor environment and help to minimize unnecessary energy consumption. At the same time, the combination of full ventilation with PCS and PV system can meet the individual thermal comfort of local environment and maximize the potential of thermal environment regulation equipment in improving individual thermal comfort and system operation energy saving.

4.5 CONCLUSIONS AND FUTURE DIRECTIONS

This chapter first introduces the relationship between occupant behavioral characteristics and HVAC system operations, followed by a detailed discussion of demand-driven regulation case studies for full ventilation, PV, PCS, and natural ventilation based on occupant information. The regulation performance is evaluated in terms of thermal comfort, air quality, and energy consumption. Application examples include the following: (1) adjusting fresh air supply volume and HVAC operating parameters based on occupancy and activity levels. (2) Regulating window openings and equipment settings according to occupancy and environmental parameters. (3) Regulating HVAC operation parameters and localized environmental control devices (e.g., fans) using physiological parameters (e.g., skin temperature) and environmental data. The chapter also introduces advanced control methods for HVAC systems that integrate detected environmental parameters and occupant information. Finally, it provides a comprehensive review of existing data-driven thermal comfort prediction models, covering research methodologies, model categories, and predictive performance. By examining occupant behavior types relevant to demand-driven indoor thermal environments, regulation strategies for HVAC systems and localized environmental control devices, and thermal comfort prediction approaches, this chapter offers insights for implementing occupant-centric HVAC control strategies.

In the field of building HVAC system operation, current research integrating occupant information with system regulation remains exploratory. Future studies could focus on the following directions: (1) thermal comfort prediction models should account for geographical and environmental diversity, as well as occupant demographics, to enhance practical applicability. (2) PV systems and individual thermal comfort systems offer greater flexibility. There are few studies related to the comprehensive adjustment of PV systems or PCS, and full ventilation systems. (3) More field studies are needed to quantify control system performance, with analysis of variations in regulation performance across regions and building types to validate real-world feasibility. (4) Current demand-driven regulation relies primarily on occupancy data, activity levels, physiological parameters, and thermal comfort votes, with minimal consideration of psychological factors. (5) Further exploration is required for group thermal sensation prediction methods that reconcile diverse individual thermal preferences within the same thermal zone.

REFERENCES

1. M. Pal, A.A. Alyafi, S. Ploix, P. Reignier, S. Bandyopadhyay, Unmasking the causal relationships latent in the interplay between occupant's actions and indoor ambience: A building energy management outlook, Applied Energy 238 (2019) 1452–1470.
2. W. O'Brien, A. Wagner, M. Schweiker, A. Mahdavi, J. Day, M.B. Kjærgaard, S. Carlucci, B. Dong, F. Tahmasebi, D. Yan, others, Introducing IEA EBC Annex 79: Key challenges and opportunities in the field of occupant-centric building design and operation, Building and Environment 178 (2020) 106738.
3. ANSI/ASHRAE Standard 62.1-2019, Ventilation for Acceptable Indoor Air Quality, ASHRAE, Atlanta, Georgia, 2019.
4. ASHRAE, ASHRAE Handbook, Fundamentals, Atlanta, Georgia, 2017.
5. W. Jung, F. Jazizadeh, Human-in-the-loop HVAC operations: A quantitative review on occupancy, comfort, and energy-efficiency dimensions, Applied Energy 239 (2019) 1471–1508.
6. J. Xie, H. Li, C. Li, J. Zhang, M. Luo, Review on occupant-centric thermal comfort sensing, predicting, and controlling, Energy and Buildings 226 (2020) 110392.
7. X. Zhou, Y. Lu, S. Hu, Z. Yang, D. Yan, New perspectives on temporal changes in occupancy characteristics of residential buildings, Journal of Building Engineering 64 (2023) 105590.
8. J.R. Dobbs, B.M. Hencey, Model predictive HVAC control with online occupancy model, Energy and Buildings 82 (2014) 675–684.
9. Z. Li, B. Dong, A new modeling approach for short-term prediction of occupancy in residential buildings, Building and Environment 121 (2017) 277–290.
10. B. Dong, B. Andrews, K.P. Lam, M. Höynck, R. Zhang, Y.-S. Chiou, D. Benitez, An information technology enabled sustainability test-bed (ITEST) for occupancy detection through an environmental sensing network, Energy and Buildings 42 (2010) 1038–1046.
11. B. Dong, K.P. Lam, Building energy and comfort management through occupant behaviour pattern detection based on a large-scale environmental sensor network, Journal of Building Performance Simulation 4 (2011) 359–369.
12. B. Dong, K.P. Lam, A real-time model predictive control for building heating and cooling systems based on the occupancy behavior pattern detection and local weather forecasting, Building Simulation 7 (2014) 89–106.
13. F. Haldi, D. Robinson, Interactions with window openings by office occupants, Building and Environment 44 (2009) 2378–2395.
14. S. Shi, B. Zhao, Occupants' interactions with windows in 8 residential apartments in Beijing and Nanjing, China, Building Simulation 9 (2016) 221–231.
15. G.Y. Yun, K. Steemers, Time-dependent occupant behaviour models of window control in summer, Building and Environment 43 (2008) 1471–1482.
16. S. Khadka, H.B. Rijal, K. Amano, T. Saito, H. Imagawa, T. Uno, K. Genjo, H. Takata, K. Tsuzuki, T. Nakaya, others, Development of adaptive behaviour models for thermal comfort: Longitudinal investigation in Japanese office buildings and literature review, Journal of Building Engineering 89 (2024) 109220.
17. M. Vellei, S. Martinez, J. Le Dréau, Agent-based stochastic model of thermostat adjustments: A demand response application, Energy and Buildings 238 (2021) 110846.
18. V. Fabi, R.V. Andersen, S.P. Corgnati, Influence of occupant's heating set-point preferences on indoor environmental quality and heating demand in residential buildings, HVAC&R Research 19 (2013) 635–645.
19. M. He, H. Pen, M. Li, Y. Huang, D. Yan, S. Lou, L. Wen, Investigation on typical occupant behavior in air-conditioned office buildings for South China's Pearl River Delta, Architectural Intelligence 1 (2022) 8.

20. X. Ren, D. Yan, C. Wang, Air-conditioning usage conditional probability model for residential buildings, Building and Environment 81 (2014) 172–182.
21. P. Correia da Silva, V. Leal, M. Andersen, Occupants' behaviour in energy simulation tools: Lessons from a field monitoring campaign regarding lighting and shading control, Journal of Building Performance Simulation 8 (2015) 338–358.
22. Y. Ding, X. Ma, S. Wei, W. Chen, A prediction model coupling occupant lighting and shading behaviors in private offices, Energy and Buildings 216 (2020) 109939.
23. V. Inkarojrit, Balancing Comfort: Occupants' Control of Window Blinds in Private Offices, University of California, Berkeley, 2005.
24. G.R. Newsham, C. Arsenault, A camera as a sensor for lighting and shading control, Lighting Research & Technology 41 (2009) 143–163.
25. W. Li, S. Wang, A multi-agent based distributed approach for optimal control of multi-zone ventilation systems considering indoor air quality and energy use, Applied Energy 275 (2020) 115371.
26. X. Tian, Y. Cheng, Z. Lin, Modelling indoor environment indicators using artificial neural network in the stratified environments, Building and Environment 208 (2022) 108581.
27. X. Kong, C. Xi, H. Li, Z. Lin, A comparative experimental study on the performance of mixing ventilation and stratum ventilation for space heating, Building and Environment 157 (2019) 34–46.
28. Y. Cheng, J. Niu, X. Liu, N. Gao, Experimental and numerical investigations on stratified air distribution systems with special configuration: Thermal comfort and energy saving, Energy and Buildings 64 (2013) 154–161.
29. B. Yang, A.K. Melikov, A. Kabanshi, C. Zhang, F.S. Bauman, G. Cao, H. Awbi, H. Wigö, J. Niu, K.W.D. Cheong, others, A review of advanced air distribution methods-theory, practice, limitations and solutions, Energy and Buildings 202 (2019) 109359.
30. I. Mutis, A. Ambekar, V. Joshi, Real-time space occupancy sensing and human motion analysis using deep learning for indoor air quality control, Automation in Construction 116 (2020) 103237.
31. J. Wang, J. Huang, Q. Fu, E. Gao, J. Chen, Metabolism-based ventilation monitoring and control method for COVID-19 risk mitigation in gymnasiums and alike places, Sustainable Cities and Society 80 (2022) 103719.
32. J. Zhao, L. Shi, J. Li, H. Li, Q. Han, A model predictive control regulation model for radiant air conditioning system based on delay time, Journal of Building Engineering 62 (2022) 105343.
33. B. Chenari, J.D. Carrilho, M.G. Da Silva, Towards sustainable, energy-efficient and healthy ventilation strategies in buildings: A review, Renewable and Sustainable Energy Reviews 59 (2016) 1426–1447.
34. T. Ekwevugbe, N. Brown, V. Pakka, D. Fan, Real-time building occupancy sensing using neural-network based sensor network, in: 2013 7th IEEE International Conference on Digital Ecosystems and Technologies (DEST), IEEE, 2013: pp. 114–119.
35. C. Ren, H. Yu, J. Wang, H.-C. Zhu, Z. Feng, S.-J. Cao, Zonal demand-controlled ventilation strategy to minimize infection probability and energy consumption: A coordinated control based on occupant detection, Environmental Pollution 345 (2024) 123550.
36. L. Mansson, S. Svennberg, M. Liddament, Technical synthesis report: A summary of IEA ECBCS Annex 18-Demand controlled ventilating systems, (1997).
37. G. Guyot, M.H. Sherman, I.S. Walker, Smart ventilation energy and indoor air quality performance in residential buildings: A review, Energy and Buildings 165 (2018) 416–430.
38. W. Zhang, W. Zhang, Y. Bai, S. Wen, Enhancing indoor environmental quality: Personalized recommendation method for demand-oriented indoor ventilation strategy, Sustainable Cities and Society 101 (2024) 105061.

39. N. Nassif, S. Kajl, R. Sabourin, Ventilation control strategy using the supply CO_2 concentration setpoint, Hvac&R Research 11 (2005) 239–262.
40. T. Kusuda, Control of Ventilation to Conserve Energy While Maintaining Acceptable Indoor Air Quality, National Bureau of Standards, Department of Commerce, 1976.
41. A. Hesaraki, J.A. Myhren, S. Holmberg, Influence of different ventilation levels on indoor air quality and energy savings: A case study of a single-family house, Sustainable Cities and Society 19 (2015) 165–172.
42. B.J. Wachenfeldt, M. Mysen, P.G. Schild, Air flow rates and energy saving potential in schools with demand-controlled displacement ventilation, Energy and Buildings 39 (2007) 1073–1079.
43. M. Zuraimi, A. Pantazaras, K. Chaturvedi, J. Yang, K. Tham, S. Lee, Predicting occupancy counts using physical and statistical CO_2-based modeling methodologies, Building and Environment 123 (2017) 517–528.
44. C. Jiang, M.K. Masood, Y.C. Soh, H. Li, Indoor occupancy estimation from carbon dioxide concentration, Energy and Buildings 131 (2016) 132–141.
45. J. Zou, Q. Zhao, W. Yang, F. Wang, Occupancy detection in the office by analyzing surveillance videos and its application to building energy conservation, Energy and Buildings 152 (2017) 385–398.
46. J. Yang, A. Pantazaras, K.A. Chaturvedi, A.K. Chandran, M. Santamouris, S.E. Lee, K.W. Tham, Comparison of different occupancy counting methods for single system-single zone applications, Energy and Buildings 172 (2018) 221–234.
47. J. Wang, J. Huang, Z. Feng, S.-J. Cao, F. Haghighat, Occupant-density-detection based energy efficient ventilation system: Prevention of infection transmission, Energy and Buildings 240 (2021) 110883.
48. T. Lipinski, D. Ahmad, N. Serey, H. Jouhara, Review of ventilation strategies to reduce the risk of disease transmission in high occupancy buildings, International Journal of Thermofluids 7 (2020) 100045.
49. H. Wang, J. Wang, Z. Feng, F. Haghighat, S.-J. Cao, An intelligent anti-infection ventilation strategy: From occupant-centric control and computer vision perspectives, Energy and Buildings 296 (2023) 113403.
50. F.I. Eyiokur, A. Kantarcı, M.E. Erakın, N. Damer, F. Ofli, M. Imran, J. Križaj, A.A. Salah, A. Waibel, V. Štruc, others, A survey on computer vision based human analysis in the COVID-19 era, Image and Vision Computing 130 (2023) 104610.
51. P. Kaviya, P. Chitra, B. Selvakumar, A unified framework for monitoring social distancing and face mask wearing using deep learning: An approach to reduce COVID-19 risk, Procedia Computer Science 218 (2023) 1561–1570.
52. S. Verbeke, A. Audenaert, Thermal inertia in buildings: A review of impacts across climate and building use, Renewable and Sustainable Energy Reviews 82 (2018) 2300–2318.
53. Y. Meng, T. Li, G. Liu, S. Xu, T. Ji, Real-time dynamic estimation of occupancy load and an air-conditioning predictive control method based on image information fusion, Building and Environment 173 (2020) 106741.
54. K. Sun, Q. Zhao, Z. Zhang, X. Hu, Indoor occupancy measurement by the fusion of motion detection and static estimation, Energy and Buildings 254 (2022) 111593.
55. F. Wang, Q. Feng, Z. Chen, Q. Zhao, Z. Cheng, J. Zou, Y. Zhang, J. Mai, Y. Li, H. Reeve, Predictive control of indoor environment using occupant number detected by video data and CO_2 concentration, Energy and Buildings 145 (2017) 155–162.
56. H. Wang, C. Liang, G. Wang, X. Li, Energy-saving potential of fresh air management using camera-based indoor occupancy positioning system in public open space, Applied Energy 356 (2024) 122358.
57. M. Kong, B. Dong, R. Zhang, Z. O'Neill, HVAC energy savings, thermal comfort and air quality for occupant-centric control through a side-by-side experimental study, Applied Energy 306 (2022) 117987.

58. B. Yang, Y. Liu, P. Liu, F. Wang, X. Cheng, Z. Lv, A novel occupant-centric stratum ventilation system using computer vision: Occupant detection, thermal comfort, air quality, and energy savings, Building and Environment 237 (2023) 110332.
59. Z. Zhang, R. Gao, Y. Liu, M. Liu, Y. Wang, W. Zhu, L. Zhou, A. Li, Smart air supply terminal for floor-standing room air conditioners based on the identification of human positions, Building and Environment 202 (2021) 108041.
60. J. Rodriguez, N. Fumo, Simultaneous thermal zoning and demand control ventilation, Energy Reports 11 (2024) 4412–4422.
61. Z. Cui, Y. Sun, D. Gao, J. Ji, W. Zou, Computer-vision-assisted subzone-level demand-controlled ventilation with fast occupancy adaptation for large open spaces towards balanced IAQ and energy performance, Building and Environment 239 (2023) 110427.
62. H. Lan, H.C. Hou, Z. Gou, M.S. Wong, Z. Wang, Computer vision-based smart HVAC control system for university classroom in a subtropical climate, Building and Environment 242 (2023) 110592.
63. P. Zhou, G. Huang, Z. Li, Demand-based temperature control of large-scale rooms aided by wireless sensor network: Energy saving potential analysis, Energy and Buildings 68 (2014) 532–540.
64. E.P. Ijjina, K.M. Chalavadi, Human action recognition in RGB-D videos using motion sequence information and deep learning, Pattern Recognition 72 (2017) 504–516.
65. P.W. Tien, S. Wei, J.K. Calautit, J. Darkwa, C. Wood, A vision-based deep learning approach for the detection and prediction of occupancy heat emissions for demand-driven control solutions, Energy and Buildings 226 (2020) 110386.
66. E.J. Choi, B.R. Park, N.H. Kim, J.W. Moon, Effects of thermal comfort-driven control based on real-time clothing insulation estimated using an image-processing model, Building and Environment 223 (2022) 109438.
67. S. Jung, J. Jeoung, T. Hong, H. Jang, Vision-based multi-label detection framework for capturing occupant action and clothing information using large-scale dataset, Building and Environment 257 (2024) 111537.
68. A.K. Melikov, Personalized ventilation, Indoor Air 14 (2004) 157–167.
69. S. Schiavon, A.K. Melikov, Energy-saving strategies with personalized ventilation in cold climates, Energy & Buildings 41 (2009) 543–550.
70. A.K. Melikov, Indoor air quality assessment by a breathing thermal manikin, in: Proceedings of Japan-Denmark Workshop on Indoor Air Chemical Pollution and Indoor Environment, 2000: pp. 31–36.
71. D. Al Assaad, C. Habchi, K. Ghali, N. Ghaddar, Effectiveness of intermittent personalized ventilation in protecting occupant from indoor particles, Building and Environment 128 (2017) 22–32.
72. C. Xu, P.V. Nielsen, L. Liu, R.L. Jensen, G. Gong, Impacts of airflow interactions with thermal boundary layer on performance of personalized ventilation, Building and Environment 135 (2018) 31–41.
73. H. Alsaad, C. Voelker, Qualitative evaluation of the flow supplied by personalized ventilation using schlieren imaging and thermography, Building and Environment 167 (2020) 106450.1–106450.11.
74. G. Song, Z. Ai, Z. Liu, G. Zhang, A systematic literature review on smart and personalized ventilation using CO_2 concentration monitoring and control, Energy Reports 8 (2022) 7523–7536.
75. R. Gao, H. Li, H. Zhang, C. Wang, B. Deng, Research on a personalized targeted air supply device based on body movement capture, Indoor Air 31 (2020) 206-219.
76. E. Katramiz, N. Ghaddar, K. Ghali, D. Al-Assaad, S. Ghani, Effect of individually controlled personalized ventilation on cross-contamination due to respiratory activities, Building and Environment 194 (2021) 107719.

77. J. Wang, C. Zhang, Y. Jing, Hybrid CMAC-PID controller in heating ventilating and air-conditioning system, in: 2007 International Conference on Mechatronics and Automation, 2007: pp. 3706–3711.
78. G. Liu, S. Liang, S. Hu, Calculation method of mean skin temperature weighted by temperature sensitivity of various parts of human body, Journal of Thermal Biology 100 (2021) 102995.
79. Y. Wu, B. Cao, Recognition and prediction of individual thermal comfort requirement based on local skin temperature, Journal of Building Engineering 49 (2022) 104025.
80. Z. Wang, R. Matsuhashi, H. Onodera, Towards wearable thermal comfort assessment framework by analysis of heart rate variability, Building and Environment 223 (2022) 109504.
81. M. Ning, M. Zaheeruddin, Neuro-optimal operation of a variable air volume HVAC&R system, Applied Thermal Engineering 30 (2010) 385–399.
82. Y. Li, Y. Fan, C. Zhi, W. Ye, X. Zhang, Multi-objective optimization of mechanical ventilation with the aid of purifiers in two scenarios: Regular operation and mitigating the spread of respiratory infectious diseases, Building Simulation 16 (2023) 795–811.
83. K.R. Subramaniam, C.-T. Cheng, T.Y. Pang, Fuzzy logic controlled simulation in regulating thermal comfort and indoor air quality using a vehicle heating, ventilation, and air-conditioning system, Sensors 23 (2023) 1395.
84. Y. Cui, X. Hong, Design and implementation of the air conditioning fuzzy controller based on the genetic algorithms, in: 2020 Chinese Automation Congress (CAC), IEEE, 2020: pp. 4642–4646.
85. Z. Jiang, Z. Deng, X. Wang, B. Dong, PANDEMIC: Occupancy driven predictive ventilation control to minimize energy consumption and infection risk, Applied Energy 334 (2023) 120676.
86. S. Yang, M.P. Wan, W. Chen, B.F. Ng, D. Zhai, An adaptive robust model predictive control for indoor climate optimization and uncertainties handling in buildings, Building and Environment 163 (2019) 106326.
87. Z. Li, Z. Sun, Q. Meng, Y. Wang, Y. Li, Reinforcement learning of room temperature set-point of thermal storage air-conditioning system with demand response, Energy and Buildings 259 (2022) 111903.
88. L. Yu, Y. Sun, Z. Xu, C. Shen, D. Yue, T. Jiang, X. Guan, Multi-agent deep reinforcement learning for HVAC control in commercial buildings, IEEE Transactions on Smart Grid 12 (2020) 407–419.
89. M. Veselý, W. Zeiler, Personalized conditioning and its impact on thermal comfort and energy performance – A review, Renewable and Sustainable Energy Reviews 34 (2014) 401–408.
90. R. Rawal, M. Schweiker, O.B. Kazanci, V. Vardhan, Q. Jin, L. Duanmu, Personal comfort systems: A review on comfort, energy, and economics, Energy and Buildings 214 (2020) 109858.
91. B. Yang, X. Ding, F. Wang, A. Li, A review of intensified conditioning of personal micro-environments: Moving closer to the human body, Energy and Built Environment 2 (2021) 260–270.
92. W. Song, Z. Zhang, Z. Chen, F. Wang, B. Yang, Thermal comfort and energy performance of personal comfort systems (PCS): A systematic review and meta-analysis, Energy and Buildings 256 (2022) 111747.
93. Y. Ke, F. Wang, P. Xu, B. Yang, On the use of a novel nanoporous polyethylene (nanoPE) passive cooling material for personal thermal comfort management under uniform indoor environments, Building and Environment 145 (2018) 85–95.
94. F. Wang, D. Pang, X. Liu, M. Liu, W. Du, Y. Zhang, X. Cheng, Progress in application of phase-change materials to cooling clothing, Journal of Energy Storage 60 (2023) 106606.

95. H. Oi, K. Yanagi, K. Tabata, Y. Tochihara, Effects of heated seat and foot heater on thermal comfort and heater energy consumption in vehicle, Ergonomics 54 (2011) 690–699.
96. H. Yang, Y. Deng, B. Cao, Y. Zhu, Study on the local and overall thermal perceptions under nonuniform thermal exposure using a cooling chair, Building and Environment 176 (2020) 106864.
97. M. Alain, G. Kamel, G. Nesreen, A simplified combined displacement and personalized ventilation model, Hvac&R Research 18 (2012) 737–749.
98. J. Kaczmarczyk, A. Melikov, D. Sliva, Effect of warm air supplied facially on occupants' comfort, Building and Environment 45 (2010) 848–855.
99. H. Zhang, E. Arens, D. Kim, E. Buchberger, F. Bauman, C. Huizenga, Comfort, perceived air quality, and work performance in a low-power task–ambient conditioning system, Building and Environment 45 (2010) 29–39.
100. D.R. Vissers, The Human Body as Sensor for Thermal Comfort Control, Eindhoven University of Technology, Eindhoven, The Netherlands, 2012.
101. H. Zhang, E. Arens, M. Taub, D. Dickerhoff, F. Bauman, M. Fountain, W. Pasut, D. Fannon, Y. Zhai, M. Pigman, Using footwarmers in offices for thermal comfort and energy savings, Energy and Buildings 104 (2015) 233–243.
102. C. Du, H. Liu, C. Li, J. Xiong, B. Li, G. Li, Z. Xi, Demand and efficiency evaluations of local convective heating to human feet and low body parts in cold environments, Building and Environment 171 (2020) 106662.
103. Y. He, N. Li, N. Li, J. Li, J. Yan, C. Tan, Control behaviors and thermal comfort in a shared room with desk fans and adjustable thermostat, Building and Environment 136 (2018) 213–226.
104. A. Aryal, B. Becerik-Gerber, G.M. Lucas, S.C. Roll, Intelligent agents to improve thermal satisfaction by controlling personal comfort systems under different levels of automation, IEEE Internet of Things Journal 8 (2021) 7089–7100.
105. L. Wang, D.A. Dalgo, N. Mattise, S. Zhu, J. Srebric, Physiological responses and data-driven thermal comfort models with personal conditioning devices (PCD), Building and Environment 236 (2023) 110290.
106. Y. Wu, B. Cao, Y. Zhu, Development of an automatic personal comfort system (APCS) based on real-time thermal sensation prediction, Building and Environment 246 (2023) 110958.
107. A. Aryal, B. Becerik-Gerber, Thermal comfort modeling when personalized comfort systems are in use: Comparison of sensing and learning methods, Building and Environment 185 (2020) 107316.
108. Y. Wu, J. Fan, B. Cao, Comparison among different modeling approaches for personalized thermal comfort prediction when using personal comfort systems, Energy and Buildings 285 (2023) 112873.
109. S.S. Shetty, D.C. Hoang, M. Gupta, S. Panda, Learning desk fan usage preferences for personalised thermal comfort in shared offices using tree-based methods, Building and Environment 149 (2019) 546–560.
110. H.D. Chinh, S.S. Shetty, M. Gupta, S. Panda, A wireless sensor and actuator network (WSAN) framework for personalized thermal comfort in office buildings, in: 2016 IEEE International Conference on Sustainable Energy Technologies (ICSET), IEEE, 2016: pp. 42–47.
111. I.J. Al-Rikabi, J. Karam, H. Alsaad, K. Ghali, N. Ghaddar, C. Voelker, The impact of mechanical and natural ventilation modes on the spread of indoor airborne contaminants: A review, Journal of Building Engineering 85 (2024) 108715.
112. S. Irving, D.J. Clements-Croome, Natural ventilation in non-domestic buildings, CIBSE, London, 2005.
113. H. Liu, H. Zheng, F. Li, H. Cai, A hybrid model for predicting window opening state in buildings based on non-intrusive monitoring, Indoor and Built Environment 30 (2021) 1400–1410.

114. M. Yao, B. Zhao, Window opening behavior of occupants in residential buildings in Beijing, Building and Environment 124 (2017) 441–449.
115. F. Stazi, F. Naspi, M. D'Orazio, A literature review on driving factors and contextual events influencing occupants' behaviours in buildings, Building and Environment 118 (2017) 40–66.
116. R. Andersen, V. Fabi, J. Toftum, S.P. Corgnati, B.W. Olesen, Window opening behaviour modelled from measurements in Danish dwellings, Building and Environment 69 (2013) 101–113.
117. Y. Zhang, P. Barrett, Factors influencing the occupants' window opening behaviour in a naturally ventilated office building, Building and Environment 50 (2012) 125–134.
118. V. Fabi, R.V. Andersen, S. Corgnati, B.W. Olesen, Occupants' window opening behaviour: A literature review of factors influencing occupant behaviour and models, Building and Environment 58 (2012) 188–198.
119. R. Fritsch, A. Kohler, M. Nygård-Ferguson, J.-L. Scartezzini, A stochastic model of user behaviour regarding ventilation, Building and Environment 25 (1990) 173–181.
120. Z. Shi, H. Qian, X. Zheng, Z. Lv, Y. Li, L. Liu, P.V. Nielsen, Seasonal variation of window opening behaviors in two naturally ventilated hospital wards, Building and Environment 130 (2018) 85–93.
121. X. Dai, J. Liu, X. Zhang, A review of studies applying machine learning models to predict occupancy and window-opening behaviours in smart buildings, Energy and Buildings 223 (2020) 110159.
122. F. Haldi, D. Robinson, On the behaviour and adaptation of office occupants, Building and Environment 43 (2008) 2163–2177.
123. B. Yang, S. Yang, X. Zhu, M. Qi, H. Li, Z. Lv, X. Cheng, F. Wang, Computer vision technology for monitoring of indoor and outdoor environments and HVAC equipment: A review, Sensors 23 (2023) 6186.
124. L. Bourikas, E. Costanza, S. Gauthier, P. James, J. Kittley-Davies, C. Ornaghi, A. Rogers, E. Saadatian, Y. Huang, Camera-based window-opening estimation in a naturally ventilated office, Building Research & Information 46 (2018) 148–163.
125. H. Zheng, F. Li, H. Cai, K. Zhang, Non-intrusive measurement method for the window opening behavior, Energy and Buildings 197 (2019) 171–176.
126. D. Luong, R. Richman, M. Touchie, Towards window state detection using image processing in residential and office building facades, Building and Environment 207 (2022) 108486.
127. J. Donges, F. Morandi, A. Prada, F. Cappelletti, A. Gasparella, Occupants' interaction with building services: Development of a camera-based method for detailed monitoring of windows, shadings, and lights, Building and Environment 248 (2024) 111078.
128. R. Markovic, E. Grintal, D. Wölki, J. Frisch, C. van Treeck, Window opening model using deep learning methods, Building and Environment 145 (2018) 319–329.
129. P.W. Tien, S. Wei, T. Liu, J. Calautit, J. Darkwa, C. Wood, A deep learning approach towards the detection and recognition of opening of windows for effective management of building ventilation heat losses and reducing space heating demand, Renewable Energy 177 (2021) 603–625.
130. P.W. Tien, S. Wei, J.K. Calautit, J. Darkwa, C. Wood, Real-time monitoring of occupancy activities and window opening within buildings using an integrated deep learning-based approach for reducing energy demand, Applied Energy 308 (2022) 118336.
131. C. Sun, X. Guo, T. Zhao, Y. Han, Real-time detection method of window opening behavior using deep learning-based image recognition in severe cold regions, Energy and Buildings 268 (2022) 112196.
132. Z. Wang, J. Calautit, P.W. Tien, S. Wei, W. Zhang, Y. Wu, L. Xia, An occupant-centric control strategy for indoor thermal comfort, air quality and energy management, Energy and Buildings 285 (2023) 112899.

133. X. Chen, Z. Zou, F. Hao, Y. Wang, C. Mei, Y. Zhou, D. Wang, X. Yang, Remote sensing of indoor thermal environment from outside the building through window opening gap by using infrared camera, Energy and Buildings 286 (2023) 112975.
134. H. Al Jebaei, A. Aryal, Quantifying the impact of personal comfort systems on thermal satisfaction and energy consumption in office buildings under different US climates, Energy and Buildings 274 (2022) 112448.
135. L.T. Graham, T. Parkinson, S. Schiavon, Lessons learned from 20 years of CBE's occupant surveys, Buildings & Cities 2 (2021).
136. P.O. Fanger, Thermal Comfort. Analysis and Applications in Environmental Engineering, Danish Technical Press, Copenhagen, 1970.
137. R. De Dear, G.S. Brager, Developing an adaptive model of thermal comfort and preference, ASHRAE Transactions 104 (1998) 145–167.
138. ANSI/ASHRAE Standard 55-2020, Thermal Environmental Conditions for Human Occupancy, ASHRAE, Atlanta, Georgia, 2020.
139. GB/T 50785-2012, Evaluation standard for indoor thermal environment in civil buildings, Standards Press of China, Beijing, 2012.
140. J. Miura, M. Demura, K. Nishi, S. Oishi, Thermal comfort measurement using thermal-depth images for robotic monitoring, Pattern Recognition Letters 137 (2020) 108–113.
141. P. Zheng, Y. Liu, H. Wu, H. Wang, Non-invasive infrared thermography technology for thermal comfort: A review, Building and Environment 248 (2023) 111079.
142. Z.Q. Fard, Z.S. Zomorodian, S.S. Korsavi, Application of machine learning in thermal comfort studies: A review of methods, performance and challenges, Energy and Buildings 256 (2022) 111771.
143. S. Li, X. Zhang, Y. Li, W. Gao, F. Xiao, Y. Xu, A comprehensive review of impact assessment of indoor thermal environment on work and cognitive performance – Combined physiological measurements and machine learning, Journal of Building Engineering 71 (2023) 106417.
144. S. Liu, S. Schiavon, H.P. Das, M. Jin, C.J. Spanos, Personal thermal comfort models with wearable sensors, Building and Environment 162 (2019) 106281.
145. M. Abdallah, C. Clevenger, T. Vu, A. Nguyen, Sensing occupant comfort using wearable technologies, in: Construction Research Congress 2016, 2016: pp. 940–950.
146. S. Lu, W. Wang, S. Wang, E. Cochran Hameen, Thermal comfort-based personalized models with non-intrusive sensing technique in office buildings, Applied Sciences 9 (2019) 1768.
147. Y. Zhou, X. Wang, Z. Xu, Y. Su, T. Liu, C. Shen, X. Guan, A model-driven learning approach for predicting the personalized dynamic thermal comfort in ordinary office environment, in: 2019 IEEE 15th International Conference on Automation Science and Engineering (CASE), IEEE, 2019: pp. 739–744.
148. Y. Feng, S. Liu, J. Wang, J. Yang, Y.-L. Jao, N. Wang, Data-driven personal thermal comfort prediction: A literature review, Renewable and Sustainable Energy Reviews 161 (2022) 112357.
149. J. Ren, R. Zhang, X. Cao, X. Kong, Experimental study on the physiological parameters of occupants under different temperatures and prediction of their thermal comfort using machine learning algorithms, Journal of Building Engineering 84 (2024) 108676.
150. K. Katić, R. Li, W. Zeiler, Machine learning algorithms applied to a prediction of personal overall thermal comfort using skin temperatures and occupants' heating behavior, Applied Ergonomics 85 (2020) 103078.
151. J. Wu, X.-Y. Chen, H. Zhang, L.-D. Xiong, H. Lei, S.-H. Deng, Hyperparameter optimization for machine learning models based on Bayesian optimization, Journal of Electronic Science and Technology 17 (2019) 26–40.
152. M. Luo, J. Xie, Y. Yan, Z. Ke, P. Yu, Z. Wang, J. Zhang, Comparing machine learning algorithms in predicting thermal sensation using ASHRAE Comfort Database II, Energy and Buildings 210 (2020) 109776.

153. B. Yang, X. Li, Y. Liu, L. Chen, R. Guo, F. Wang, K. Yan, Comparison of models for predicting winter individual thermal comfort based on machine learning algorithms, Building and Environment 215 (2022) 108970.
154. B. Yang, X. Cheng, D. Dai, T. Olofsson, H. Li, A. Meier, Real-time and contactless measurements of thermal discomfort based on human poses for energy efficient control of buildings, Building and Environment 162 (2019) 106284.
155. L. Kang, G. Huan, Z. Xuejin, Z. Hua, D. Binlin, L. Ni, Z. Yi, T. Ran, H. Qize, S. Lin, Deep learning and thermographic imaging method for thermal comfort prediction in different genders, International Journal of Thermal Sciences 197 (2024) 108804.
156. Z. Hou, J. Lyu, D. Wu, J. Chen, J. Shi, Z. Lian, Occupant-centric cabin thermal sensation assessment system based on low-cost thermal imaging, Building and Environment 261 (2024) 111692.
157. X. Zhou, Z. Miao, M. Yuan, K. Li, H. Guo, X. Lin, Y. Zeng, R. Tu, J. Zhong, Dual-phase prediction model of passenger thermal sensation using facial thermal imaging and environmental factors, Case Studies in Thermal Engineering 58 (2024) 104439.

5 Outdoor Thermal Environmental Monitoring

Ruiqi Guo, Luting Bai, and Zhihan Lyu

5.1 URBAN MICROCLIMATE AND THERMAL COMFORT FOR PEDESTRIANS

Urban meteorology [1] categorizes urban climate into three layers: the urban canopy layer, the urban boundary layer, and the urban plume layer. The urban canopy layer climate is the most relevant to human life, as it directly impacts the comfort of outdoor spaces in cities. Urban microclimate, a subset of the urban canopy layer, is deeply intertwined with urban activities and human life. It's defined as "the temperature, humidity, pressure, wind speed, sunlight, and radiation at specific spots around buildings, like the ground, roofs, walls, and windowsills" [2]. This microclimate operates within a 1 km range and is particularly important for architects and researchers, as it aligns with the scale of their work. In urban climatology, urban microclimate is unique because it's directly shaped by the surfaces of buildings and the surrounding environment, such as the layout of the city and local heat conditions. These factors not only affect the amount of energy buildings use but also influence the heat levels and comfort of people walking outdoors.

As defined by ASHRAE Standard 55, thermal comfort is essentially "how satisfied we feel with the temperature around us". This satisfaction is a mental state shaped by various factors. When we talk about outdoor thermal comfort for pedestrians, it's about "how comfortable people feel with the heat and conditions outside". Assessing this comfort level is tricky because it's influenced by a mix of physical, physiological, and psychological elements [3]. Among these, physical factors play the most significant role. Specifically, the comfort of people walking outdoors is affected by key microclimate factors like air temperature, humidity, sunlight, and wind speed [4].

Cities, with their dense populations and bustling human activities, create unique microclimates that are distinct from surrounding areas [5, 6]. Over the past few decades, China's rapid urbanization has led to significant challenges in maintaining the quality of these urban microclimates. Large-scale construction, urban renewal, and population surges have particularly impacted big cities. Changes in the urban landscape, combined with heat emissions from traffic and buildings, have intensified the "urban heat island" (UHI) effect. Additionally, air pollution struggles to disperse, further degrading the livability and outdoor environmental quality of cities. The rising demand for air conditioning has also increased building energy consumption,

DOI: 10.1201/9781003611455-5

worsening the energy crisis and triggering a cascade of economic, social, and environmental issues [7]. While current technology doesn't allow us to directly control the broader climate, we can still influence local urban microclimates through careful planning and design. This makes monitoring urban microclimates an essential task in today's urban development.

Monitoring urban microclimate today primarily involves two methods: deploying sensors for on-site measurements and using computers for numerical simulations [8]. Traditionally, designers have relied on meteorological data from similar climate zones to guide new projects and updates. While this approach is efficient, it often lacks precision, potentially leading to resource wastage [9]. However, with the rapid advancements in computer technology, numerical simulations have become both more accurate and efficient. Combining these simulations with traditional design standards is now seen as a highly effective way to optimize urban planning. For instance, the U.S. Green Building Design Guidelines (LEED) suggests specific strategies to improve neighborhood microclimates and reduce heat island effects. Similarly, Japan's CASBEE standards emphasize the importance of thermal and wind environments in neighborhoods, explicitly recommending the use of computer simulations to refine designs. In developed countries like those in Europe and America, it's common practice to conduct computer simulations or numerical calculations before finalizing neighborhood plans, ensuring designs are guided by precise data rather than just traditional experience [10]. China's Ministry of Construction and National Environmental Protection Agency also encourage the use of numerical simulations in their green building and ecological neighborhood standards [11]. Leveraging computer tools to simulate urban microclimates, analyze impacts, and guide scientific planning is not only practical but also essential for the future of sustainable, data-driven design.

Pedestrian thermal comfort is a critical measure of urban environmental quality and residents' daily experiences, with significant implications for urban planning and public health [12]. It refers to how comfortable people feel in outdoor environments, particularly in terms of temperature. This comfort level affects not only the ease of daily activities but also the overall health and well-being of city dwellers [12]. During hot summers, extreme heat and humidity can lead to serious health issues like heatstroke and dehydration, especially for vulnerable groups such as the elderly, children, and those with chronic illnesses. By incorporating thoughtful urban design, such as adding more green spaces, installing shading structures, and optimizing building layouts, cities can significantly improve pedestrian thermal comfort, enhancing residents' quality of life and making outdoor spaces more enjoyable [13].

Additionally, pedestrian thermal comfort plays a vital role in creating sustainable cities [14]. High temperatures often lead to increased energy use, particularly for air conditioning, which in turn worsens UHI effects and boosts carbon emissions. By improving thermal comfort, cities can reduce energy consumption, lower their carbon footprint, and contribute to environmental protection. Accurate monitoring and analysis of thermal comfort data also help urban planners and policymakers make informed decisions, optimizing the design and maintenance of public spaces to enhance a city's livability and sustainability. Ultimately, improving pedestrian thermal comfort not only boosts public health and residents' quality of life but also fosters greener, healthier, and more livable urban environments, providing safe and comfortable outdoor spaces that support social harmony and economic growth [15].

Urban microclimate and pedestrian thermal comfort are closely intertwined. Urban microclimate describes the localized weather patterns within a city, shaped by factors like dense buildings, heavy traffic, and heat buildup. Pedestrian thermal comfort, on the other hand, refers to how people perceive temperature in such environments [16]. A key feature of urban microclimate is the UHI effect, where cities are significantly warmer than surrounding rural areas, especially in summer. This intensifies discomfort for people outdoors, making them feel hotter and more uneasy. Beyond affecting daily life and productivity, this discomfort can lead to health risks like heatstroke and dehydration. To tackle this, urban planners and designers must consider microclimate factors, such as adding green spaces, optimizing building arrangements, and using reflective materials to cool cities and enhance pedestrian comfort [17]. Other microclimate elements like wind speed, humidity, and solar radiation also play a crucial role [18]. A gentle breeze can help sweat evaporate, cooling the body, while high humidity makes it harder for sweat to evaporate, increasing discomfort. Thoughtful urban design can improve these conditions, creating more pleasant outdoor spaces. For instance, adding parks and water features can regulate temperature and humidity, improve air quality, and boost pedestrian comfort [19]. In short, understanding and improving urban microclimate is vital for creating healthier, more livable cities.

Monitoring urban microclimate and pedestrian thermal comfort is essential for enhancing urban living conditions. Urban microclimate, influenced by dense buildings, paved roads, and human activity, leads to heat buildup and the heat island effect, altering temperature, humidity, and wind patterns within cities [20, 21]. Pedestrian thermal comfort directly impacts how comfortable and healthy people feel in their daily lives. By accurately monitoring microclimate conditions, urban planners can identify the extent and distribution of heat islands and implement solutions like increasing greenery, using better building materials, and improving ventilation to reduce the impact of high temperatures. Monitoring thermal comfort also helps design better urban spaces, such as adding shade structures, misting systems, and optimizing public areas to make them more enjoyable. Additionally, thermal comfort data can provide early warnings for extreme heat, ensuring timely measures to prevent heat-related illnesses. Accurate microclimate and thermal comfort insights not only guide long-term urban planning and infrastructure but also improve daily life, making cities more livable and sustainable. With these tools, city leaders can develop effective strategies to address climate challenges and protect public health and well-being.

5.1.1 Measurement Methods

5.1.1.1 Urban Microclimate Monitoring

Urban microclimate monitoring typically involves three main approaches: on-site local monitoring, thermal comfort surveys, and computer simulations. Key microclimate parameters include air temperature, relative humidity (RH), wind speed, wind direction, globe temperature, and solar radiation intensity. Currently, local monitoring is done by manually placing sensors at selected points [22]. The data collected from these sensors helps calculate thermal indices, which are used to evaluate how comfortable people feel in a given environment [23]. Below is an illustration of the commonly used monitoring tools and weather stations.

In past research, several thermal indices have been widely used to assess comfort levels. These include predicted mean vote (PMV) [24], standard effective temperature (SET) [25], outdoor SET (OUT_SET*) [26], wet bulb globe temperature (WBGT) [27], physiological equivalent temperature (PET) [28], and universal thermal climate index (UTCI) [29]. SET and PMV are often used for indoor comfort evaluation, while OUT_SET*, WBGT, UTCI, and PET are more suited for outdoor environments [3]. The OUT_SET* model is rarely used in practice because its calculations are complex, relying on factors like clothing insulation and environmental variables. WBGT was originally designed to measure heat stress in high-temperature workplaces, using a simple four-level scale based on work intensity. However, this method is too basic for assessing comfort during varying metabolic activities. On the other hand, PET and UTCI are the most popular outdoor thermal indices today, as they take into account both environmental conditions and human physiological responses [3, 9].

5.1.1.2 Thermal Comfort Questionnaire

When designing a thermal comfort questionnaire, it's essential to adhere to the ISO10551 standard, which typically divides the survey into two main sections [9]. The first part collects general information about the respondent, such as their gender, age, height, weight, what they're wearing, their activity level, why they're visiting the park, how often they come, how long they've been there, and their current mood. The second part focuses on thermal comfort, asking respondents to rate their thermal sensation vote (TSV), temperature acceptable (TA), and overall comfort level (TCV, thermal comfort vote) in various environments. The scales for these ratings are chosen based on ASHRAE 55 standard and local weather conditions [30]. For thermal sensation, a 7-point scale is most commonly used, ranging from "Very cold" (–3) to "Very hot" (+3). Thermal acceptability is usually measured on a 5-point scale, from "Very unacceptable" (–2) to "Very acceptable" (+2). Lastly, thermal comfort is often assessed using a 7-point scale, from "Very uncomfortable" (–3) to "Very comfortable" (+3).

5.1.1.3 Thermal Environment Simulation

Several current computational theoretical models: energy balance equation model, CTTC and its improved models, computational fluid dynamics (CFD)-based models [31–33], and the corresponding simulation tools developed based on these theoretical models are introduced. This chapter mainly classifies and details the corresponding simulation tools based on different computational theoretical models according to existing research results.

Simulation tools based on the energy balance equation model. The energy balance equation model is an important part of the lumped parameter method, focusing on energy balance considerations, and is a key theoretical model when considering long-wave and short-wave radiation. This model effectively avoids the influence of various complex factors, simplifies a large amount of data analysis, and has the characteristics of fast calculation, simplicity, and convenience. However, there is currently a lack of commercial software based on this model, with commonly used tools including RayMan and SOLWEIG [34]. RayMan is a microclimate software focused on thermal comfort and human radiation load calculation, developed by the University of Freiburg, Germany. It is mainly used to evaluate the impact of outdoor microclimate on human thermal comfort. The RayMan model can calculate the total short-wave and long-wave radiation received by the human body based on input meteorological data (such as temperature, humidity, wind speed) and geographic data (such as terrain, vegetation, and three-dimensional (3D) building information), and then evaluate outdoor thermal comfort indicators (such as PET, UTCI). The advantage of RayMan lies in its precise solar radiation calculation method, which can consider complex factors such as building shadows, tree canopy shading, and terrain undulations, and generate hourly radiation load data. This precise radiation calculation is particularly important in complex environments such as urban blocks or parks, helping researchers and urban planners better understand the impact of different designs on outdoor thermal environments. Additionally, RayMan supports simple interface operations, making it easy for users to quickly import meteorological data and perform calculations, suitable for projects requiring rapid analysis and thermal comfort evaluation [30]. SOLWEIG (Solar and LongWave Environmental Irradiance Geometry) is a microclimate model developed by the University of Gothenburg, Sweden, mainly used to evaluate the distribution of solar radiation in urban environments and its impact on human thermal comfort. SOLWEIG simulates surface temperature, PET of pedestrians, and other indicators by calculating the transmission of solar radiation and surface long-wave radiation. Its model structure is relatively simple, usually requiring only the input of 3D building and terrain data, combined with sunlight and meteorological conditions, to predict radiation intensity and shadow distribution at different time points. This model is particularly suitable for short-term and small-scale analysis, helping researchers and urban planners evaluate the impact of urban design on thermal comfort and providing a basis for the layout of greening or shading facilities [35].

The impact of human activities on climate has become increasingly significant with the development of urbanization. People have begun to re-examine the relationship between cities and climate, and between humans and climate. In this context, the application of CFD technology in environmental science has developed rapidly, with ENVI-met being the most famous software for urban microclimate simulation. ENVI-met is a 3D microclimate model developed by the University of Mainz, Germany, commonly used to simulate air temperature, humidity, wind speed, and pollutant dispersion in urban environments. The main feature of ENVI-met is its detailed simulation of urban elements such as vegetation, water bodies, and buildings, especially excelling in handling the interaction between atmosphere and surface. Based on CFD principles, this model can simulate high-resolution urban microclimates and is typically used for longer time scales (such as 24 hours or more) [36].

The SOLWEIG model performs well in calculating solar radiation distribution and pedestrian thermal comfort, effectively capturing the distribution and changes of shadows, especially in complex urban building environments. Its calculations are relatively simple, suitable for small-scale, rapid analysis. Application scenarios include thermal comfort evaluation: used to analyze the impact of urban design and building layout on pedestrian thermal comfort, suitable for street-scale thermal environment evaluation; short-term solar radiation studies: can quickly provide radiation distribution and shadow simulation, suitable for evaluating the immediate effects of urban shading measures; optimization of greening and shading design: helps plan the location of greening or shading facilities and evaluate their impact on improving the thermal environment.

ENVI-met has high computational requirements but can provide very detailed climate simulations, suitable for long-term microclimate simulations of larger areas. Application scenarios include UHI effect research: suitable for large-scale urban microclimate analysis, such as evaluating the impact of urban green spaces and water bodies on local temperature; air quality and pollutant dispersion simulation: simulates pollutant dispersion, suitable for studying the effectiveness of air quality improvement measures; impact of the built environment on climate: used to simulate the impact of building layout and materials on microclimate, suitable for thermal environment analysis of building groups; high-precision research in complex environments: when studying multi-dimensional climate characteristics of streets, squares, or urban areas, ENVI-met can provide detailed microclimate data.

RayMan is relatively simple to operate and does not require very detailed terrain data, making it suitable for rapid analysis. Application scenarios include outdoor thermal comfort evaluation: commonly used to study pedestrian thermal comfort under different meteorological conditions, suitable for human thermal comfort analysis in micro-environments (such as urban parks and blocks); research on the impact of climate change on humans: evaluates the impact of heat exposure on human health under climate change scenarios, suitable for public health and heat exposure risk assessment; small-scale space heat load calculation: particularly suitable for small outdoor areas, helping urban planners evaluate the thermal comfort of outdoor public spaces (such as squares and pedestrian streets).

These three models can be used complementarily according to specific research needs, leveraging their respective advantages in different scenarios to provide comprehensive and detailed support for urban microclimate monitoring.

5.1.2 Case Studies

When it comes to monitoring microclimates, there are two main approaches. The first involves measuring microclimate parameters and conducting surveys, while the second relies on numerical simulations, using measured data to fine-tune the models. To give you a clearer picture, let's take a look at a study by Bo Hong, published in 2020 in *Building and Environment*. The paper, titled "Outdoor thermal benchmarks and their application to climate–responsive designs of residential open spaces in a cold region of China" [37], combines microclimate measurements with surveys, as shown in Figure 5.1. By walking through the experiment, we hope to make outdoor thermal comfort research more accessible to beginners and encourage more people to dive into this field. The study is broken down into five key sections: choosing the study site, designing the experiment and collecting data, analyzing the data, highlighting the main findings, and applying the results to design. If this piques your interest, feel free to explore other journal articles on the topic.

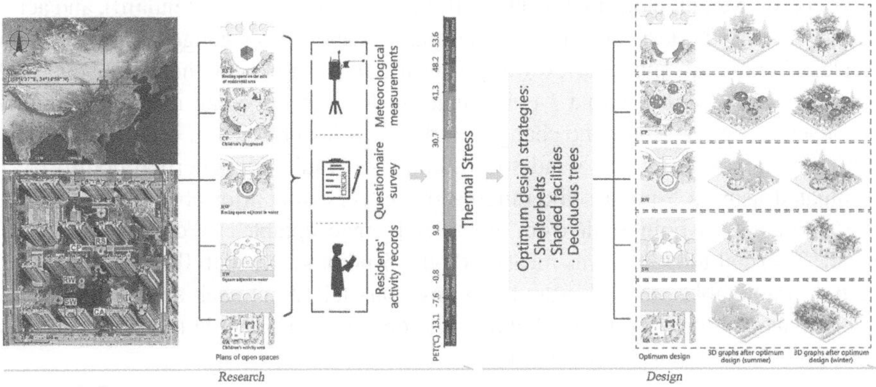

FIGURE 5.1 Overall research framework. (Source: Reference [37], Copyright 2020 by Elsevier.)

5.1.2.1 Study Site Selection

The study took place in Xi'an, a city in China's cold region. Xi'an experiences a mix of semi-arid and humid subtropical climates, characterized by chilly, dry winters and sweltering summers. The research focused on a community spanning about 13 ha, featuring 32-story buildings surrounded by plenty of greenery and diverse open spaces.

The study focused on five distinct spaces to explore various activities and how they interact with climate factors: the resting space sits at the heart of the residential zone, characterized by hard surfaces and pavilions. The children's play area is packed with fun equipment for kids and is conveniently located near fitness amenities. The waterfront rest area offers a serene spot by artificial water features, complete with circular pavilions for relaxation. The waterfront square area boasts artistic sculptures and stone seating, all with a view of the water. Lastly, the children's activity area is designed for toddlers under five, featuring safe, rubberized flooring and granite surfaces.

5.1.2.2 Experimental Design and Data Collection

5.1.2.2.1 Meteorological Monitoring

Five automatic weather stations were strategically placed, with their sensors positioned 1.2 m above the ground to meet the guidelines of ASHRAE 55 and ISO 7726 standards. These stations tracked several key weather factors: air temperature (T_a) was monitored using a HOBO U23-001 temperature and humidity logger, which is precise to within ±0.21°C. RH was also logged by the same device, with a precision of ±2.5%. Wind speed (V_a) was measured with a Kestrel 5500 anemometer, accurate to 0.1 m/s. Solar radiation (G) was captured by a TBQ-2 pyranometer, with a margin of error of ±2%. Globe temperature (T_g) was recorded using a Delta OHM HD2107.2 instrument, accurate to ±0.25°C. Finally, the mean radiant temperature (T_{mrt}) was derived from the globe temperature and wind speed, following the ISO 7726 standard.

5.1.2.2.2 Questionnaire Survey

While conducting weather measurements, a survey was simultaneously carried out among residents, covering several key areas: demographic details such as gender, age, height, weight, clothing insulation (following ASHRAE 55 standard), and activity levels over the last 20 minutes. Thermal perception was gauged using a 7-point scale, where –3 meant "cold", 0 was "neutral", and 3 indicated "hot". Thermal preferences were assessed with a 3-point scale: –1 for wanting conditions to be "higher, stronger, or wetter", 0 for "no change", and +1 for "lower, weaker, or drier", focusing on temperature, humidity, wind speed, and solar radiation. Thermal comfort was evaluated using a 5-point scale ranging from "very uncomfortable" to "very comfortable", and a 4-point scale from "completely unacceptable" to "completely acceptable". Thermal tolerance was measured on a 5-point scale, with 0 being "tolerable" and 4 "intolerable", to understand how residents felt about each space. Two volunteers managed the survey: one handled the questionnaire, while the other recorded the types of activities every 30 minutes.

5.1.2.2.3 Resident Activity Record

In five different areas, volunteers tracked the activity levels and age groups of residents every half hour. The intensity of these activities was categorized into four levels according to the World Health Organization's Metabolic Equivalent of Task (MET) standards: sedentary, light, moderate, and vigorous. The age groups recorded included infants and toddlers (0–2 years), children (3–17 years), adults (18–55 years), and seniors (over 55 years).

5.1.2.3 Data Analysis Method

5.1.2.3.1 Descriptive Analysis

This study used ANOVA to examine weather data from different locations and seasons, and then applied Tukey's post hoc test to check for significant differences. The findings revealed that, while air temperature and RH didn't vary much, other factors like wind speed, solar radiation, globe temperature, and PET showed notable differences across the five areas.

5.1.2.3.2 Thermal Comfort Standard

By analyzing the linear regression between residents' TSV and PET, researchers identified Xi'an's "neutral PET" as 19.7°C. The range where people feel thermally comfortable (with TSV between −0.5 and 0.5) spans from 12.4°C to 26.9°C. Additionally, 90% of residents find temperatures between 9.8°C and 30.7°C acceptable.

5.1.2.3.3 The Relationship Between Activity and Thermal Comfort

By applying linear regression, we examined how the number of people in different spaces relates to PET. Interestingly, during winter, more people tend to gather in areas with higher PET values, suggesting they seek warmth. Conversely, in summer, people avoid spaces with high PET, opting instead for cooler, more comfortable environments. This behavior highlights a clear preference for thermal comfort in daily activities.

5.1.2.4 Key Findings

Spatial Microclimate Variations: The research highlights notable differences in microclimate factors like temperature, wind speed, and solar radiation across various spaces. For instance, the axis rest area experiences much higher wind speeds compared to other areas, while the south-facing children's play zone, lacking shade, sees elevated levels of solar radiation and globe temperature.

Seasonal Preferences and Comfort Zones: During winter, people gravitate toward sunny spots, whereas in summer, shaded areas are more appealing. Interestingly, residents cope better with winter's "cold stress" than summer's "heat stress". The study identified Xi'an's "neutral PET" as 19.7°C, with a comfortable range spanning from 9.8°C to 30.7°C.

Design Solutions: To enhance thermal comfort, the study suggests practical design strategies such as installing windbreaks and planting deciduous trees for shade. These measures aim to mitigate summer heat and winter cold, ultimately creating open spaces that cater to residents' seasonal needs.

5.1.2.5 Design and Application

The study used thermal environment analysis to come up with the best designs for each space. For instance, in the axis rest area, a sunshade pavilion is recommended for winter when there's plenty of sunlight, while deciduous plants are suggested for summer to create shade. In the children's activity area, mushroom-shaped sunshades and large-canopy deciduous trees are proposed to provide extra shade during the summer. For the waterfront rest area, adjustable sunshade facilities are recommended to manage solar radiation and wind effects based on the season.

In summary, the research methodology took into account weather data, residents' personal experiences, and their activities to understand how microclimates affect comfort and behavior. By measuring weather conditions in different spaces and analyzing how people feel about the heat, the study offers valuable data and design tips for creating climate-friendly spaces in cold areas.

5.1.3 Practical Applications

Urban microclimate and pedestrian thermal comfort monitoring technologies have a wide range of practical uses. Here are some specific examples and real-world applications:

1. *Urban Planning and Design, Application*: Leveraging microclimate data to enhance urban planning and boost the quality of the urban environment. Case: Singapore: Singapore's urban greening initiative employs microclimate data to choose the right types of plants and their placement, resulting in numerous "green lung" zones. For instance, Gardens by the Bay and Changi Airport not only add to the city's greenery but also enhance air quality and mitigate the UHI effect by fine-tuning the microclimate within these areas.
2. *Tackling Urban Heat Islands, Application*: Keeping an eye on UHIs and implementing strategies to lessen their effects. Case: Los Angeles: By analyzing heat island data, Los Angeles has encouraged the adoption of reflective roof paints and green spaces. In hotspots where the UHI effect is most pronounced, increasing the number of trees and plants has led to a noticeable drop in city temperatures.
3. *Smart Building Energy Use, Application*: Fine-tuning heating and cooling systems in buildings using detailed microclimate information to boost

energy efficiency. Case: Boston Museum: The museum employs a microclimate tracking system to perfect its indoor climate control. Real-time data helps adjust heating and cooling, cutting down on energy use while ensuring a stable environment to safeguard valuable artifacts.

4. *Traffic Management, Application*: By leveraging climate data, cities can fine-tune traffic signals and management systems to ease congestion. Take Copenhagen, for example; the city taps into microclimate monitoring data to tweak traffic light timings, especially during harsh weather like snowstorms. This smart adjustment not only keeps traffic moving smoothly but also boosts safety on the roads.
5. *Health Warning System, Application*: Keeping an eye on thermal comfort in real-time allows cities to send out health alerts, particularly to vulnerable groups. Tokyo is a prime example: during scorching summer heatwaves, the city's thermal comfort monitoring system kicks in, sending out alerts to help residents stay cool. This is especially crucial for the elderly and those with chronic conditions, as it helps cut down on heat-related illnesses.
6. *Public Space Design, Application*: Cities are now using thermal comfort data to craft public spaces that feel just right. Melbourne's city center and Xi'an's Yongxingfang are leading the way: in Melbourne, sunshades are strategically placed to keep things cool, while in Xi'an, a misting system adds a refreshing touch during hot spells (Figure 5.2). These thoughtful designs not only make urban spaces more inviting but also encourage people to get out and enjoy them.

FIGURE 5.2 Yongxingfang spray system in Xi'an, China.

These real-world examples highlight how urban microclimate and pedestrian thermal comfort monitoring technologies play a crucial role in making cities more livable and improving residents' quality of life.

5.1.4 Conclusions and Future Directions

In an era marked by rapid urbanization and climate change, the urban microclimate is playing an increasingly crucial role in shaping the quality of life for city dwellers. Among these concerns, pedestrian thermal comfort has emerged as a central focus in urban planning and environmental management. This book explores the mechanisms behind urban microclimate formation, the factors that influence it, and its effects on how people experience heat while walking. The findings reveal that elements like building density, types of land cover, and the presence of green spaces and water bodies significantly shape local microclimates, affecting how pedestrians feel and behave in different seasons and weather conditions. Drawing on scientific data, the book proposes practical strategies to enhance thermal comfort, such as expanding green areas, improving street layouts, and adding shade structures. These measures not only help cool down urban areas but also make cities more livable.

Looking to the future, research on urban microclimate and pedestrian thermal comfort is set to become more sophisticated, multi-scale, and interdisciplinary. Advances in technologies like remote sensing, Internet of Things, and artificial intelligence (AI) will enable more precise and real-time monitoring of microclimates, capturing subtle changes across various urban spaces. By leveraging diverse data sources and high-resolution climate models, future studies will better understand how microclimates dynamically affect pedestrian comfort, particularly during extreme weather events. Additionally, this research will increasingly integrate with urban planning, architectural design, and environmental conservation efforts, fostering interdisciplinary solutions to optimize urban microclimates and thermal comfort. Ultimately, improving pedestrian thermal comfort is not just about enhancing well-being; it's also a vital step toward building greener, more sustainable cities.

5.2 REMOTE SENSING IMAGE MEASUREMENT

Remote sensing image measurement technology is a method that allows us to observe and analyze objects on the Earth's surface from afar using specialized equipment. This equipment, often mounted on satellites, airplanes, or drones, captures signals from electromagnetic waves, like visible light, infrared, and microwaves, that bounce off or are emitted by the ground. These signals are then turned into detailed images [38]. This technology has a wide range of uses, from mapping and environmental monitoring to agriculture, city planning, and even predicting natural disasters. At its core, the technology works by detecting how different materials on the ground reflect or emit electromagnetic waves across various wavelengths. Sensors pick up these signals and convert them into images, providing us with extensive, high-resolution data about the Earth's surface [39]. This data is crucial for many applications, helping us make informed decisions in various fields.

Today's advanced remote sensing sensors offer exceptional spatial and spectral resolution, delivering detailed data on temperature, humidity, vegetation cover, and more. This high-accuracy information is crucial for precisely analyzing and understanding shifts in urban heat patterns. Remote sensing satellites provide consistent, long-term data collection, enabling researchers to track and analyze trends in thermal environments over time. By examining and comparing satellite images from various periods, scientists can uncover the underlying patterns and causes of these thermal changes. By analyzing remote sensing data, city managers and planners can gain a complete picture of the thermal environment, allowing them to develop effective strategies to manage and improve it. For instance, adding more green spaces and water bodies, optimizing building arrangements, and enhancing urban airflow can all help reduce the UHI effect and make outdoor spaces more comfortable for people.

Remote sensing technology is a game-changer in monitoring outdoor thermal conditions. It offers a fast, wide-ranging, and accurate way for scientists and urban planners to study how urban heat patterns change over time. Let's dive deeper into how this technology works and why it's so important. Remote sensing images use infrared bands to measure surface temperatures, which helps track and analyze the UHI effect (a phenomenon where cities are noticeably hotter than surrounding rural areas). These images provide a clear visual of temperature variations across different parts of a city, pinpointing hotspots and measuring the intensity of the heat island effect [40, 41].

Beyond temperature, researchers also use remote sensing to study land use, water bodies, vegetation, and air quality. For example, it helps monitor changes in urban landscapes, like shrinking green spaces or increasing building density, which directly impact the thermal environment. Long-term remote sensing data can reveal how urban growth and land use changes affect heat patterns [42, 43]. Additionally, remote sensing is great for tracking water bodies and vegetation, both of which play a key role in cooling cities. By studying these natural features, planners can develop targeted strategies to improve the thermal environment [44, 45]. Some researchers even use remote sensing to monitor air pollution, as air quality and heat are closely linked. Better thermal conditions can lead to cleaner air [46, 47].

With climate change becoming a growing concern, remote sensing provides valuable long-term and large-scale data to assess its impact on urban heat. This information helps cities adapt to a changing climate [48, 49]. During extreme heatwaves, remote sensing can quickly deliver real-time thermal data, supporting emergency responses and public safety measures [50]. In short, remote sensing technology not only broadens and sharpens our ability to monitor outdoor thermal conditions but also equips city planners with the tools they need to create more livable and sustainable urban environments.

5.2.1 Measurement Methods

Many large cities have undergone rapid urbanization, drastically changing the landscape and the surfaces beneath our feet. Vast areas of natural spaces like forests, lakes, farmlands, and wetlands have been replaced by concrete and asphalt, altering the way heat interacts with the air and land in urban areas. This disruption in the energy balance between the ground and the atmosphere has led to the UHI effect, where cities tend to be significantly warmer than their rural surroundings [51]. The UHI effect can be categorized into four types based on how heat behaves at different levels: canopy

UHI (near the ground), boundary layer UHI (in the air above), surface UHI (on the ground), and subsurface UHI (below the ground) [52–55]. Each type operates at a different scale, is driven by distinct processes, and requires specific methods to measure. For example, remote sensing techniques can extract land surface temperature (LST) from satellite images to study surface UHI [56]. Thanks to the availability, high resolution, and ease of use of remote sensing data, scientists can now monitor UHI over entire cities, regions, or even the globe for extended periods at a low cost.

Remote sensing is a key tool for studying UHIs, as it allows researchers to calculate LSTs from satellite images. LST is a crucial indicator of urban heat, revealing how intense and widespread the heat island effect is. Traditional methods of measuring LST include using thermometers that touch the surface or infrared devices that measure heat from a distance [57]. However, these methods only capture temperatures at specific points and can be influenced by surrounding heat or even alter the surface's thermal state. For larger areas or long-term studies, satellite-based methods are more practical. Most ground observation stations are located in flat areas for easier maintenance [58], but measuring LST in rugged terrain is more complex. Remote sensing works by detecting the heat energy emitted by the Earth's surface and using mathematical models to calculate temperatures.

Remote sensing techniques for estimating temperature vary based on the type of spectrum used, primarily falling into two categories: thermal infrared and passive microwave methods. Thermal infrared remote sensing measures the radiation temperature of an object's surface, often referred to as skin temperature. On the other hand, microwaves, with their longer wavelengths, can penetrate the surface to some extent, providing an equivalent temperature for a certain depth within the object [59]. Over time, the algorithms for temperature retrieval using thermal infrared and microwaves have become more sophisticated. Popular thermal infrared methods include single-channel [60–63], multi-channel [64–67], multi-angle [68–71], and multi-temporal algorithms [72, 73]. For passive microwave temperature retrieval, models are generally categorized into statistical [74–76], physical [77–80], and neural network models [81–83]. As sensors and algorithms continue to advance, the variety of surface temperature data products has expanded, finding applications in climate change monitoring, disaster assessment, and land cover analysis. Spatially, MODIS products with a 1-km resolution are widely used, while the Landsat series offers higher resolution at 100 m. Temporally, thanks to geostationary satellites, we can now obtain surface temperature data that captures changes from as frequently as every 15 minutes to hourly intervals.

Whether we're talking about thermal infrared or passive microwave models, figuring out the temperature from the data is always tricky. The problem is that there are more unknowns than equations, making it a "sick" problem in math terms. To solve it, scientists have to make some educated guesses and set certain rules, but these guesses and rules can also introduce errors. When it comes to thermal infrared temperature inversion, we need to know things like how much the ground emits heat (emissivity), how much the atmosphere lets through (transmittance), and the radiation coming from the atmosphere. But here's the catch: surface temperature, emissivity, and atmospheric radiation are all linked, making it harder to figure out. The atmosphere's effect on temperature errors grows with the surface's brightness temperature, and the wetter the air, the bigger the impact. That's why most methods work best in dry conditions [84]. To estimate atmospheric transmittance, scientists often look at how it relates to water vapor, using models like MODTRAN, 6S, and LOWTRAN. One of the toughest parts of thermal infrared inversion is separating temperature from emissivity. Depending on whether we know the emissivity or not, we either figure them out step by step or try to get both at once if we know the atmospheric profile. There are several ways to estimate emissivity, like the reference channel method, emissivity normalization, and the two-temperature method [85, 86].

For passive microwave temperature inversion, the science isn't as solid. Most methods rely on statistical models or neural networks, and physical models often simplify things with assumptions or empirical data to reduce unknowns. To get better results, researchers focus on making the input data more reliable, like correcting for atmospheric effects on emissivity and water vapor. Over the years, many improved models have been developed, getting the temperature within 1K of accuracy. But for complex landscapes, the environment messes with the thermal infrared signals, making it hard to separate temperature from emissivity. In rugged areas, factors like nearby temperatures, mixed pixels, and the direction of radiation add to the challenge. Thermal infrared is also heavily affected by clouds and the atmosphere, while passive microwaves can work day and night, even though clouds. This makes microwaves useful for filling in the gaps in cloudy, mountainous areas. However, microwaves measure an average temperature over a depth, and for rough surfaces, emissivity can vary, leading to lower resolution. Combining microwave and thermal infrared data brings its own set of problems, like converting temperatures and scaling them up or down. For different cloud thicknesses, we need to figure out if the atmosphere's effect on microwave readings is significant and how to measure it better.

Water vapor is a big player in the atmosphere, absorbing radiation in ways that change with frequency, temperature, pressure, and density. Clouds, full of water droplets and particles, also absorb and scatter radiation. On clear days, water vapor and oxygen are the main absorbers, but when it's cloudy, thermal infrared signals get distorted, making it impossible to get ground temperatures. Microwaves can cut through clouds, but we still have to account for how water droplets scatter the signal. Clouds affect different surfaces in different ways, and pixels at the edge of a cloud are less affected than those in the middle. The longer the cloud cover, the more it changes the surface temperature, and cloud shadows have their unique effects. Current temperature inversion methods mostly work for clear skies, but figuring out temperatures under clouds is still a puzzle. Some researchers use interpolation or combine thermal infrared and microwave data to estimate it. Water vapor and clouds can't be ignored, especially in hilly areas, where good correction methods can help fill in the missing data. While we have decent ways to fill in cloud-covered pixels in visible and near-infrared images, these methods assume that emissivity doesn't change quickly. The same can't be said for thermal infrared under clouds, where the methods' effectiveness isn't well understood.

Combining microwave and thermal infrared data to cover gaps is tough, with challenges in temperature and scale conversion. Most passive microwave temperature inversion methods either ignore the atmosphere or treat it as a constant, but we need to better understand its impact. Water vapor and clouds are major hurdles in getting accurate temperatures for complex landscapes. Reconstructing thermal infrared data, identifying different cloud types, and filling in data for cloud-covered areas are all areas that need more research.

5.2.2 Case Studies

In this section, we'll explore a 2019 study by Zhaowu Yu, published in *Science of The Total Environment*, titled "Spatiotemporal patterns and characteristics of remotely sensed region heat islands during the rapid urbanization (1995–2015) of Southern China". This study serves as a great example of how remote sensing technology can be used to monitor environmental changes, especially for those new to the field [87]. The paper is broken down into four key sections: the research background, data sources, methods used, and the main findings. The research framework is shown in Figure 5.3. For those who want to dive deeper, there are plenty of other journal articles to explore on this topic.

5.2.2.1 Research Background

The UHI effect is a well-known phenomenon where cities are significantly warmer than their rural surroundings, primarily due to human activities. This issue is considered one of the most serious environmental challenges brought about by urbanization. The intense UHI effect leads to a host of problems, including poorer air quality, higher energy and water usage, and negative impacts on the health and comfort of city dwellers. Over the last decade, urban agglomerations, clusters of cities that are closely packed and economically linked, have emerged as a defining feature of global urbanization. These agglomerations typically consist of at least one megacity and several major cities within a specific area. Unlike individual cities, urban agglomerations can dramatically change the thermal landscape of a region because the distances between cities are much smaller or even nonexistent. This means that the thermal environment of a region can't be understood by looking at just one city. Unfortunately, most past research has focused on individual cities or specific areas within large cities, leaving the broader patterns and characteristics of UHI in urban

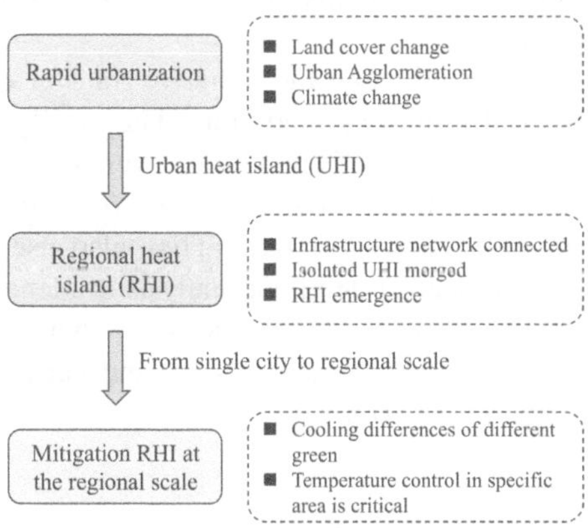

FIGURE 5.3 Research framework.

agglomerations largely unexplored. This oversight not only risks underestimating the UHI effect in key cities but also misses the bigger picture of how the thermal environment evolves across entire regions. Understanding these patterns is essential for policymakers to develop effective strategies to mitigate and adapt to these changes.

5.2.2.2 Data Sources

5.2.2.2.1 Land Cover Classification

Given the high resolution and extensive time frame of this research, they utilized Landsat series images to map the land cover and surface temperature of the PRDR. To ensure clarity, they applied a straightforward cloud-free algorithm within Google Earth Engine (GEE) to filter the Landsat images. Next, they prepared the image set for land cover classification in the code editor interface. They then manually created approximately 400 polygons to serve as training data, each representing a specific land cover type. For the classification process, they employed the classification and regression tree (CART) classifier, a widely used supervised classification method, to generate the final results.

5.2.2.2.2 Land Surface Temperature Data

To map the entire study area, they needed at least five Landsat image scenes. Since summer Landsat images in this region are often of poor quality, they selected images taken between June and September from 1988 to 2017. However, historical climate data from PRDR shows that June and September are, on average, 2°C cooler than July and August, so this should be factored into further analysis. All Landsat images were sourced from USGS Earth Explorer.

5.2.2.3 Research Methodology

This study focuses on analyzing remote sensing data and conducting multi-scale spatiotemporal analysis to explore the regional heat island (RHI) effect in China's Pearl River Delta from 1995 to 2015. First, they used Landsat satellite images and the GEE platform to gather LST data. Then, they classified the images into five land cover types (built-up areas, forests, water bodies, grasslands, and bare land) using a supervised method called CART. To ensure accuracy, they validated the classification with 100 points each year.

The LST data was calculated using the radiative transfer equation (RTE) algorithm, which performs well in humid conditions. They analyzed the relative value of LST (RLST) to understand how different land cover types influence the thermal environment. By categorizing RLST into 2°C, 4°C, 6°C, and 8°C intervals, they explored the spatiotemporal patterns of RHI. They also used tools like standard deviation ellipse (SDE) and nearest neighbor analysis to identify spatial trends and clustering patterns. Finally, they quantified the area, shape complexity, and spatial connectivity of different RHI categories using landscape indices, revealing the multi-scale spatial distribution of RHI.

5.2.2.4 Key Findings

Heat Island Expansion Trend: As urbanization progressed, the RHI in the Pearl River Delta region expanded, especially in the Pearl River Estuary area. Isolated

UHIs began to merge, forming a larger RHI, highlighting the strong clustering effect of RHI. The changes in LST and RLST in the Qinghai-Tibet Plateau and the Pearl River Delta between 1995 and 2015 show how the heat island effect grew over time, particularly in the Pearl River Estuary, where isolated heat islands combined into a single RHI, demonstrating the significant clustering of RHI.

Cooling Effects Vary: Forests and water bodies outperform grasslands in cooling, and the impact of grasslands remains unclear. This insight highlights the need for urban planners to prioritize the heat-reducing benefits of forests and water bodies in future city designs. By analyzing the RLST values and their yearly variations across different land cover types, the study arrives at the following results. Urban areas and bare land tend to have higher RLST values, ranging from 3°C to 8°C, while forests and water bodies show a cooling effect, with temperatures dropping to between −3°C and 0°C. Grasslands, on the other hand, have a less predictable cooling impact, with RLST values fluctuating between 0°C and 2°C. The study highlights regions with RLST values consistently between 4°C and 8°C as high-risk zones for heat exposure. These areas, which remain stable over time, should be a top priority in future strategies for climate adaptation. By analyzing the nearest neighbor ratio (NNR) for four categories of RHI ($2<\text{RLST}\leq 4$, $4<\text{RLST}\leq 6$, $6<\text{RLST}\leq 8$, and $\text{RLST}>8$), the study arrives at the following results. Regions with RLST values ranging from 4°C to 8°C show a higher degree of clustering and maintain this pattern consistently over time. These regions are flagged as high-risk zones for heat exposure.

Spatiotemporal Evolution: Over the years, the spatial distribution of RHI has expanded outward from the urban center, shifting from an east-west alignment to a southwest-northeast orientation. The SDE analysis highlights this trend from 1995 to 2015. Initially, RHI was concentrated in the central area during 1995–2000, but it gradually extended southwestward, forming a distinct southwest-northeast pattern.

Landscape Features: Analysis of landscape indices reveals that areas with high RLST values (6°C–8°C) are more tightly clustered and have more regular shapes, indicating strong spatial clustering and stability in high-temperature RHI zones. Figure 5.4 presents key landscape indices for each RHI category, such as PLAND (percentage of landscape area), MPS (mean patch area), AWMSI (area-weighted shape index), and AWMFD (area-weighted fractal dimension index). Regions with higher RLST values ($4<\text{RLST}\leq 8$ and $\text{RLST}>8$) demonstrate greater clustering and regularity, underscoring the spatial stability and concentration of high-temperature areas.

This study delves into the changing patterns of the RHI effect in China's Pearl River Delta from 1995 to 2015, a period marked by rapid urban growth. By analyzing remote sensing data at various scales, the research highlights how different land types, like buildings, forests, and water bodies, affect the area's heat dynamics. The findings show that as cities expanded, smaller heat islands in the Pearl River Estuary grew and merged into larger, more persistent heat zones. These areas, with temperatures consistently ranging between 4°C and 8°C, pose a significant risk of prolonged heat exposure. Interestingly, forests and water bodies were found to cool the environment more effectively than grasslands, offering valuable insights for managing urban heat. This study serves as a crucial resource for planning climate-resilient cities and promoting sustainable urban development.

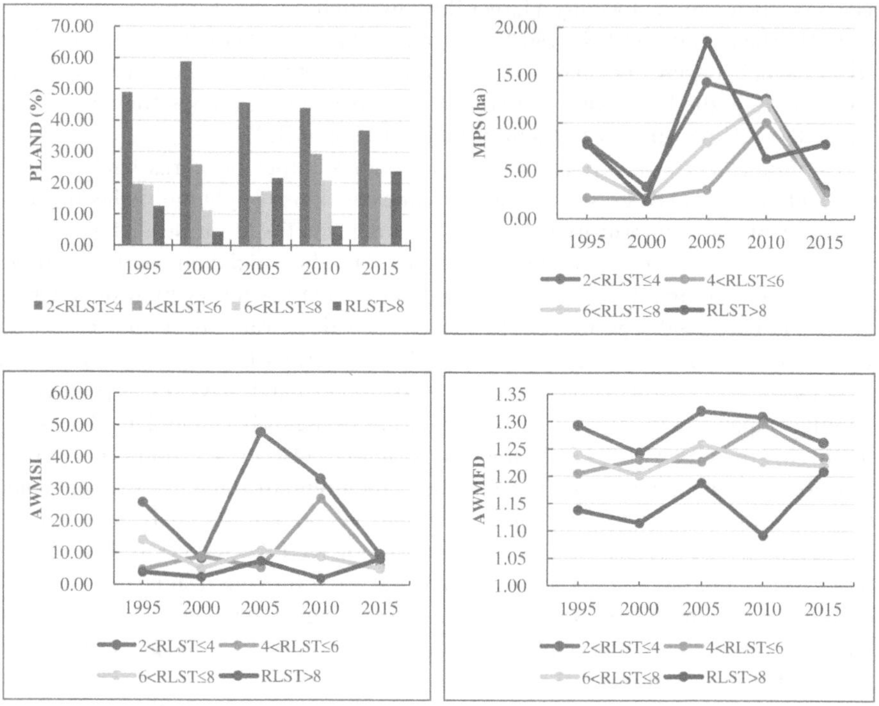

FIGURE 5.4 Landscape index values for the four RHI grades in 1995, 2000, 2005, 2010, and 2015 (Source: Reference [87], Copyright 2019 by Elsevier.)

5.2.3 Practical Applications

Remote sensing image technology plays a crucial role in detecting outdoor thermal environments, with a wide range of practical applications in engineering. Here are some key areas where it's making a difference:

1. *Urban Heat Island Analysis*: Remote sensing helps identify and monitor heat islands in cities. By analyzing thermal infrared images from satellites or drones, it pinpoints high-temperature zones, giving urban planners the data they need to implement cooling strategies like adding green spaces or water features.
2. *Green Space Monitoring*: This technology tracks urban green areas and their changes over time. By comparing images from different seasons or years, it evaluates how effectively green spaces reduce city temperatures and improve comfort, aiding in the planning of urban greening projects.
3. *Building Material Analysis*: Remote sensing monitors the surface temperatures of various building materials, helping choose those with better reflectivity and insulation. This reduces buildings' heat load and optimizes designs for better indoor and outdoor comfort.
4. *Meteorological Data Integration*: Combining remote sensing images with ground weather station data offers a detailed microclimate analysis. Integrating

satellite images with temperature, humidity, and wind speed data creates precise models to support urban planning and emergency responses.

5. *Health Risk Assessment*: Remote sensing assesses thermal stress levels across urban areas, predicting the health impacts of heatwaves. By combining this with population and medical resource data, it helps craft emergency plans to protect public health.
6. *Infrastructure Monitoring*: It monitors the temperature of large-scale infrastructure like roads and bridges, spotting potential heat damage early for preventive maintenance, ensuring safety and stability.
7. *Policy Support*: Remote sensing data provides a scientific foundation for governments to develop and implement effective environmental management policies for future smart cities.

5.2.4 Conclusions and Future Directions

This chapter delves into how remote sensing technology is revolutionizing the way we monitor urban heat environments. It highlights the technology's ability to gather detailed, large-scale surface temperature data across different times and spaces, and its crucial role in tracking UHIs and understanding how heat patterns change over time. Studies reveal that remote sensing data can precisely map out where and when cities heat up, shedding light on how various urban layouts and land types influence the thermal landscape. This information is invaluable for tackling urban heat issues. Moreover, the combination of high-resolution satellite images with cloud computing and big data analytics allows for real-time, detailed monitoring of heat in fast-growing urban areas, making urban heat management more effective and science-driven.

Looking to the future, remote sensing in urban heat monitoring is set to become more precise, diverse, and smart. Advances in satellite and drone technology will deliver even finer thermal data, enabling us to pinpoint heat hotspots in cities with greater accuracy. By blending data from different sources like Landsat, MODIS, and Sentinel satellites, we'll achieve a more continuous and comprehensive view of urban heat patterns. The integration of AI and machine learning will speed up data processing and analysis, paving the way for automated, intelligent heat monitoring. As cities face the twin challenges of climate change and rapid growth, remote sensing will increasingly inform urban planning and eco-friendly designs, offering a scientific backbone for policies aimed at managing heat and fostering sustainable urban development. In essence, remote sensing is poised to be a cornerstone in building smarter, more climate-resilient cities.

5.3 STREET VIEW IMAGE MEASUREMENT

The expansion of urban scale and the acceleration of urbanization have brought tremendous pressure to urban development. The original natural environmental conditions have been altered and replaced to varying degrees by a large number of artificial environments, such as roads, buildings, and urban green landscapes. The natural spatial structure of the city has undergone significant changes, evolving from a completely natural environment to a complex and multifunctional artificial environment.

Therefore, conducting an objective evaluation of the environment holds significant importance. It can effectively adjust urban functions, improve urban environmental quality, assist in formulating principles and plans for environmental protection, and ensure the sustainable and coordinated development of the urban economy and environment. Traditional urban environmental evaluation research has predominantly employed methods based on field surveys, which are challenging to implement on a large scale and at a fine-grained level.

Street view images (SVIs) provide big data in the form of real-scene images for urban environmental evaluation research, which offers new opportunities for urban studies at a large-scale level. Compared with traditional field-survey data, SVIs not only have wide coverage and a large amount of data, but also map providers offer API development interfaces that allow the free download of SVIs. More importantly, the images contain rich urban infrastructure information, including both man-made and natural landscapes at the street level, and they intuitively and accurately reflect the urban facade information. These advantages make SVIs an important data source for urban environmental evaluation research, while also bringing new methods and opportunities to this field of study.

5.3.1 Measurement Methods

In recent years, SVIs have gained considerable momentum in urban studies. SVIs are an innovative type of geospatial data used for perceiving the physical environment of cities. By providing urban spatial information and observational images of cities from a pedestrian's perspective, SVIs enable users to remotely explore authentic street landscapes, including both natural and man-made scenes. Owing to the continuous progress in machine learning, computer vision (CV), and the computational technologies for processing large-scale images, SVIs have swiftly emerged as a significant data source for geospatial data collection and urban analysis [88, 89]. The vast data sources provide unprecedented opportunities for digitalization, enabling researchers to conduct large-scale studies on urban environments and human activities.

SVIs record the detailed 3D profiles of urban streets from a micro and pedestrian perspective, creating panoramic views and more effectively capturing micro-scale street elements [90]. Currently, SVIs have covered half of the global population [91], providing a valuable, large-scale data source for urban data. Under certain conditions, they can even replace field surveys [92]. Image data are characterized by their extensive coverage, high resolution, large volume, and low cost [93, 94], which can provide data support for the quantitative study of the urban physical environment. Based on the sources of data, SVI can be categorized into panoramic images and crowdsourced images [89].

5.3.1.1 Panoramic Images

Panoramic images are primarily collected by map service providers, using street view vehicles to traverse urban road networks and gather 360-degree panoramic visual information. Currently, dozens of street view services serve as sources of SVI data, with the majority covering one or several countries. The primary services include Google Street View (GSV), Tencent Street View (TSV), and Baidu Street

TABLE 5.1
APIs of Street View Images

Street View API	Coverage Area	The Maximum Resolution of the Image (Width × Height)
Google	114 countries and regions	2048×2048
Baidu	372 cities in China	1024×512
Tencent	296 cities in China	960×640

View (BSV) (descriptions and comparisons of the three APIs of SVIs are presented in Table 5.1). The high coverage of SVIs not only provides rich information for a comprehensive description of urban environments but also enables researchers to assess and measure diverse landscapes across countries on a global scale [95, 96].

GSV is a free image service and is considered the most well-known and widely used service for providing SVI, with image retrieval sourced from the GSV API (Figure 5.5). Since its launch in 2007, GSV has covered over 90 countries and has even extended to indoor spaces. With the exception of a few cases where wearable cameras are used to survey narrow roads, panoramic images are predominantly obtained through a standardized approach that involves mounting multiple cameras on the roof of a vehicle, along with various sensors such as LiDAR [97]. With the exception of some landmark buildings and unconventional locations, the vast majority of images are omnidirectionally captured and taken from public roads, closely resembling what pedestrians would see [98]. The panoramic images provided by GSV allow users to drag the view to inspect different parts of the surrounding environment. The service can be accessed through a web interface integrated with Google Maps, smartphone applications, and application programming interfaces.

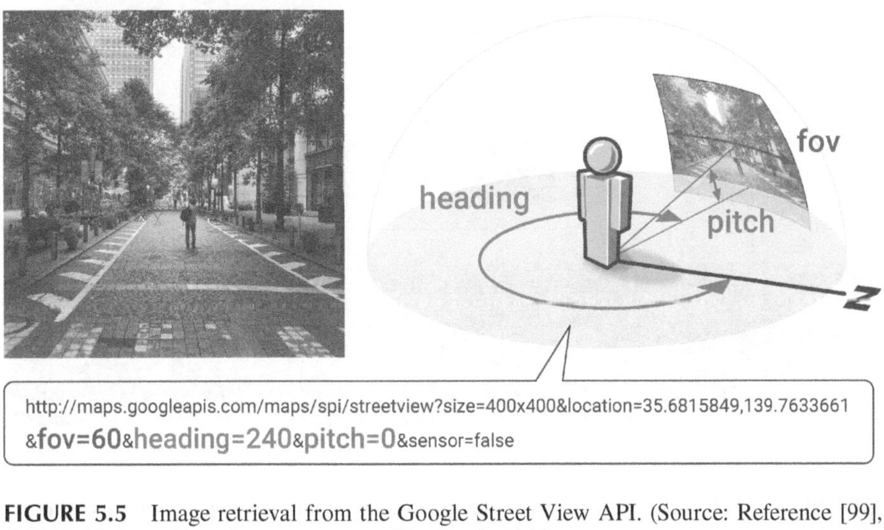

FIGURE 5.5 Image retrieval from the Google Street View API. (Source: Reference [99], CC BY 4.0, https://creativecommons.org/licenses/by/4.0/.)

However, the application programming interface does not permit the retrieval of historical images, and the image resolution is relatively low [88].

Given that the operations of GSV are restricted in China and thus cannot be used normally, TSV and BSV can be considered as alternative services to Google Maps. BSV is a web-based map service provided by Baidu, and its SVIs are only used in China. To comprehensively reflect the spatial characteristics of street views at a given measurement point, BSV collects a sample approximately every 40 m. Each measurement point includes four SVIs, which together provide a panoramic view of the surrounding environment (Figure 5.6).

5.3.1.2 Crowdsourced Images

With the popularization of network information technology, location-based spatial services have entered a period of rapid development, generating a large amount of

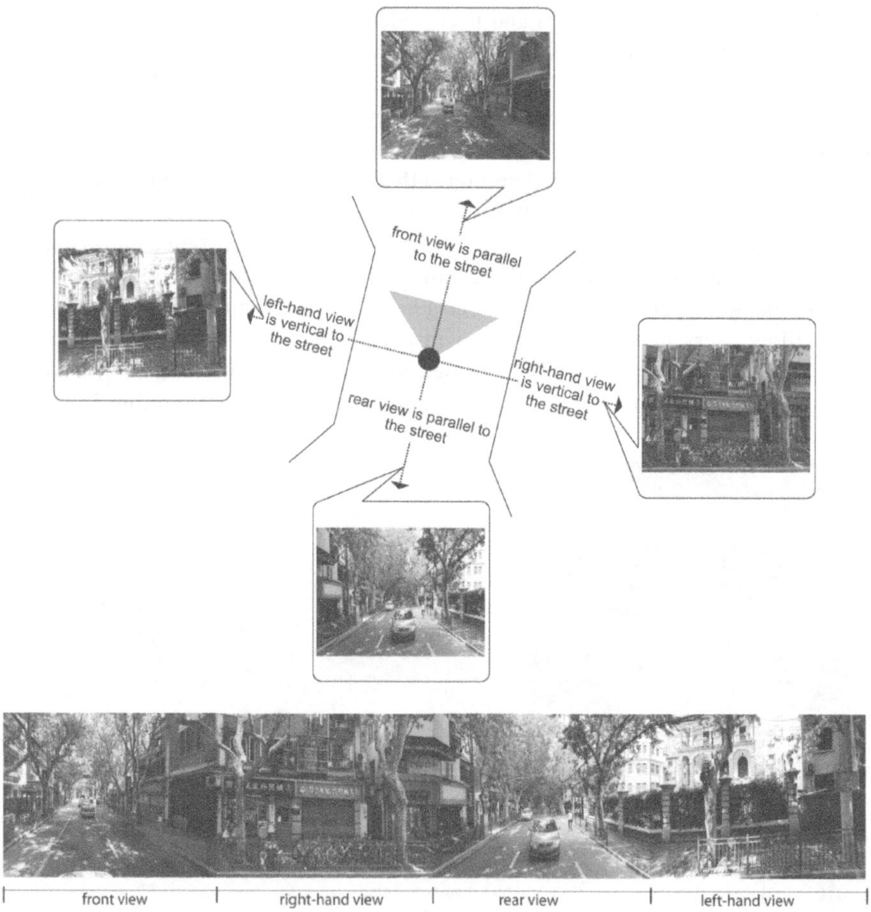

FIGURE 5.6 The collection of street view images. (Source: Reference [100], CC BY 4.0, https://creativecommons.org/licenses/by/4.0/.)

crowdsourced geographic data every day. Crowdsourced geographic spatial data can better reflect various aspects of information related to urban development, including human activities, socioeconomics, environmental pollution, and more [101]. Owing to the high quality and good perspective of images captured by users, they are frequently employed as a supplement to panoramic images. Compared with panoramic images, a significant portion of crowdsourced images is collected by users while walking and cycling. Spatial sampling is one of the main factors that distinguish crowdsourcing service platforms from commercial services such as panoramic services, which often provide comprehensive coverage of cities with relatively uniform sampling. Mapillary, KartaView, and Apple Maps are all widely popular crowdsourcing service platforms. These services are mostly applicable to any type of SVI captured by any device from different mobile platforms.

Mapillary is a crowdsourced street view platform that allows users to upload street photos to create a public, searchable database of street images [102, 103]. These photos can be used for a variety of purposes, including mapmaking, urban planning, and traffic management. Mapillary also provides tools and APIs that enable developers and researchers to utilize this data. KartaView is an online tool provided by Mapillary that allows users to create custom maps using Mapillary's street view imagery. Users can personalize their maps by selecting different views and adding tags and annotations. KartaView can be used in a variety of scenarios, including education, tourism, and urban planning.

Apple Maps is a mapping service provided by Apple Inc., integrated into iOS devices and macOS systems. In addition to offering basic map and navigation functions, Apple Maps also includes a street view service known as "Look Around". This service provides high-resolution 360-degree panoramic images, allowing users to explore city streets. Apple Maps also includes additional features such as real-time traffic information, satellite imagery, and 3D views.

These platforms all leverage the concept of crowdsourcing, which involves utilizing the participation of a large number of users to collect and update map data. This approach can help them rapidly expand their map coverage and enhance the accuracy and timeliness of the data. However, each platform has its own specific user base, data collection methods, and use cases, which means that they may differ in terms of functionality and user experience. The advantage of such crowdsourced services may be higher temporal resolution and the ability to cover areas not served by GSV. On a micro scale, crowdsourced services may include images from sidewalks and bike lanes. On a macro scale, they can provide images of cities/countries that are not available through GSV and other commercial services. Crowdsourced services allow for the download of images of the same location captured by different users at different times using various devices. Therefore, in some locations, the temporal resolution of crowdsourced images is finer than that of panoramic images. Panoramic images are typically captured every few years, which limits access to historical imagery. Compared with panoramic images, crowdsourced images can be downloaded at a higher resolution. Although crowdsourced images have the advantages of flexibility and resistance to the limitations of commercial services, they also have significant drawbacks, such as the field of view and coverage. Most volunteers use non-panoramic cameras, such as dashcams mounted on

FIGURE 5.7 An example of image segmentation for a street scene: Street view images before (top) and after the segmentation (bottom). (Source: Reference [105], CC BY 4.0, https://creativecommons.org/licenses/by/4.0/.)

cars and smartphones, which can only provide a limited (single-frame) perspective [104], resulting in a relatively narrow field of view. Given the substantial differences among image takers and the devices they use, the quality of the images inevitably exhibits a high degree of variability. The spatial coverage of crowdsourced services is far less comprehensive than that of panoramic images, which is a significant limitation.

CV technologies and deep learning algorithms provide technical support for the quantification of the urban physical environment [105]. The primary visual tasks of CV technologies are object detection and object segmentation. Many deep learning models, such as deep convolutional neural network (DCNN), fully convolutional networks (FCN), and SegNet, have been applied to image semantic segmentation (Figure 5.7). In the specific domain of SVI segmentation, the performance of these models may vary. DCNNs are the most commonly used image analysis models in deep learning, with representative structural frameworks including AlexNet, GoogLeNet, DenseNet, and others. Deep learning-based CV technologies describe and classify street views according to the features extracted from urban SVIs. They can also automatically extract multi-level information such as urban geometry, building facade colors, urban green space ratio, and sky view factor (SVF), solving many complex problems in image and video analysis.

Owing to global warming and rapid urbanization, the frequency and intensity of extreme heat events have increased, which greatly affect human thermal comfort and public health [106]. For pedestrians, high temperatures are a key barrier to walking activities, reducing pedestrian mobility and urban vitality [107]. The geometry of urban areas and street greening has significant impacts on the urban thermal environment. Good urban planning and design can improve the thermal comfort of pedestrians outdoors [108, 109]. The use of SVI to study urban street environments, to quantify urban geometry and levels of greening, plays an important guiding role in improving human quality of life, promoting healthy urban development, improving urban transportation, and reducing energy consumption. It has now become a popular method [106, 110].

5.3.2 Case Studies

5.3.2.1 Urban Geometry

The street canyons formed by high-density building clusters in urban centers are crucial to urban microclimate and comfort. Previous research on thermal comfort in urban outdoor spaces has shown that the aspect ratio (H/W) of street canyons and SVF are the main geometric parameters of urban form structure [111, 112].

The street H/W is defined as the ratio of the average wall height (H) to the street width (W), indicating the density of buildings relative to their height. It can categorize street canyons into three types (i.e., avenue canyons, regular canyons, and deep canyons). When the H/W is close to 1, it may be the optimal proportion for thermal comfort; when the H/W is less than 0.5, this proportion is considered shallow; when the H/W is equal to 2, this proportion is considered deep [113–115]. The traditional method for studying the H/W involves manual field surveys, which often require significant human and material resources, are inefficient, incomplete, and inaccurate. Moreover, data updates are difficult, and the data is generally not publicly available, which limits the study of the impact of street canyon geometry on the environment. SVI provides large-scale urban data, which, combined with deep learning algorithms, enables the study of the H/W of street canyons from a broader spatiotemporal perspective [116]. This method can reveal the aesthetics, spatial quality, and evolutionary process of urban form. More importantly, SVI contains more detailed and realistic information about the urban landscape, which is valuable for studying the street H/W.

Compared with the H/W, SVF is a more advantageous parameter, as it can better describe complex urban environments and is a key factor affecting urban environmental aspects such as surface heat balance, microscale air circulation, air pollution, and microclimate. Oke proposed the concept of SVF, which is defined as the ratio of the visible sky area from a given surface point to the total possible sky area in an urban street space [117]. The value is a dimensionless number between 0 and 1. A value of 0 indicates that the sky is completely obstructed, while a value of 1 indicates no obstruction at all [118]. It represents the openness of outdoor public spaces and also serves as an index for measuring the shading degree of different street canyons [119]. SVF characterizes the ratio of the radiation received (or emitted) by urban streets to the total radiation emitted (or received) by the entire hemispherical radiation environment. Increasing the height of buildings surrounding street canyons will obstruct the sky, resulting in a decrease in SVF, whereas reducing the height will lead to an increase in SVF. As the morphology of buildings along the street and street trees changes, the sky visibility from the pedestrian's perspective will also change [120].

With the increasing significance of SVF in research, over the past few decades, substantial studies have been conducted on the calculation or estimation methods of SVF to enhance its accuracy and efficiency [121–124]. A variety of computational methods and mathematical models for calculating and assessing SVF have been developed, which can be primarily divided into five categories: geometric methods [125], global positioning system (GPS) methods [126–128], simulation methods based on 3D urban models or digital surface models (DSMs) [129–132], and image-based methods [124, 133–136]. There may be some overlap among these methods, with image-based methods being more accurate and direct.

The fisheye imaging method uses a circular fisheye lens to shoot directly at the sky, projecting the hemispherical environment onto a circular plane to produce a two-dimensional circular fisheye image, from which the SVF is derived. Figure 5.8 shows an example of fisheye image generation used for measuring SVF. To segment the sky and obstruction regions, image processing software is used to perform binarization, contrast adjustment, and brightness adjustment on fisheye images. SVF is calculated as the percentage of visible sky area [137, 138], and the height at which fisheye photos are taken may vary [139]. The fisheye imaging method takes into account the obstruction of vegetation and other urban infrastructures, thus providing more accurate estimation results. However, the segmentation process of different types of image processing software (such as RayMan and SkyView) relies on manual operation and parameter settings, which is time-consuming and labor-intensive. The method of calculating SVF using fisheye photos can show how obstacles such as trees and buildings enter the hemispherical view, but fisheye imaging requires on-site photography [140] and is also affected by lighting and weather conditions, making it unsuitable for large-scale studies [135].

SVI covers most urban areas in the world, especially GSV images, which are increasingly being used in scientific research. The wide availability of SVI has expanded the scope of research. The SVI method is an open-source visualization method based on SVI for estimating SVF, which is low-cost and highly efficient [134]. Middel et al. [142] proposed utilizing fisheye images generated by Google Earth to calculate SVF in large urban areas at high spatial resolution. This method can calculate SVF for the full 3D environment of large urban areas, including vegetation, and is independent of existing 3D building databases or DSMs. Carrasco-Hernandez et al. [143] used the open-source software Hugin for image merging and projection, as well as the Rayman software tool to calculate the SVF of street canyons. These studies have demonstrated the feasibility of estimating SVF based on SVI. Although this approach significantly reduces the time required for field surveys and photography, the time needed for manual processing of large-scale urban image data remains substantial. With the advancement of computer technology, some researchers have attempted to utilize the efficiency of deep learning and image recognition technologies to extract information from a large number of SVIs and conduct relevant analyses (Figure 5.9). Xia et al. [144] proposed a DeepLabV3+ semantic segmentation model and a deep learning-based automatic estimation method for SVF values. The method employs a pre-trained DeepLabV3+ deep learning model to perform semantic segmentation on SVI, subsequently generating fisheye images and automatically calculating SVF based on the results. The method achieved a recognition rate of up to 98.62% for sky regions in SVI when comparing the semantic segmentation results with manually labeled sky regions. Zeng et al. [134] developed an SVF estimation tool based on SVI using Python programming language and OpenCV software library. The tool is capable of handling various outdoor weather conditions, and

FIGURE 5.8 Street views of the three investigated intersections (a–c), and their sky view factor values (calculated by RayMan 1.2). (Source: Reference [141], CC BY 4.0, https://creativecommons.org/licenses/by/4.0/.)

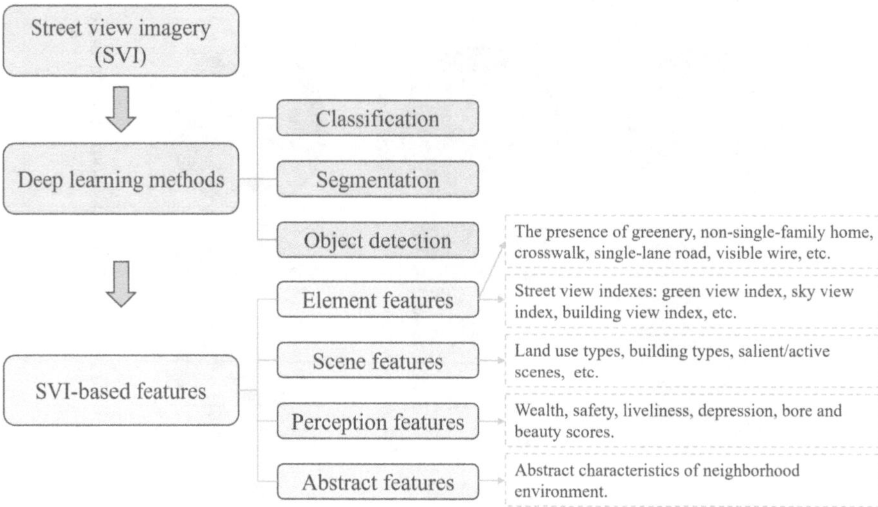

FIGURE 5.9 Street view imagery (SVI)-based features and deep learning methods.

its performance in sky region segmentation and SVF calculation has been verified through photos taken with fisheye lenses. However, the method is susceptible to vegetation obstruction, and seasonal variations in vegetation can also introduce errors in SVF estimation. Gong et al. [135] proposed a method that utilizes the Pyramid Scene Parsing Network (PSPNet) to extract street features from GSV images and estimate all street view factors, including SVF, tree view factor (TVF), and building view factor (BVF) for street canyons. Taking the urban area of Hong Kong as the research object, the accuracy and effectiveness of this method were compared with the SVF estimated from hemispherical photography reference data. Nice et al. [145] proposed an automatic sky region detection system based on an adaptive algorithm and demonstrated that it can produce accurate results using a wide range of SVI types. The method employs mean shift segmentation, K-means clustering, and Sobel filtering to process the images and identify sky pixels. The algorithm for specific images is selected by a convolutional neural network trained on 25,000 images from the Skyfinder dataset, which can adaptively choose the optimal image processing method to improve detection accuracy. These studies demonstrate the feasibility of low-cost, automated, and efficient SVF estimation.

The further quantitative research on the urban environment using SVI mainly relies on the extraction of data from various urban elements such as buildings, bridges, street facilities, and roads [146], including building names, locations, uses, heights, number of floors, years, sizes, types, and so on [147]. For any city, a rapid and universal method for determining building data based on open-source data is a better choice, which greatly enriches the building dataset. Li et al. [148] first attempted to use deep learning technology to estimate the age of buildings from images. The algorithm mainly includes three steps: downloading GSV images, extracting image features, and estimating the age of buildings. By using a DCNN, the age of buildings can be accurately predicted. The precise measurement of urban facade colors is a necessary condition for urban color planning. Zhong et al. [149] used a deep learning algorithm to automatically extract the dominant colors of urban facades (DCUF) from BSV panoramic images, proving that SVIs can serve as an important data source for urban color research. With the development of measurement tools and data sources, numerous automatic building classification methods have emerged [150]. The classification methods combined with SVIs and deep learning are open-source and favored by researchers in the fields of urban analysis and urban computing, showing great potential in many applications, such as urban population mapping [151], density analysis [135], and urban public facility planning [152]. The nature of cities is complex, and classical urban modeling may not be able to capture the actual complexity of urban systems. However, the combination of CV technologies, deep learning algorithms, and data sources such as SVI, points of interest (POI), satellite remote sensing, and social media has made progress in handling a variety of complex physical and non-physical visual tasks [153].

5.3.2.2 Urban Greening

Under the influence of global warming and urbanization, the urban thermal environment is gradually deteriorating, posing a threat to human health. Therefore, effectively regulating the urban thermal environment has become a key challenge [154]. Green spaces are an important component of urban landscapes. Research has shown that green spaces can cool the urban environment and mitigate the heat island effect, which is crucial for improving the urban thermal environment and pedestrian thermal comfort [155–158].

Traditional green landscape extraction often employs satellite or drone remote sensing methods. By utilizing the strong reflectance characteristics of green landscapes in the near-infrared band, these methods can accurately identify and extract vegetation types such as forests and grasslands. However, this approach often fails to detect natural or artificial vegetation beneath tree canopies, which affects more detailed green landscape planning. In recent years, the emergence of SVIs has to some extent overcome the aforementioned issues. SVI has unparalleled advantages over traditional data sources such as satellites or drones. It can perceive green landscape features such as the distribution of vegetation on street facades from a human-level perspective and has become an important supplementary data source for green landscape extraction. The use of SVI for extracting urban green landscapes can more accurately and three-dimensionally reflect the spatial distribution of green landscapes. It has significant application value in fields such as green landscape visibility assessment [159], urban economic development [160], measurement of residents' physical and mental health [161, 162], and urban planning interventions [163].

In most cases, studies primarily measure the quantity of roadside greenery at the urban scale or across multiple cities (e.g., a relative measure of vegetation visible at the street level at a given location), for urban planning interventions, street tree maintenance, planting efforts, and greenway networks. Traditional urban greening indices, such as green coverage [164], leaf area index (LAI) [165], and normalized difference vegetation index (NDVI) [19], are mostly measured and calculated from an aerial perspective using remote sensing images or geographic information systems. Due to the angle of capture, remote sensing images often overlook shrubs and lawns beneath tree canopies, as well as green vegetation on building walls [166].

Green view index (GVI) is a horizon-based index used to assess the level of street greening. This value represents the proportion of green pixels in the pedestrian's field of view, reflecting the extent to which pedestrians perceive the surrounding green environment [167]. The GVI is a commonly used indicator for quantifying urban greening and assessing the visibility of urban forests, focusing primarily on pedestrians' views of green spaces. There are three methods for measuring GVI: on-site measurement, remote sensing measurement, and SVI measurement. There are three methods for measuring GVI: on-site measurement, remote sensing measurement, and SVI measurement. The accuracy of remote sensing measurement is limited [168], and remote sensing images and LiDAR data are often used in conjunction with SVI data to assess urban greening [169]. However, LiDAR data is typically collected at specific points, with high acquisition costs, and there is a lack of methods that can be universally or automatically applied for high-precision processing over large areas [170]. We need novel, easily accessible methods and data sources to achieve economical, continuous, and frequent data collection and rapid processing [171]. The availability of low-cost, detailed, and increasingly popular SVIs is growing [93, 172]. Researchers have utilized SVI, such as GSV and TSV images, as a new data source and combined them with CV technologies to automatically extract green areas and calculate GVI values, thereby compensating for these shortcomings [170, 173, 174]. The GVI focuses on green landscapes at the pedestrian level and is primarily extracted using image semantic segmentation techniques [88].

Color space transformation, band operations, and other commonly used methods for extracting green spaces are also continuously evolving [175]. Ki et al. [174] used GSV and deep learning to calculate GVI through semantic segmentation. The results showed that there are differences between traditional greening variables and GVI in terms of specific greening forms and perspectives. Dong et al. [176] used Beijing as the study area, extracted green vegetation areas from TSV images, and calculated GVI values using image segmentation algorithms. This demonstrated the feasibility of using SVI to assess the complex street greening levels in megacities. Since GVI is based on point estimates, it is sensitive to the location of sampling points when aggregated at the regional level using means or medians, potentially over-weighting values from densely sampled locations. Kumakoshi et al. [99] proposed a summarization method called standardized GVI (sGVI). By using Voronoi grids, the weights of GVI locations are not proportional to the density of locations within the aggregation area. This method takes into account the positional density of GVI and improves the accuracy of street greening evaluation. Chiang et al. [177] established the consistency between deep learning and manual classification methods for urban environmental features and investigated the impact of street-level SVF and GVI on thermal comfort, particularly the differences in their effects during hot and cold seasons. Zhang et al. [178] employed a semantic segmentation model to segment SVIs, analyzed the GVI of Osaka City, and utilized an adjacency matrix and the Floyd–Warshall algorithm to calculate the optimal GVI path, thereby achieving visualization and integration of urban greening at the neighborhood level. To fully describe the complexity of the lateral distribution and vertical structure of street greening, researchers have studied new greening evaluation indices from different perspectives. For example, Tong et al. [179] combined remote sensing and street view perspectives to propose a new index for street vegetation structural diversity (VSD). Many studies on the quantification of urban greening involve satellite images. However, if the goal is to assess the side view of street greening and understand what people see on the ground, SVI has unparalleled advantages that most remote sensing methods cannot achieve. In fact, SVI is commonly used in combination with remote sensing data, such as airborne LiDAR and satellite observations, in research to obtain a comprehensive picture of urban green spaces, including roadside areas, parks, and off-road courtyards.

The GVI is also continuously developing and improving (see Table 5.2). In terms of data sources, some researchers have used BSV images to obtain Baidu GVI (BGVI) [180], while others have used panoramic SVIs to obtain panoramic GVI (PGVI) [181], further expanding the data sources for calculating GVI. In terms of the observational perspective, Dong et al. [176] expanded the range of GVI perspectives by adding upward and downward views to the horizontal view, making it more consistent with the visual range of pedestrians. Some scholars have used fisheye cameras to calculate TVF and have standardized and compared various GVI results horizontally. They found a strong correlation among different GVI results. Therefore, in studies assessing the visibility of green landscapes using SVIs as the base data, it is difficult to achieve revolutionary results solely through innovations in the calculation methods of indicators. However, some researchers have innovated indicators from other dimensions. Wu et al. [161] proposed urban traffic GVI (GVI-UT) from a temporal perspective, which quantifies the average GVI along streets during

TABLE 5.2
Commonly Used Evaluation of Urban Forest Visibility Indicator

Indicator	Calculation Method	The Meaning of the Variable
Green view index (GVI)	$GVI = \frac{\sum_{i=0}^{4(or6)} \sum_{j=0}^{n} Pix_{Green}}{\sum_{i=0}^{4(or6)} \sum_{j=0}^{n} Pix_{All}}$	i is the number of SVIs obtained in the horizontal direction, which can be 4 or 6; j is the number of pixels in a single image; Pix_{Green} is the number of pixels identified as green landscape; Pix_{All} is the total number of pixels.
Baidu green view index (BGVI)	$BGVI = \frac{\sum_{i=1}^{4} \sum_{j=1}^{3} area_{green_{ij}}}{\sum_{i=1}^{4} \sum_{j=1}^{3} area_{all_{ij}}}$	i represents the four Baidu Street View images obtained in the horizontal direction; j represents the three images obtained in the vertical direction.
Panoramic green view index (PGVI)	$PGVI = \frac{area_{green}}{area_{all}} \times 100\%$	$area_{green}$ is the number of green landscape pixels in the panoramic image; $area_{all}$ is the total number of pixels in the image.
Tree view factor (TVF)	$TVF = \frac{1}{2\pi} \sin\frac{\pi}{2n} \sum_{i=1}^{n} \sin\left[\frac{\pi(2i-1)}{2n}\right] \alpha_{i,x}$	Convert SVI to fisheye images and divide them into several concentric circular rings of equal width; n is the total number of rings; i is the index of the ring; $a_{i,x}$ is the angular width of the tree pixels in the ith ring.

residents' travel. This can provide residents with more comfortable route choices for their travel.

SVI provides rich urban micro-data, which is an important extension of traditional remote sensing data sources. However, due to the limitations of its field of view, it needs to be integrated with the advantages of large-scale observation from traditional remote sensing data sources. Currently, research on assessing urban green landscapes using the integration of SVI and remote sensing data is relatively scarce. Therefore, to fully express the level of urban street greening, future efforts are still needed to explore the fusion methods of these two heterogeneous data sources, providing reference indicators for the refined planning of green landscapes that balance human-scale and accuracy.

The segmentation accuracy of SVI has made significant progress with the rapid development of machine learning, providing possibilities for detailed research on green landscapes. As urban greening policies shift from increasing quantity to improving quality, the focus is now on the quality of urban greening. The spatial distribution of green landscapes should not be limited to a two-dimensional perspective. Instead, it should involve enriching the types of vegetation, expanding greening beneath tree canopies, and configuring vertical greening on street-side buildings or fences. The existing SVI segmentation datasets have a relatively simple classification system for green landscapes and cannot distinguish between types of vegetation. More precise classifications, such as those between trees, shrubs, and grasses or between different tree species, can provide support for more refined green landscape configurations. SVI has also been proven to be useful for location estimation [92]. In addition, it can be combined with oblique photogrammetry data and LiDAR data to more accurately depict the location, height, radius, and other information of green landscape points, thereby providing richer vegetation parameter information for the fine classification of green landscapes.

5.3.3 Practical Applications

The widespread adoption of large-scale image platforms, advances in CV and machine learning technologies, and the availability of computing resources have propelled SVIs to become an important data source for geospatial data collection and urban analysis. Since the initial promotion of large-scale SVI services, researchers have realized that they are very suitable for assessing the characteristics of the built environment [182]. In recent years, SVI has been used in various aspects of urban research, such as collecting pedestrian count data [183], analyzing accessibility [184], demographic studies [185], mapping infrastructure [186], creating new geospatial data [187], expanding existing datasets [188], and inferring building types [88], among others. SVI provides unprecedented opportunities for digitalization, enabling researchers to conduct large-scale studies on urban environments and human activities.

Exploring the quantification and visualization of urban elements using SVIs and AI technologies, mining additional geospatial information in cities, and using more complex or specific indicators to achieve large-scale and quantitative urban environment evaluation research has become a hot topic in urban environment evaluation research in the big data era. SVIs are related to the scene information directly captured and directly perceived by humans, as well as the social, economic, psychological, and physical feelings associated with them. A review of the application of SVIs in urban environment evaluation is conducted according to the evaluation of urban physical environment, urban social environment, and urban aesthetic environment.

5.3.3.1 Quantification of Urban Physical Environment

The urban physical environment includes natural environments such as topography, geology, soil, hydrology, climate, vegetation, and animals, as well as artificial environments like buildings, roads, pipelines, infrastructure, different types of land use, exhaust gases, and wastewater. The urban natural environment forms the foundation of the urban environment, while the urban artificial environment constitutes the essential material infrastructure for realizing various urban functions. The urban physical environment is substantially linked to various health outcomes and comfort levels of residents. SVI enables researchers to capture landscapes at eye level and serves as an important data source for observing and analyzing the urban physical environment. Given that SVIs mainly capture urban elements related to the built environment, such as buildings, roads, infrastructure, and street trees, current research on urban physical environment evaluation based on SVIs primarily focuses on the assessment of the built environment. SVI data is an important and publicly available data source that can be used to create indicators for the physical environment through machine learning techniques. The use of SVI, CV techniques, and deep learning algorithms to explore the quantification and visualization of urban elements, to mine detailed geographical and spatial information in cities, and to identify complex or specific indicators is gradually becoming a hot topic in urban environment evaluation research. This approach enables large-scale and quantitative assessments of urban physical environments.

SVI can be used for thermal environment simulation, acoustic and light environment monitoring, as well as air quality assessment. Combine meteorological data calculations, numerical model simulations, and the shooting location and geometric feature information of the SVI system for radiation and temperature simulations [143]. Using deep learning techniques, the SVI system can infer the SVF in the environment and assess the UHI effect and thermal comfort [189]. Using SVI can also extract different types of roadside vegetation to analyze the spatial relationship between urban layout and thermal environment. Projecting the solar trajectory onto SVI can be used to calculate the duration of solar irradiation and quantify the total shortwave radiation intensity at the street level [142]. However, this requires the integration of weather conditions and analysis of the separated radiation patterns. Compared to using expensive and limited 3D building models to calculate solar radiation, SVI is an ideal complementary data source for simulating solar radiation within street canyons. SVI presents the urban physical environment in the form of 3D cross-sectional views, calculates the various impacts of different elements on people's behavior and perception, and conveys detailed visual content. Using SVI can quantify the impact of building facades, courtyards, and street landscapes on noise disturbance and stress levels, and it can be used to monitor traffic noise in urban environments. Deep learning methods can be used to analyze features extracted from SVI data and apply SVI to assess air quality. Mobile monitoring has been employed to collect real-time air quality measurement data to evaluate local air quality and exposure to air pollutants. CV techniques can also be employed to extract relevant features from SVIs, such as color, texture, and shape. These features can reflect the distribution of particulate matter, haze, and other pollutants in the air, thereby enabling the assessment of air quality [190, 191].

SVIs, as a type of geographic big data for perceiving the urban physical environment, are characterized by their wide coverage, high density, detailed content expression, and high acquisition efficiency. The application of SVIs to the quantification of urban physical environments is a complex and multifaceted process that integrates various technologies and methods, providing rich data support for urban studies. With the continuous development and improvement of computer technology, there will be more breakthroughs and progress in using SVI to study urban physical environments.

5.3.3.2 Urban Social Environment Analysis

The urban social environment includes population distribution and dynamics, service facilities, recreational facilities, and social life. It reflects the characteristics of urban areas as human settlements distinct from rural areas and other forms of agglomeration, providing the conditions necessary to meet the various activity needs of people in cities. By identifying urban elements in SVIs and integrating the social environmental attributes and meanings that these elements reflect, it is possible to conduct evaluations and research on the urban social environment.

Based on the identification of elements in SVIs, the use of deep learning techniques for object detection and image classification has demonstrated potential at the street level for maintaining spatial data on infrastructure. At the street level, the focus of element recognition is often on smaller and easily overlooked street infrastructure elements, such as manhole covers [192], road surface damage [193], streetlights [194], and traffic signs [195]. At the building level, based on the recognition of repetitive elements and composition within them, it is possible to determine information such as architectural style [196], building age [197], and window-to-wall ratio [198]. In addition, analyzing and inferring non-visual characteristics of the social environment based on specific elements identified at the visual level is another research focus. These characteristics can be linked to social phenomena such as crime [199], social equity [200], and gentrification [201], or combined with spatial socioeconomic data to test, predict, and assess the economic environment of an area.

Currently, the quality of the social environment is mainly explored through themes such as environmental perception, accessibility, urban vitality, public health, and physical exercise. The subjective evaluation of SVIs often continues to use a combination of public scoring and machine learning. The measurement of objective variables mostly utilizes semantic segmentation from deep learning to quantify visual elements in images, thereby constructing relevant indicators for assessing spatial quality. A spatial environment evaluation model integrating multiple data sources has been formed, including POI, location-based services (LBS), and social media data.

SVI is also used in disaster risk assessment. In pre-disaster vulnerability assessments of exposed elements, data is collected through on-site surveys using multi-sensor fusion systems that include panoramic cameras, or by leveraging SVIs combined with object detection and image classification techniques from the field of CV to investigate the exposure information of buildings and infrastructure. For example, assessing tilted utility poles susceptible to tsunami disasters [202], identifying soft-story buildings vulnerable to earthquake disasters [203], detecting architectural features of buildings prone to flood damage [204], and categorizing building types susceptible to earthquake disasters [205, 206]. In addition, studies have shown that SVIs captured on-site can support post-disaster assessments of damage to the built environment [207].

Given that panoramic images can reflect urban micro-characteristics and the functional types of buildings, they can be used either independently or in combination with data from social media, POI, and other sources to provide more precise classification of urban land use as a supplement to traditional remote sensing data. In terms of research methods, image classification is commonly used to categorize the land-use functions in SVIs by scene type [208], or a correlation between land-use functions and visual elements quantified through semantic segmentation is established [105]. In terms of classification scale, multi-level classification tasks can be achieved, ranging from neighborhood scale, multiple parcels to city-wide levels.

5.3.3.3 Urban Aesthetic Environment Evaluation

The urban aesthetic environment includes landscapes, urban fabric, architectural features, and cultural relics. It is the external manifestation and reflection of a city's image, temperament, and charm. SVIs record the 3D cross-sectional views of urban streets at a large-scale and fine-grained level, allowing for detailed evaluation of the urban aesthetic environment. People develop unique feelings about a place due to the visual environment, their experiences, and the activities of residents within it. Their emotional and perceptual responses to the environment reflect their views on the surroundings, such as whether they find it safe, beautiful, and vibrant. These emotional perceptions can also play an important role in people's mental health. SVI can represent street space from a human perspective and has been used in a large number of urban perception studies. SVI is an important source for deriving various indicators of environmental characteristics. Deep learning models trained on datasets can simulate people's emotional responses to the environment, thereby assessing health, quality of life, and sense of security.

Many urban perception studies focus on aspects that are less tangible and less measurable, such as inferring the function, vitality, and form of a city, which are more likely to be subjective. The core theme is to measure the perceived quality of street landscapes. Researchers use SVI to measure urban perception attributes, such as perceived safety and affluence, vitality, comfort, and attitudes toward greenery. The purpose of these studies is largely to contribute to urban planning and design. However, many studies also have more specific applications in urban perception, such as analyzing physical activity [209], the impact of architectural form on human physiological responses [210], understanding residents' color preferences [211], the symmetry of urban blocks [212], and commercial hotspots [213]. SVI data is an important and publicly available data source that can be used to create indicators describing the characteristics of the physical environment through machine learning techniques. Studies have shown that characteristics of the built environment are correlated with mental health and chronic diseases. Most studies on residents' health rely on quantifying green exposure. For example, residential greenness is a unique and modifiable exposure structure that can alleviate physiological stress, reduce the risk of depression and anxiety, and improve people's health [214]. People's perception of community safety can alleviate psychological stress and enhance life satisfaction, thereby affecting their mental health. In addition, architectural features may also indirectly affect residents' mental health through factors such as accessibility [94], greenery [215], and public open spaces [216]. Stress and mental health are the primary research focuses. Research on infectious diseases is also crucial for health and well-being, as disease transmission is directly related to environmental variables. SVI provides a great opportunity to study environments that harbor pathogens of infectious diseases. Current research has covered potential breeding environments for dengue fever, high-risk areas for COVID-19, as well as environmental factors that cause diseases and their transmission pathways.

SVIs record the detailed 3D profiles of urban streets from a micro and pedestrian perspective, close to the human visual angle. They create panoramic views that can effectively capture micro-scale street elements. The architectural form, function, and human-scale features in the built environment are the main factors that influence street vitality. SVI can be used to extract environmental features, including recreational facilities [217], sidewalk quality [218], and street enclosure characteristics [219]. The analysis of residents' quality of life in the built environment using SVI also includes identifying potential urban congestion points [220], understanding measures to mitigate nearby road pollution [221], predicting the difficulty of driving [222], and identifying dumpsites [223]. In addition, the built environment can also influence people's behavior in engaging in physical activities [223]. SVI can be used to measure the residential environment related to walking infrastructure and traffic safety, such as the impact of greenery on walking behavior [224] and walking infrastructure [225]. Cycling is a form of physical activity that offers both health and environmental benefits [226]. Images collected through SVI can be used to assess environmental factors that influence cycling behavior and to determine the usability of roads. The application of SVIs in emotional perception not only helps to understand people's perceptions and emotional responses to urban spaces but also provides decision-making support for urban planning and design, promoting the development of urban spaces that better align with human behavioral patterns and needs.

In the era of big data, the widespread use of navigation and positioning devices, mobile devices, and mapping services has given rise to a new type of geographic big data—SVIs. The widespread use of SVI tools, along with the rapid development of CV, machine learning, and deep learning technologies in the field of AI, has driven the large-scale and automated identification and interpretation of SVIs in urban research. SVIs are a novel big data source for representing urban environments, with an observational perspective closer to that of urban residents and richly detailed content. SVIs not only can depict the visible environment of urban physical space in detail, such as buildings, roads, and natural features, but also can implicitly express the invisible environment. The application fields of SVIs are rapidly expanding and diversifying.

In urban research utilizing SVIs, virtual audits that are accurate and comprehensive can be conducted based on the extraction of elements or scene information from the images; a range of public health and perception assessments can be carried out by analyzing the quality of the built environment and the level of greenery landscapes in SVIs; in assessments of greenery and thermal environments, the quantification of greenery and sky information in SVIs using AI algorithms can rapidly and cost-effectively calculate SVF and GVI; in the fields of 3D modeling and virtual reality, SVIs provide an alternative to the previously expensive 3D modeling; in the field of autonomous driving, SVIs can help vehicles achieve comprehensive perception of the entire semantic scene; as a supplement to traditional data (such as GIS data and remote sensing imagery), SVIs can refine the results of land classification in land use and air quality assessments based on land use analysis. Compared with traditional methods such as field observation, SVIs offer a lower cost and higher efficiency for studying urban environments at the resident level.

5.3.4 Conclusions and Future Directions

SVIs focus more on recording urban street-level scenes from a human perspective. With a wide data coverage and low data acquisition costs, they have become an important data source in current urban application research. Before the advent of SVIs, there was a lack of a low-cost and efficient way to comprehensively record urban physical space from a human perspective on a large scale, which limited most studies related to urban physical space and the built environment. With the support of AI technologies, SVIs can be used to conduct more in-depth studies on the characteristics, patterns, evolution, impacts on the socio-economic environment, and mechanisms of interaction with human activities of urban physical space.

Compared with traditional data sources, SVIs, with their high coverage, relatively low data bias, cost-effectiveness, time-efficiency, and pedestrian-level perspective, have become an important data source for urban environmental research. In terms of coverage and volume, global map service providers have already generated and stored a large-scale collection of SVIs. For example, GSV has covered most cities in nearly 200 countries, providing a wealth of information for urban environment research. All SVIs essentially follow the same standard rules and production processes, which to some extent ensures the quality of the images and minimizes data bias when using SVIs. By using the API services provided by map companies, users can download available images for free in a short period of time. When downloading images, users can customize image parameters, including location, time, camera angle, and image size. It is also important that SVIs can capture the objective scenes of cities from a human perspective. Unlike aerial photographs and remote sensing images taken from a bird's-eye view, SVIs are more in line with human perception. They study the urban environment from a human perspective and more comprehensively capture the details of objects in the built environment. In addition, SVIs can also be used to study the intangible aspects of urban life and people's perceptions of the environment.

First, SVI can capture elements of the built environment at the visual level of residents, allowing for the assessment of urban environments at a lower cost. With the development of SVI-related services, using SVI for urban analysis has become a trend. It will continue to evolve and receive more research and applications in the future. Second, with the support of AI technologies, representative evaluation elements such as roads, pedestrians, buildings, and trees can be extracted from the images provided by SVI to analyze and evaluate specific or integrated environmental elements within spatial scopes such as streets, neighborhoods, and cities. This makes it possible to quantify the urban environment and perceive environmental qualities. SVI provides practical assistance for the spatiotemporal coverage of urban environmental perception, making it possible to describe urban forms more finely, efficiently, and on a larger scale from both physical and social spaces.

While SVIs bring new data support and research ideas to urban environment evaluation studies, they also have some limitations. First, there is temporal variability. Some characteristics of the street view environment change over time, including the number, characteristics, and activities of pedestrians, as well as parked or moving vehicles, and many signs of physical disorder, such as trash. The conditions and characteristics of these elements may simultaneously exhibit random variations and regular fluctuations related to daily, seasonal, or weather patterns, which can lead to higher measurement errors. In addition, inconsistencies in the timing, season, and weather between field observations and SVI capture may also lead to biases. For example, collecting street view videos in the early morning may lead to lower levels of observed social and pedestrian activities, which may (depending on the timing of garbage collection) affect the measurement of physical disorder in the street view. The second limitation is temporal consistency. SVIs are not all captured at the same time; some images may be taken in winter, while others are captured in summer, spring, or autumn. In studies assessing urban green spaces, images captured during the same period are required. To maintain temporal consistency, it is necessary to check the images and exclude those that are not from the same period. Finally, there is the issue of privacy. To protect the privacy of individuals captured in SVIs, images depicting personal attributes need to be deleted or blurred. This may lead to an underestimation of elements related to neighborhood physical disorder and urban safety studies, resulting in biased research outcomes.

Given the advantages of urban SVIs, they have become a new data source in urban research, and using SVIs to study urban environment evaluation has also become a trend. Currently, urban environment evaluation studies based on SVIs are still in their infancy. On the one hand, the development of AI technologies, represented by deep learning, has made it possible to identify patterns that cannot be explicitly expressed. On the other hand, in the era of big data, the integration of urban application research data, represented by SVIs, remote sensing imagery, and geotagged social media data, poses challenges for the development of new theories and methods in urban environment evaluation research. Future urban environment evaluation research based on SVIs should focus on the following three areas of work: (1) integrating various data sources (such as SVIs, remote sensing images, geotagged social media data, and public transit cards) and new methods (deep learning, data fusion, big data analysis, and visualization) to conduct multi-angle, multi-level, and fine-grained studies of the urban environment, thereby enhancing the reliability and precision of urban environment evaluations; (2) comparative studies of current street environments and historical environments based on the time series of SVIs. SVIs from different time points are comparable, allowing the observation of environmental characteristics from historical SVIs and aiding in the study of the temporal dynamics of urban landscapes. In addition to using SVIs alone, one can also carry mobile devices and visit the site, enabling simultaneous virtual audits based on street views and on-site observations. The current urban environment can be directly compared with the historical environment; (3) comparative studies of urban environment evaluations based on SVIs across different cities. Given the wide coverage of SVIs, and with BSV and TSV gradually expanding to cover more cities nationwide, a unified data source for urban environment evaluation on a larger national scale is provided. Since SVIs focus on recording the built environment and do not include natural environment features such as urban atmosphere, soil, and water environment, future research needs to address how to expand the analysis of key elements based on local-scale SVIs to larger scales, such as entire cities or multiple cities nationwide. This can be achieved by integrating diverse spatiotemporal data that record natural environment elements like urban atmosphere, soil, and water environment to conduct comprehensive urban environment analysis and enable environmental evaluations across different cities. The continuous development of deep learning technologies and computational resources also provides technical support for conducting environmental evaluations across cities nationwide. This will provide more objective reference data for urban environment evaluation policies. In future research, various data sources should be integrated, such as remote sensing images and social media data. Utilize deep learning and big data analytics to conduct multi-perspective and multi-level research, thereby enhancing the reliability and accuracy of assessments based on SVI. With the advancement of CV technology, in addition to using static images, it is also possible to extract more detailed information about people's movements and activities from dynamic videos. Well-trained neural networks will play a significant role in helping us understand how urban environments influence residents' behaviors, perceptions, and health.

In summary, SVIs offer the advantages of wide coverage, the ability to provide street-level landscape information, and low data acquisition costs. These features make SVI an important large-sample data source and provide new research ideas for urban environment evaluation studies. The continuous breakthroughs in AI technologies and their applications in various fields have made it possible to conduct urban environment evaluation studies based on SVIs at large spatial scales.

REFERENCES

1. J.T. Snow, X. Zeng, P. Klein, S.E. Sarnat, J.M. Shepherd, E.M. Stanley, Needs Defining the, Urban Meteorology: Forecasting, Monitoring, and Meeting Users' Needs, National Academies Press, Washington, USA, 2012.
2. E. Erell, D. Pearlmutter, T. Williamson, Urban Microclimate: Designing the Spaces Between Buildings, Routledge, London, 2010.
3. D. Lai, Z. Lian, W. Liu, C. Guo, W. Liu, K. Liu, Q. Chen, A comprehensive review of thermal comfort studies in urban open spaces, Science of The Total Environment 742 (2020) 140092.

4. D. Lai, W. Liu, T. Gan, K. Liu, Q. Chen, A review of mitigating strategies to improve the thermal environment and thermal comfort in urban outdoor spaces, Science of The Total Environment 661 (2019) 337–353.
5. M. Salmanian, N. Ujang, Emerging need for micro-climatic considerations in urban design process: A REVIEW, Jurnal Teknologi (Sciences & Engineering) 84 (2022) 129–148.
6. Florida LTAP Center, Urban Heat Island and Pavement Preservation Explained [Webinar], Florida Pavement Preservation Council, 2021, https://floridaltap.org/urban-heat-island-and-pavement-preservation-explained/.
7. W. Zhang, J. Lian, C.-Y. Chang, K. Kalsi, Aggregated modeling and control of air conditioning loads for demand response, IEEE Transactions on Power Systems 28 (2013) 4655–4664.
8. I. Kousis, M. Manni, A.L. Pisello, Environmental mobile monitoring of urban microclimates: A review, Renewable and Sustainable Energy Reviews 169 (2022) 112847.
9. R. Guo, B. Yang, Y. Guo, H. Li, Z. Li, B. Zhou, B. Hong, F. Wang, Machine learning-based prediction of outdoor thermal comfort: Combining Bayesian optimization and the SHAP model, Building and Environment 254 (2024) 111301.
10. Y. Toparlar, B. Blocken, B. Maiheu, G.J.F. van Heijst, A review on the CFD analysis of urban microclimate, Renewable and Sustainable Energy Reviews 80 (2017) 1613–1640.
11. N.H. Wong, S.K. Jusuf, C.L. Tan, Integrated urban microclimate assessment method as a sustainable urban development and urban design tool, Landscape and Urban Planning 100 (2011) 386–389.
12. M. Taleghani, U. Berardi, The effect of pavement characteristics on pedestrians' thermal comfort in Toronto, Urban Climate 24 (2018) 449–459.
13. M. Taleghani, D. Sailor, G.A. Ban-Weiss, Micrometeorological simulations to predict the impacts of heat mitigation strategies on pedestrian thermal comfort in a Los Angeles neighborhood, Environmental Research Letters 11 (2016) 024003.
14. A. Mochida, I.Y.F. Lun, Prediction of wind environment and thermal comfort at pedestrian level in urban area, Journal of Wind Engineering and Industrial Aerodynamics 96 (2008) 1498–1527.
15. B.M. de Quadros, M.G.O. Mizgier, Urban green infrastructures to improve pedestrian thermal comfort: A systematic review, Urban Forestry & Urban Greening 88 (2023) 128091.
16. E. Jamei, P. Rajagopalan, Urban development and pedestrian thermal comfort in Melbourne, Solar Energy 144 (2017) 681–698.
17. L. Kleerekoper, M. van Esch, T.B. Salcedo, How to make a city climate-proof, addressing the urban heat island effect, Resources, Conservation and Recycling 64 (2012) 30–38.
18. M. Taleghani, L. Kleerekoper, M. Tenpierik, A. van den Dobbelsteen, Outdoor thermal comfort within five different urban forms in the Netherlands, Building and Environment 83 (2015) 65–78.
19. Z. Yu, G. Yang, S. Zuo, G. Jørgensen, M. Koga, H. Vejre, Critical review on the cooling effect of urban blue-green space: A threshold-size perspective, Urban Forestry & Urban Greening 49 (2020) 126630.
20. S. Yang, L. (Leon) Wang, T. Stathopoulos, A.M. Marey, Urban microclimate and its impact on built environment – A review, Building and Environment 238 (2023) 110334.
21. D. Kunaka, How urban climate differs from the surrounding climates (The Urban Heat Island Phenomenon), The Geo Room, 2020, https://thegeoroom.co.zw/climatology/how-urban-climate-differs-from-the-surrounding-climates-the-urban-heat-island-phenomenon/.
22. J. Niu, B. Hong, Y. Geng, J. Mi, J. He, Summertime physiological and thermal responses among activity levels in campus outdoor spaces in a humid subtropical city, Science of The Total Environment 728 (2020) 138757.

23. K. Blazejczyk, Y. Epstein, G. Jendritzky, H. Staiger, B. Tinz, Comparison of UTCI to selected thermal indices, International Journal of Biometeorology 56 (2012) 515–535.
24. P. Ole Fanger, J. Toftum, Extension of the PMV model to non-air-conditioned buildings in warm climates, Energy and Buildings 34 (2002) 533–536.
25. R.R. Gonzalez, Y. Nishi, A.P. Gagge, Experimental evaluation of standard effective temperature a new biometeorological index of man's thermal discomfort, International Journal of Biometeorology 18 (1974) 1–15.
26. J. Pickup, R. de Dear, An outdoor thermal comfort index (Out_SET*)–Part 1–The model and its assumptions, in: Biometeorology and Urban Climatology at the Turn of the Millennium, WMO, Geneva, 2000: pp. 279–283.
27. G.M. Budd, Wet-bulb globe temperature (WBGT)—Its history and its limitations, Journal of Science and Medicine in Sport 11 (2008) 20–32.
28. P. Höppe, The physiological equivalent temperature – A universal index for the biometeorological assessment of the thermal environment, International Journal of Biometeorology 43 (1999) 71–75.
29. G. Jendritzky, R. de Dear, G. Havenith, UTCI—Why another thermal index? International Journal of Biometeorology 56 (2012) 421–428.
30. A. Matzarakis, F. Rutz, H. Mayer, Modelling radiation fluxes in simple and complex environments—Application of the RayMan model, International Journal of Biometeorology 51 (2007) 323–334.
31. J. Bouyer, C. Inard, M. Musy, Microclimatic coupling as a solution to improve building energy simulation in an urban context, Energy and Buildings 43 (2011) 1549–1559.
32. L. Shashua-Bar, M.E. Hoffman, The Green CTTC model for predicting the air temperature in small urban wooded sites, Building and Environment 37 (2002) 1279–1288.
33. N. Antoniou, H. Montazeri, M. Neophytou, B. Blocken, CFD simulation of urban microclimate: Validation using high-resolution field measurements, Science of The Total Environment 695 (2019) 133743.
34. D. Liu, S. Hu, J. Liu, Contrasting the performance capabilities of urban radiation field between three microclimate simulation tools, Building and Environment 175 (2020) 106789.
35. F. Lindberg, B. Holmer, S. Thorsson, SOLWEIG 1.0 – Modelling spatial variations of 3D radiant fluxes and mean radiant temperature in complex urban settings, International Journal of Biometeorology 52 (2008) 697–713.
36. S. Tsoka, A. Tsikaloudaki, T. Theodosiou, Analyzing the ENVI-met microclimate model's performance and assessing cool materials and urban vegetation applications – A review, Sustainable Cities and Society 43 (2018) 55–76.
37. J. Mi, B. Hong, T. Zhang, B. Huang, J. Niu, Outdoor thermal benchmarks and their application to climate–responsive designs of residential open spaces in a cold region of China, Building and Environment 169 (2020) 106592.
38. T. Lillesand, R.W. Kiefer, J. Chipman, Remote Sensing and Image Interpretation, John Wiley & Sons, 2015.
39. J.G. Liu, P.J. Mason, Essential Image Processing and GIS for Remote Sensing, John Wiley & Sons, 2013.
40. X.-L. Chen, H.-M. Zhao, P.-X. Li, Z.-Y. Yin, Remote sensing image-based analysis of the relationship between urban heat island and land use/cover changes, Remote Sensing of Environment 104 (2006) 133–146.
41. M.L. Imhoff, P. Zhang, R.E. Wolfe, L. Bounoua, Remote sensing of the urban heat island effect across biomes in the continental USA, Remote Sensing of Environment 114 (2010) 504–513.
42. J.R. Anderson, A Land Use and Land Cover Classification System for Use with Remote Sensor Data, U.S. Government Printing Office, 1976.

43. J. Rogan, D. Chen, Remote sensing technology for mapping and monitoring land-cover and land-use change, Progress in Planning 61 (2004) 301–325.
44. M. Govender, K. Chetty, H. Bulcock, A review of hyperspectral remote sensing and its application in vegetation and water resource studies, Water SA 33 (2007) 145–151.
45. J. Luo, X. Li, R. Ma, F. Li, H. Duan, W. Hu, B. Qin, W. Huang, Applying remote sensing techniques to monitoring seasonal and interannual changes of aquatic vegetation in Taihu Lake, China, Ecological Indicators 60 (2016) 503–513.
46. P. Gupta, S.A. Christopher, J. Wang, R. Gehrig, Y. Lee, N. Kumar, Satellite remote sensing of particulate matter and air quality assessment over global cities, Atmospheric Environment 40 (2006) 5880–5892.
47. R.V. Martin, Satellite remote sensing of surface air quality, Atmospheric Environment 42 (2008) 7823–7843.
48. M.E. Brown, K.M. de Beurs, M. Marshall, Global phenological response to climate change in crop areas using satellite remote sensing of vegetation, humidity and temperature over 26 years, Remote Sensing of Environment 126 (2012) 174–183.
49. J. Yang, P. Gong, R. Fu, M. Zhang, J. Chen, S. Liang, B. Xu, J. Shi, R. Dickinson, The role of satellite remote sensing in climate change studies, Nature Climate Change 3 (2013) 875–883.
50. T.M. Lenton, J.F. Abrams, A. Bartsch, et al., Remotely sensing potential climate change tipping points across scales, Nature Communications 15 (2024) 343.
51. K. Deilami, M. Kamruzzaman, Y. Liu, Urban heat island effect: A systematic review of spatio-temporal factors, data, methods, and mitigation measures, International Journal of Applied Earth Observation and Geoinformation 67 (2018) 30–42.
52. Y. Hu, M. Hou, G. Jia, C. Zhao, X. Zhen, Y. Xu, Comparison of surface and canopy urban heat islands within megacities of eastern China, ISPRS Journal of Photogrammetry and Remote Sensing 156 (2019) 160–168.
53. T.R. Oke, The heat island of the urban boundary layer: Characteristics, causes and effects, in: J.E. Cermak, A.G. Davenport, E.J. Plate, D.X. Viegas (Eds.), Wind Climate in Cities, Springer Netherlands, Dordrecht, 1995: pp. 81–107.
54. I.D. Stewart, E.S. Krayenhoff, J.A. Voogt, J.A. Lachapelle, M.A. Allen, A.M. Broadbent, Time evolution of the surface urban heat island, Earth's Future 9 (2021) e2021EF002178.
55. C. Liu, B. Shi, C. Tang, L. Gao, A numerical and field investigation of underground temperatures under urban heat island, Building and Environment 46 (2011) 1205–1210.
56. S.W. Kim, R.D. Brown, Urban heat island (UHI) variations within a city boundary: A systematic literature review, Renewable and Sustainable Energy Reviews 148 (2021) 111256.
57. B. Baker, Temperature Sensing Technologies, AN679, Microchip Technology Inc, 1998.
58. H. Li, Q. Xia, C. Yin, W. Wan, The current status of research on GNSS-R remote sensing technology in China and future development (In Chinese), Journal of Radars 2 (2013) 389–399.
59. J.D. Kalma, T.R. McVicar, M.F. McCabe, Estimating land surface evaporation: A review of methods using remotely sensed surface temperature data, Surveys in Geophysics 29 (2008) 421–469.
60. J.C. Price, Estimating surface temperatures from satellite thermal infrared data—A simple formulation for the atmospheric effect, Remote Sensing of Environment 13 (1983) 353–361.
61. J. Susskind, J. Rosenfield, D. Reuter, M.T. Chahine, Remote sensing of weather and climate parameters from HIRS2/MSU on TIROS-N, Journal of Geophysical Research: Atmospheres 89 (1984) 4677–4697.
62. Z.-L. Li, B.-H. Tang, H. Wu, H. Ren, G. Yan, Z. Wan, I.F. Trigo, J.A. Sobrino, Satellite-derived land surface temperature: Current status and perspectives, Remote Sensing of Environment 131 (2013) 14–37.

63. Z. Qin, A. Karnieli, P. Berliner, A mono-window algorithm for retrieving land surface temperature from Landsat TM data and its application to the Israel-Egypt border region, International Journal of Remote Sensing 22 (2001) 3719–3746.
64. P.Y. Deschamps, T. Phulpin, Atmospheric correction of infrared measurements of sea surface temperature using channels at 3.7, 11 and 12 Mm, Boundary-Layer Meteorology 18 (1980) 131–143.
65. F. Becker, Z.-L. Li, Towards a local split window method over land surfaces, International Journal of Remote Sensing 11 (1990) 369–393.
66. Y.-G. Qian, N. Wang, L.-L. Ma, Y.-K. Liu, H. Wu, B.-H. Tang, L.-L. Tang, C.-R. Li, Land surface temperature retrieved from airborne multispectral scanner mid-infrared and thermal-infrared data, Optics Express 24 (2016) A257–A269.
67. B.-H. Tang, Nonlinear split-window algorithms for estimating land and sea surface temperatures from simulated Chinese Gaofen-5 satellite data, IEEE Transactions on Geoscience and Remote Sensing 56 (2018) 6280–6289.
68. A. Chedin, N.A. Scott, A. Berroir, A single-channel, double-viewing angle method for sea surface temperature determination from coincident METEOSAT and TIROS-N radiometric measurements, Journal of Applied Meteorology and Climatology 21 (1982) 613–618.
69. A.J. Prata, Land surface temperatures derived from the advanced very high resolution radiometer and the along-track scanning radiometer: 1. Theory, Journal of Geophysical Research: Atmospheres 98 (1993) 16689–16702.
70. A.J. Prata, Land surface temperatures derived from the advanced very high resolution radiometer and the along-track scanning radiometer: 2. Experimental results and validation of AVHRR algorithms, Journal of Geophysical Research: Atmospheres 99 (1994) 13025–13058.
71. L. Zhaoliang, M.P. Stoll, Z. Renhua, J. Li, S. Zhongbo, On the separate retrieval of soil and vegetation temperatures from ATSR data, Science in China Series D-Earth Sciences 44 (2001) 97–111.
72. K. Watson, Spectral ratio method for measuring emissivity, Remote Sensing of Environment 42 (1992) 113–116.
73. Z. Wan, New refinements and validation of the MODIS land-surface temperature/emissivity products, Remote Sensing of Environment 112 (2008) 59–74.
74. M.J. McFarland, R.L. Miller, C.M.U. Neale, Land surface temperature derived from the SSM/I passive microwave brightness temperatures, IEEE Transactions on Geoscience and Remote Sensing 28 (1990) 839–845.
75. E.G. Njoku, L. Li, Retrieval of land surface parameters using passive microwave measurements at 6–18 GHz, IEEE Transactions on Geoscience and Remote Sensing 37 (1999) 79–93.
76. X.-J. Han, S.-B. Duan, C. Huang, Z.-L. Li, Cloudy land surface temperature retrieval from three-channel microwave data, International Journal of Remote Sensing 40 (2019) 1793–1807.
77. X. Xiang, E.A. Smith, Feasibility of simultaneous surface temperature-emissivity retrieval using SSM/I measurements from HAPEX-Sahel, Journal of Hydrology 188–189 (1997) 330–360.
78. A. Basist, N.C. Grody, T.C. Peterson, C.N. Williams, Using the special sensor microwave/imager to monitor land surface temperatures, wetness, and snow cover, Journal of Applied Meteorology and Climatology 37 (1998) 888–911.
79. M. Fily, A. Royer, K. Goïta, C. Prigent, A simple retrieval method for land surface temperature and fraction of water surface determination from satellite microwave brightness temperatures in sub-arctic areas, Remote Sensing of Environment 85 (2003) 328–338.
80. X. Zhang, J. Zhou, F.-M. Göttsche, W. Zhan, S. Liu, R. Cao, A method based on temporal component decomposition for estimating 1-km all-weather land surface temperature

by merging satellite thermal infrared and passive microwave observations, IEEE Transactions on Geoscience and Remote Sensing 57 (2019) 4670–4691.
81. F. Aires, C. Prigent, W.B. Rossow, M. Rothstein, A new neural network approach including first guess for retrieval of atmospheric water vapor, cloud liquid water path, surface temperature, and emissivities over land from satellite microwave observations, Journal of Geophysical Research: Atmospheres 106 (2001) 14887–14907.
82. C. Prigent, C. Jimenez, F. Aires, Toward "all weather," long record, and real-time land surface temperature retrievals from microwave satellite observations, Journal of Geophysical Research: Atmospheres 121 (2016) 5699–5717.
83. C. Jiménez, C. Prigent, S.L. Ermida, J.-L. Moncet, Inversion of AMSR-E observations for land surface temperature estimation: 1. Methodology and evaluation with station temperature, Journal of Geophysical Research: Atmospheres 122 (2017) 3330–3347.
84. J.A. Sobrino, G. Sòria, A.J. Prata, Surface temperature retrieval from Along Track Scanning Radiometer 2 data: Algorithms and validation, Journal of Geophysical Research: Atmospheres 109 (2004).
85. Z.-L. Li, H. Wu, N. Wang, S. Qiu, J.A. Sobrino, Z. Wan, B.-H. Tang, G. Yan, Land surface emissivity retrieval from satellite data, International Journal of Remote Sensing 34 (2013) 3084–3127.
86. M. Mira, E. Valor, V. Caselles, E. Rubio, C. Coll, J.M. Galve, R. Niclos, J.M. Sanchez, R. Boluda, Soil moisture effect on thermal infrared (8–13-μm) emissivity, IEEE Transactions on Geoscience and Remote Sensing 48 (2010) 2251–2260.
87. Z. Yu, Y. Yao, G. Yang, X. Wang, H. Vejre, Spatiotemporal patterns and characteristics of remotely sensed region heat islands during the rapid urbanization (1995–2015) of Southern China, Science of The Total Environment 674 (2019) 242–254.
88. F. Biljecki, K. Ito, Street view imagery in urban analytics and GIS: A review, Landscape and Urban Planning 215 (2021) 104217.
89. B. Yang, S. Yang, X. Zhu, M. Qi, H. Li, Z. Lv, X. Cheng, F. Wang, Computer vision technology for monitoring of indoor and outdoor environments and HVAC equipment: A review, Sensors 23(13) (2023) 6186.
90. Y.C. Li, L. Peng, C.W. Wu, J.Z. Zhang, Street view imagery (SVI) in the built environment: A theoretical and systematic review, Buildings 12(8) (2022) 1167.
91. R. Goel, L.M. Garcia, A. Goodman, R. Johnson, R. Aldred, M. Murugesan, S. Brage, K. Bhalla, J. Woodcock, Estimating city-level travel patterns using street imagery: A case study of using Google Street View in Britain, PLoS One 13(5) (2018) e0196521.
92. A. Berland, D.A. Lange, Google Street View shows promise for virtual street tree surveys, Urban Forestry & Urban Greening 21 (2017) 11–15.
93. X.J. Li, C.R. Zhang, W.D. Li, R. Ricard, Q.Y. Meng, W.X. Zhang, Assessing street-level urban greenery using Google Street View and a modified green view index, Urban Forestry & Urban Greening 14(3) (2015) 675–685.
94. R.Y. Wang, Y. Lu, J.B. Zhang, P.H. Liu, Y. Yao, Y. Liu, The relationship between visual enclosure for neighbourhood street walkability and elders' mental health in China: Using street view images, Journal of Transport & Health 13 (2019) 90–102.
95. F. Zhang, B. Zhou, C. Ratti, Y. Liu, Discovering place-informative scenes and objects using social media photos, Royal Society Open Science 6(3) (2019) 181375.
96. Y. Kang, F. Zhang, S. Gao, H. Lin, Y. Liu, A review of urban physical environment sensing using street view imagery in public health studies, Annals of GIS 26(3) (2020) 261–275.
97. S. Branson, J.D. Wegner, D. Hall, N. Lang, K. Schindler, P. Perona, From Google Maps to a fine-grained catalog of street trees, ISPRS Journal of Photogrammetry and Remote Sensing 135 (2018) 13–30.
98. Y. Lu, Using Google Street View to investigate the association between street greenery and physical activity, Landscape and Urban Planning 191 (2019) 103435.

99. Y. Kumakoshi, S.Y. Chan, H. Koizumi, X. Li, Y. Yoshimura, Standardized green view index and quantification of different metrics of urban green vegetation, Sustainability 12(18) (2020) 7434.
100. Y. Ye, H. Xie, J. Fang, H. Jiang, D. Wang, Daily accessed street greenery and housing price: Measuring economic performance of human-scale streetscapes via new urban data, Sustainability 11(6) (2019) 1741.
101. J.Y. He, J.B. Zhang, Y. Yao, X. Li, Extracting human perceptions from street view images for better assessing urban renewal potential, Cities 134 (2023) 104189.
102. R. Mahabir, R. Schuchard, A. Crooks, A. Croitoru, A. Stefanidis, Crowdsourcing street view imagery: A comparison of Mapillary and OpenStreetCam, ISPRS International Journal of Geo-Information 9(6) (2020) 341.
103. L. Juhász, H.H. Hochmair, User contribution patterns and completeness evaluation of Mapillary, a crowdsourced street level photo service, Transactions in GIS 20(6) (2016) 925–947.
104. F. Biljecki, T. Zhao, X. Liang, Y. Hou, Sensitivity of measuring the urban form and greenery using street-level imagery: A comparative study of approaches and visual perspectives, International Journal of Applied Earth Observation and Geoinformation 122 (2023) 103385.
105. Z.Y. Gong, Q.W. Ma, C.C. Kan, Q.Y. Qi, Classifying street spaces with street view images for a spatial indicator of urban functions, Sustainability 11(22) (2019) 6424.
106. S.W. Kim, R.D. Brown, Pedestrians' behavior based on outdoor thermal comfort and micro-scale thermal environments, Austin, TX, Science of the Total Environment 808 (2022) 152143.
107. Y.J. Kim, R.D. Brown, A multilevel approach for assessing the effects of microclimatic urban design on pedestrian thermal comfort: The high line in New York, Building and Environment 205 (2021) 108244.
108. E. Jamei, P. Rajagopalan, M. Seyedmahmoudian, Y. Jamei, Review on the impact of urban geometry and pedestrian level greening on outdoor thermal comfort, Renewable & Sustainable Energy Reviews 54 (2016) 1002–1017.
109. W. Klemm, B.G. Heusinkveld, S. Lenzholzer, B. van Hove, Street greenery and its physical and psychological impact on thermal comfort, Landscape and Urban Planning 138 (2015) 87–98.
110. J. Yang, B. Shi, G. Xia, Q. Xue, S.J. Cao, Impacts of urban form on thermal environment near the surface region at pedestrian height: A case study based on high-density built-up areas of Nanjing City in China, Sustainability 12(5) (2020) 1737.
111. M.H.H. Abdelhafez, F. Altaf, M. Alshenaifi, O. Hamdy, A. Ragab, Achieving effective thermal performance of street canyons in various climatic zones, Sustainability 14(17) (2022) 10780.
112. C.B. Hu, F. Zhang, F.Y. Gong, C. Ratti, X. Li, Classification and mapping of urban canyon geometry using Google Street View images and deep multitask learning, Building and Environment 167 (2020) 106424.
113. A. Qaid, D.R. Ossen, Effect of asymmetrical street aspect ratios on microclimates in hot, humid regions, International Journal of Biometeorology 59(6) (2015) 657–677.
114. A. Chatzidimitriou, S. Yannas, Street canyon design and improvement potential for urban open spaces; the influence of canyon aspect ratio and orientation on microclimate and outdoor comfort, Sustainable Cities and Society 33 (2017) 85–101.
115. Y. Zhang, X. Du, Y. Shi, Effects of street canyon design on pedestrian thermal comfort in the hot-humid area of China, International Journal of Biometeorology 61(8) (2017) 1421–1432.
116. W. Gao, J. Hou, Y. Gao, M. Zhao, M. Jia, Quantifying the spatial ratio of streets in Beijing based on street-view images, ISPRS International Journal of Geo-Information 12(6) (2023) 246.

117. T. Oke, Canyon geometry and the nocturnal urban heat island: Comparison of scale model and field observations, Journal of Climatology 1 (1981) 237–254.
118. M. Scarano, F. Mancini, Assessing the relationship between sky view factor and land surface temperature to the spatial resolution, International Journal of Remote Sensing 38(23) (2017) 6910–6929.
119. T.P. Lin, K.T. Tsai, R.L. Hwang, A. Matzarakis, Quantification of the effect of thermal indices and sky view factor on park attendance, Landscape and Urban Planning 107(2) (2012) 137–146.
120. T.R. Oke, Street design and urban canopy layer climate, Energy and Buildings 11(1) (1988) 103–113.
121. A. Matzarakis, D. Fröhlich, M. Gangwisch, Effect of radiation and wind on thermal comfort in urban environments—Applications of the RayMan and SkyHelios model, Urban Climate 27 (2019) 1–7.
122. T.M. Gál, M. Rzepa, B. Gromek, J. Unger, Comparison between sky view factor values computed by two different methods in an urban environment, Acta Climatologica et chorologica 40 (2007) 17–26.
123. C. Kidd, L. Chapman, Derivation of sky-view factors from LiDAR data, International Journal of Remote Sensing 33(11) (2012) 3640–3652.
124. A. Matzarakis, O. Matuschek, Sky view factor as a parameter in applied climatology – Rapid estimation by the SkyHelios model, Meteorologische Zeitschrift 20(1) (2011) 39–45.
125. D.G. Steyn, T.J. Lyons, The determination of view-factors in urban canyons – Comments, Journal of Climate and Applied Meteorology 24(4) (1985) 383–385.
126. L. Chapman, J.E. Thornes, A.V. Bradley, Sky-view factor approximation using GPS receivers, International Journal of Climatology 22(5) (2002) 615–621.
127. L. Chapman, J.E. Thornes, A.V. Bradley, Rapid determination of canyon geometry parameters for use in surface radiation budgets, Theoretical and Applied Climatology 69(1–2) (2001) 81–89.
128. L. Chapman, J.E. Thornes, Real-time sky-view factor calculation and approximation, Journal of Atmospheric and Oceanic Technology 21(5) (2004) 730–741.
129. J. Unger, Connection between urban heat island and sky view factor approximated by a software tool on a 3D urban database, International Journal of Environment and Pollution 36(1–3) (2009) 59–80.
130. X. Li, C. Ratti, I. Seiferling, Quantifying the shade provision of street trees in urban landscape: A case study in Boston, USA, using Google Street View, Landscape and Urban Planning 169 (2018) 81–91.
131. H.K.W. Cheung, D. Coles, G.J. Levermore, Urban heat island analysis of Greater Manchester, UK using sky view factor analysis, Building Services Engineering Research & Technology 37(1) (2016) 5–17.
132. T. Gal, F. Lindberg, J. Unger, Computing continuous sky view factors using 3D urban raster and vector databases: Comparison and application to urban climate, Theoretical and Applied Climatology 95(1–2) (2009) 111–123.
133. C.S.B. Grimmond, S.K. Potter, H.N. Zutter, C. Souch, Rapid methods to estimate sky-view factors applied to urban areas, International Journal of Climatology 21(7) (2001) 903–913.
134. L. Zeng, J. Lu, W. Li, Y. Li, A fast approach for large-scale Sky View Factor estimation using street view images, Building and Environment 135 (2018) 74–84.
135. F.Y. Gong, Z.C. Zeng, F. Zhang, X. Li, E. Ng, L.K. Norford, Mapping sky, tree, and building view factors of street canyons in a high-density urban environment, Building and Environment 134 (2018) 155–167.
136. F.Y. Gong, Z.C. Zeng, E. Ng, L.K. Norford, Spatiotemporal patterns of street-level solar radiation estimated using Google Street View in a high-density urban environment, Building and Environment 148 (2019) 547–566.

137. D. Steyn, The calculation of view factors from fisheye-lens photographs: Research note, Atmosphere-Ocean 18 (1980) 254–258.
138. L. Chen, E. Ng, X. An, C. Ren, M. Lee, U. Wang, Z. He, Sky view factor analysis of street canyons and its implications for daytime intra-urban air temperature differentials in high-rise, high-density urban areas of Hong Kong: A GIS-based simulation approach, International Journal of Climatology 32(1) (2012) 121–136.
139. C. Miao, S. Yu, Y. Hu, H. Zhang, X. He, W. Chen, Review of methods used to estimate the sky view factor in urban street canyons, Building and Environment 168 (2020) 106497.
140. L. Chapman, J.E. Thornes, A.V. Bradley, Sky-view factor approximation using GPS receivers, International Journal of Climatology: A Journal of the Royal Meteorological Society 22(5) (2002) 615–621.
141. F. Aram, E. Solgi, E.H. Garcia, A. Mosavi, Urban heat resilience at the time of global warming: Evaluating the impact of the urban parks on outdoor thermal comfort, Environmental Sciences Europe 32(1) (2020) 117.
142. A. Middel, J. Lukasczyk, R. Maciejewski, Sky view factors from synthetic fisheye photos for thermal comfort routing-a case study in Phoenix, Arizona, Urban Plan 2(1) (2017) 19–30.
143. R. Carrasco-Hernandez, A.R. Smedley, A.R. Webb, Using urban canyon geometries obtained from Google Street View for atmospheric studies: Potential applications in the calculation of street level total shortwave irradiances, Energy and Buildings 86 (2015) 340–348.
144. Y. Xia, N. Yabuki, T. Fukuda, Sky view factor estimation from street view images based on semantic segmentation, Urban Climate 40 (2021) 100999.
145. K.A. Nice, J.S. Wijnands, A. Middel, J. Wang, Y. Qiu, N. Zhao, J. Thompson, G.D. Aschwanden, H. Zhao, M. Stevenson, Sky pixel detection in outdoor imagery using an adaptive algorithm and machine learning, Urban Climate 31 (2020) 100572.
146. J. Urban, M. Pikl, F. Zemek, J. Novotný, Using Google Street View photographs to assess long-term outdoor thermal perception and thermal comfort in the urban environment during heatwaves, Frontiers in Environmental Science 10 (2022) 878341.
147. Z. Deng, Y. Chen, X. Pan, Z. Peng, J. Yang, Integrating GIS-based point of interest and community boundary datasets for urban building energy modeling, Energies 14(4) (2021) 1049.
148. Y. Li, Y. Chen, A. Rajabifard, K. Khoshelham, M. Aleksandrov, Estimating building age from Google Street View images using deep learning (short paper), 10th International Conference on Geographic Information Science (GIScience 2018), Schloss-Dagstuhl-Leibniz Zentrum für Informatik, 40 (2018) 1–40.
149. T. Zhong, C. Ye, Z. Wang, G. Tang, W. Zhang, Y. Ye, City-scale mapping of urban façade color using street-view imagery, Remote Sensing 13(8) (2021) 1591.
150. E.J. Hoffmann, Y. Wang, M. Werner, J. Kang, X. Zhu, Model fusion for building type classification from aerial and street view images, Remote Sensing 11(11) (2019) 1259.
151. Z. Zong, J. Feng, K. Liu, H. Shi, Y. Li, DeepDPM: Dynamic population mapping via deep neural network, Proceedings of the AAAI Conference on Artificial Intelligence 33 (2019) 1294–1301.
152. S.S. Karayazi, G. Dane, B.d. Vries, Utilizing urban geospatial data to understand heritage attractiveness in Amsterdam, ISPRS International Journal of Geo-Information 10(4) (2021) 198.
153. M.R. Ibrahim, J. Haworth, T. Cheng, Understanding cities with machine eyes: A review of deep computer vision in urban analytics, Cities 96 (2020) 102481.
154. H. Zhou, G. Tao, X. Yan, J. Sun, Influences of greening and structures on urban thermal environments: A case study in Xuzhou City, China, Urban Forestry & Urban Greening 66 (2021) 127386.

155. F. Aram, E.H. García, E. Solgi, S. Mansournia, Urban green space cooling effect in cities, Heliyon 5(4) (2019).
156. F. Aram, E. Solgi, E. Higueras García, D. Mohammadzadeh S, A. Mosavi, S. Shamshirband, Design and validation of a computational program for analysing mental maps: Aram mental map analyzer, Sustainability 11(14) (2019) 3790.
157. F. Aram, E. Solgi, E. Higueras García, A. Mosavi, A.R. Várkonyi-Kóczy, The cooling effect of large-scale urban parks on surrounding area thermal comfort, Energies 12(20) (2019) 3904.
158. F. Aram, E. Solgi, G. Holden, The role of green spaces in increasing social interactions in neighborhoods with periodic markets, Habitat International 84 (2019) 24–32.
159. J. Wu, L. Cheng, S. Chu, N. Xia, M. Li, A green view index for urban transportation: How much greenery do we view while moving around in cities? International Journal of Sustainable Transportation 14(12) (2020) 972–989.
160. J. Chen, C. Zhou, F. Li, Quantifying the green view indicator for assessing urban greening quality: An analysis based on Internet-crawling street view data, Ecological Indicators 113 (2020) 106192.
161. Y. Lap-Fu, Environmental planning and economic efficiency of green cities based on improved neural network and satellite remote sensing, Arabian Journal of Geosciences 14(14) (2021) 1399.
162. M. Subiza-Pérez, L. Vozmediano, C. San Juan, Green and blue settings as providers of mental health ecosystem services: Comparing urban beaches and parks and building a predictive model of psychological restoration, Landscape and Urban Planning 204 (2020) 103926.
163. S.J. Mooney, K. Wheeler-Martin, L.M. Fiedler, C.M. LaBelle, T. Lampe, A. Ratanatharathorn, N.N. Shah, A.G. Rundle, C.J. DiMaggio, Development and validation of a Google Street View pedestrian safety audit tool, Epidemiology 31(2) (2020) 301–309.
164. K. Gupta, P. Kumar, S.K. Pathan, K.P. Sharma, Urban Neighborhood Green Index – A measure of green spaces in urban areas, Landscape and Urban Planning 105(3) (2012) 325–335.
165. A. Shah, A. Garg, V. Mishra, Quantifying the local cooling effects of urban green spaces: Evidence from Bengaluru, India, Landscape and Urban Planning 209 (2021) 104043.
166. F. Jing, L. Liu, S. Zhou, J. Song, L. Wang, H. Zhou, Y. Wang, R. Ma, Assessing the impact of street-view greenery on fear of neighborhood crime in Guangzhou, China, International Journal of Environmental Research and Public Health 18(1) (2021) 311.
167. J. Yang, L. Zhao, J. Mcbride, P. Gong, Can you see green? Assessing the visibility of urban forests in cities, Landscape and Urban Planning 91(2) (2009) 97–104.
168. H. Riihimäki, M. Luoto, J. Heiskanen, Estimating fractional cover of tundra vegetation at multiple scales using unmanned aerial systems and optical satellite data, Remote Sensing of Environment 224 (2019) 119–132.
169. E. Barbierato, I. Bernetti, I. Capecchi, C. Saragosa, Integrating remote sensing and street view images to quantify urban forest ecosystem services, Remote Sensing 12(2) (2020) 329.
170. S. Lumnitz, T. Devisscher, J.R. Mayaud, V. Radic, N.C. Coops, V.C. Griess, Mapping trees along urban street networks with deep learning and street-level imagery, ISPRS Journal of Photogrammetry and Remote Sensing 175 (2021) 144–157.
171. Y. Ke, L.J. Quackenbush, A review of methods for automatic individual tree-crown detection and delineation from passive remote sensing, International Journal of Remote Sensing 32(17) (2011) 4725–4747.
172. S. Li, S. Dragicevic, F.A. Castro, M. Sester, S. Winter, A. Coltekin, C. Pettit, B. Jiang, J. Haworth, A. Stein, T. Cheng, Geospatial big data handling theory and methods:

A review and research challenges, ISPRS Journal of Photogrammetry and Remote Sensing 115 (2016) 119–133.
173. D. Ki, S. Lee, Analyzing the effects of Green View Index of neighborhood streets on walking time using Google Street View and deep learning, Landscape and Urban Planning 205 (2021) 103920.
174. H. Yu, Y. Zhou, R. Wang, Z. Qian, L.D. Knibbs, B. Jalaludin, M. Schootman, S.E. McMillin, S.W. Howard, L.Z. Lin, et al., Associations between trees and grass presence with childhood asthma prevalence using deep learning image segmentation and a novel green view index, Environmental Pollution 286 (2021) 117582.
175. B. Wang, L. Li, Y. Nakashima, R. Kawasaki, H. Nagahara, Y. Yagi, Noisy-LSTM: Improving temporal awareness for video semantic segmentation, IEEE Access 9 (2021) 46810–46820.
176. R. Dong, Y. Zhang, J. Zhao, How green are the streets within the sixth ring road of Beijing? An analysis based on Tencent Street View pictures and the green view index, International Journal of Environmental Research and Public Health 15(7) (2018) 1367.
177. Y.C. Chiang, H.H. Liu, D. Li, L.C. Ho, Quantification through deep learning of sky view factor and greenery on urban streets during hot and cool seasons, Landscape and Urban Planning 232 (2023) 104679.
178. J. Zhang, A. Hu, Analyzing green view index and green view index best path using Google Street View and deep learning, Journal of Computational Design and Engineering 9(5) (2022) 2010–2023.
179. M. Tong, J. She, J. Tan, M. Li, R. Ge, Y. Gao, Evaluating street greenery by multiple indicators using street-level imagery and satellite images: A case study in Nanjing, China, Forests 11(12) (2020) 1347.
180. X. Yu, G. Zhao, C. Chang, X. Yuan, F. Heng, BGVI: A new index to estimate street-side greenery using Baidu Street View image, Forests 10(1) (2018) 3.
181. X. Chen, Q. Meng, D. Hu, L. Zhang, J. Yang, Evaluating greenery around streets using Baidu panoramic street view images and the panoramic green view index, Forests 10(12) (2019) 1109.
182. C.M. Kelly, J.S. Wilson, E.A. Baker, D.K. Miller, M. Schootman, Using Google Street View to audit the built environment: Inter-rater reliability results, Annals of Behavioral Medicine 45 (2013) 108–112.
183. L. Yin, Q. Cheng, Z. Wang, Z. Shao, 'Big data' for pedestrian volume: Exploring the use of Google Street View images for pedestrian counts, Applied Geography 63 (2015) 337–345.
184. K. Hara, V. Le, J. Froehlich, Combining crowdsourcing and Google Street View to identify street-level accessibility problems, Proceedings of the SIGCHI Conference on Human Factors in Computing Systems (2013) 631–640.
185. T. Gebru, J. Krause, Y. Wang, D. Chen, J. Deng, E.L. Aiden, L. Fei-Fei, Using deep learning and Google Street View to estimate the demographic makeup of neighborhoods across the United States, Proceedings of the National Academy of Sciences 114(50) (2017) 13108–13113.
186. D.B. Abou Chacra, J.S. Zelek, Fully automated road defect detection using street view images, 2017 14th Conference on Computer and Robot Vision (CRV) (2017) 353–360.
187. I. Seiferling, N. Naik, C. Ratti, R. Proulx, Green Streets– Quantifying and mapping urban trees with street-level imagery and computer vision, Landscape and Urban Planning 165 (2017) 93–101.
188. J. Kang, M. Körner, Y. Wang, H. Taubenböck, X. Zhu, Building instance classification using street view images, ISPRS Journal of Photogrammetry and Remote Sensing 145 (2018) 44–59.
189. E. Johansson, Influence of urban geometry on outdoor thermal comfort in a hot dry climate: A stud in Fez, Morocco, Building and Environment 41(10) (2006) 1326–1338.

190. H.L. Brantley, G.S.W. Hagler, E.S. Kimbrough, R.W. Williams, S. Mukerjee, L.M. Neas, Mobile air monitoring data-processing strategies and effects on spatial air pollution trends, Atmospheric Measurement Techniques 7(7) (2014) 2169–2183.
191. S. Hankey, J.D. Marshall, On-bicycle exposure to particulate air pollution: Particle number, black carbon, PM2.5, and particle size, Atmospheric Environment 122 (2015) 65–73.
192. V. Mallikalava, S. Yuvaraj, K. Vengatesan, A. Kumar, S. Punjabi, S. Samee, Theft vehicle detection using image processing integrated digital signature based ECU, Proceedings of the 3rd International Conference on Smart Systems and Inventive Technology (2020) 913–918.
193. M. Ren, X. Zhang, X. Chen, B. Zhou, Z. Feng, YOLOv5s-M: A deep learning network model for road pavement damage detection from urban street-view imagery, International Journal of Applied Earth Observation and Geoinformation 120 (2023) 103335.
194. Y. Ao, Fully Convulutional Networks for Street Furniture Identification in Panorama Images, University of Twente, 2019.
195. A. Campbell, A. Both, Q. Sun, Detecting and mapping traffic signs from Google Street View images using deep learning and GIS, Computers, Environment and Urban Systems 77 (2019) 101350.
196. K. Siountri, C.N. Anagnostopoulos, The classification of cultural heritage buildings in Athens using deep learning techniques, Heritage 6(4) (2023) 3673–3705.
197. Y. Ogawa, C. Zhao, T. Oki, S. Chen, Y. Sekimoto, Deep learning approach for classifying the built year and structure of individual buildings by automatically linking street view images and GIS building data, IEEE Journal of Selected Topics in Applied Earth Observations and Remote Sensing 16 (2023) 1740–1755.
198. J.T. Szcześniak, Y.Q. Ang, S. Letellier-Duchesne, C.F. Reinhart, A method for using street view imagery to auto-extract window-to-wall ratios and its relevance for urban-level daylighting and energy simulations, Building and Environment 207 (2022) 108108.
199. H. Zhou, L. Liu, M. Lan, W. Zhu, G. Song, F. Jing, Y. Zhong, Z. Su, X. Gu, Using Google Street View imagery to capture micro built environment characteristics in drug places, compared with street robbery, Computers, Environment and Urban Systems 88 (2021) 101631.
200. G. Palmer, M. Green, E. Boyland, Y.S.R. Vasconcelos, R. Savani, A. Singleton, A deep learning approach to identify unhealthy advertisements in street view images, Scientific Reports 11(1) (2021) 4884.
201. J. Speake, V. Kennedy, R. Love, Visual and aesthetic markers of gentrification: Agency of mapping and tourist destinations, Tourism Geographies 25(2–3) (2023) 756–777.
202. J. Kim, M. Kamari, S. Lee, Y. Ham, Large-scale visual data-driven probabilistic risk assessment of utility poles regarding the vulnerability of power distribution infrastructure systems, Journal of Construction Engineering and Management 147(10) (2021) 04021121.
203. R. Kalfarisi, M. Hmosze, Z. Wu, Detecting and geolocating city-scale soft-story buildings by deep machine learning for urban seismic resilience, Natural Hazards Review 23(1) (2022) 04021062.
204. A.V. Vo, M. Bertolotto, U. Ofterdinger, D.F. Laefer, In search of basement indicators from street view imagery data: An investigation of data sources and analysis strategies, KI-Künstliche Intelligenz 37(1) (2023) 41–53.
205. P.A. Pelizari, C. Geiß, P. Aguirre, H. Santa María, Y.M. Peña, H. Taubenböck, Automated building characterization for seismic risk assessment using street-level imagery and deep learning, ISPRS Journal of Photogrammetry and Remote Sensing 180 (2021) 370–386.

206. F. Ghione, S. Mæland, A. Meslem, V. Oye, Building stock classification using machine learning: A case study for Oslo, Norway, Frontiers in Earth Science 10 (2022) 886145.
207. H. Stone, V. Putrino, D. D'Ayala, Earthquake damage data collection using omnidirectional imagery, Frontiers in Built Environment 4 (2018) 51.
208. X. Li, C. Zhang, W. Li, Building block level urban land-use information retrieval based on Google Street View images, GIScience & Remote Sensing 54(6) (2017) 819–835.
209. R. Wang, Y. Liu, Y. Lu, Y. Yuan, J. Zhang, P. Liu, Y. Yao, The linkage between the perception of neighbourhood and physical activity in Guangzhou, China: Using street view imagery with deep learning techniques, International Journal of Health Geographics 18 (2019) 18.
210. E. Gorgul, C. Chen, K.K. Wu, Y. Guo, Measuring street enclosure and its influence to human physiology through wearable sensors, Adjunct Proceedings of the 2019 ACM International Joint Conference on Pervasive and Ubiquitous Computing and Proceedings of the 2019 ACM International Symposium on Wearable Computers (2019) 65–68.
211. Y. Kato, S. Matsukawa, Development of generating system for architectural color icons using Google Map platform and tensorflow-segmentation, 24th International Conference on Computer-Aided Architectural Design Research in Asia: Intelligent and Informed. The Association for Computer-Aided Architectural Design Research in Asia (CAADRIA) (2019) 81–90.
212. S. Samiei, P. Rasti, H. Daniel, E. Belin, P. Richard, D. Rousseau, Toward a computer vision perspective on the visual impact of vegetation in symmetries of urban environments, Symmetry 10(12) (2018) 666.
213. W. Wang, S. Yang, Z. He, M. Wang, J. Zhang, W. Zhang, Urban perception of commercial activeness from satellite images and streetscapes, Companion Proceedings of the Web Conference 2018 (2018) 647–654.
214. J. Wang, Y. Ma, L. Tang, D. Li, J. Xie, Y. Hu, Y. Tian, Long-term exposure to residential greenness and decreased risk of depression and anxiety, Nature Mental Health 2(5) (2024) 525–534.
215. M. Helbich, Y. Yao, Y. Liu, J. Zhang, P. Liu, R. Wang, Using deep learning to examine street view green and blue spaces and their associations with geriatric depression in Beijing, China, Environment International 126 (2019) 107–117.
216. B.T. Taylor, P. Fernando, A.E. Bauman, A. Williamson, J.C. Craig, S. Redman, Measuring the quality of public open space using Google Earth, American Journal of Preventive Medicine 40(2) (2011) 105–112.
217. J. Kruse, Y. Kang, Y.-N. Liu, F. Zhang, S. Gao, Places for play: Understanding human perception of playability in cities using street view images and deep learning, Computers, Environment and Urban Systems 90 (2021) 101693.
218. J. Gustat, C.E. Anderson, Q.C. Chukwurah, M.E. Wallace, S.T. Broyles, L.A. Bazzano, Cross-sectional associations between the neighborhood built environment and physical activity in a rural setting: The Bogalusa Heart Study, BMC Public Health 20 (2020) 1426.
219. L. Meng, K.H. Wen, Z. Zeng, R. Brewin, X. Fan, Q. Wu, The impact of street space perception factors on elderly health in high-density cities in Macau—Analysis based on street view images and deep learning technology, Sustainability 12(5) (2020) 1799.
220. K. Qin, Y. Xu, C. Kang, M.P. Kwan, A graph convolutional network model for evaluating potential congestion spots based on local urban built environments, Transactions in GIS 24(5) (2020) 1382–1401.
221. C. Gabbe, E. Oxlaj, J. Wang, Residential development and near-roadway air pollution: Assessing risk and mitigation in San Jose, California, Journal of Transport & Health 13 (2019) 78–89.

222. P. Skurowski, M. Paszkuta, Saliency map based analysis for prediction of car driving difficulty in Google Street View scenes, AIP Conference Proceedings, AIP Publishing 1978(1) (2018) 110003.
223. G. Conley, S.C. Zinn, T. Hanson, K. McDonald, N. Beck, H. Wen, Using a deep learning model to quantify trash accumulation for cleaner urban stormwater, Computers, Environment and Urban Systems 93 (2022) 101752.
224. X. Li, C. Zhang, W. Li, Y.A. Kuzovkina, D. Weiner, Who lives in greener neighborhoods? The distribution of street greenery and its association with residents' socioeconomic conditions in Hartford, Connecticut, USA, Urban Forestry & Urban Greening 14(4) (2015) 751–759.
225. L. Yin, Z. Wang, Measuring visual enclosure for street walkability: Using machine learning algorithms and Google Street View imagery, Applied Geography 76 (2016) 147–153.
226. Y. Lu, Y. Yang, G. Sun, Z. Gou, Associations between overhead-view and eye-level urban greenness and cycling behaviors, Cities 88 (2019) 10–18.

6 Building Construction Safety Monitoring

He Li, Ruiqi Guo, Luting Bai, and Bin Yang

6.1 MONITORING METHODS

6.1.1 UAV Flight and Information Collection Technology

Unmanned aerial vehicles (UAVs) are unmanned aircraft controlled by wireless remote devices and flight control chips. UAV technology is renowned for its flexibility, accessibility, and efficiency, allowing it to perform tasks in environments that are either inaccessible or extremely hazardous for humans. The initial application of UAV technology in the field of construction engineering was to replace manual labor in measuring various types of data for engineering projects, which is known as UAV mapping technology. With continuous advancements in technology, UAV platforms can now be equipped with various advanced, high-precision, high-resolution sensors with stable transmission capabilities, fundamentally improving the level and quality of detection.

The application of UAV technology in construction safety management addresses many of the limitations inherent in traditional safety inspection models, such as limited personnel, high time costs, small coverage areas, and significant blind spots in detection. This technology integrates UAV aerial photography with computer vision (CV) techniques, utilizing deep learning algorithms like YOLO [1] to automatically identify key elements on construction sites, such as machinery and workers, forming the basis for analyzing safety conditions. UAVs are not only employed for routine monitoring and safety inspections but are also used for mapping and emergency response, effectively improving construction efficiency and safety levels [2]. For example, UAVs can complete large-scale geographic data collection within a few hours, providing support for emergency operations while also being used for topographic mapping in complex terrains and hard-to-reach areas, significantly enhancing the precision and efficiency of data collection. The research framework for monitoring construction safety using UAVs is shown in Figure 6.1.

By capturing aerial footage, UAVs provide a wide perspective and large-scale data coverage, significantly enhancing the efficiency and scope of safety inspections. Based on these data, researchers have developed automated systems, such as human–machine distance detection and alarm systems, to prevent collisions between workers and machinery [3]. Additionally, image stitching techniques are used to create panoramic images covering entire construction sites [4]. The flight safety of UAVs is also a key research area, involving UAV anti-interference algorithms, flight path planning, and obstacle avoidance technologies. For instance, the proportional-integral-derivative (PID) control algorithm is widely used in UAV control systems due to its simplicity and stable control performance. Moreover, depth imaging [5] and machine learning

DOI: 10.1201/9781003611455-6

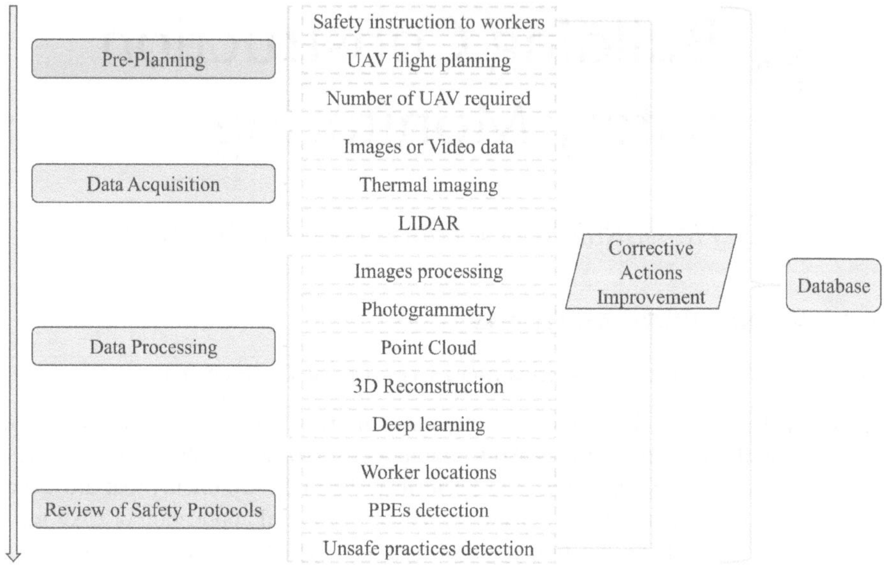

FIGURE 6.1 A process map for UAV-based construction-site monitoring.

technologies [6] have been applied to improve UAVs' obstacle avoidance capabilities, enhancing their safety and stability in complex environments. Overall, UAV technology not only improves the efficiency of safety monitoring in construction but also modernizes site management through advanced data analysis and automation features.

6.1.2 Behavior Recognition Technology

In the field of construction safety, CV-based human–machine safety management technology is transforming traditional monitoring systems. These techniques collect visual data, such as point clouds and images, from construction sites using laser scanning, radar scanning, and other monitoring systems. After processing this data through manual feature extraction and machine learning algorithms (e.g., neural networks and support vector machines (SVM)), they can accurately identify the positions, postures, and relative distances of workers and construction machinery, thereby enabling effective proactive safety control and risk management.

Behavior recognition technology places particular emphasis on the use of image-based information because images contain rich data on human movements and impose relatively low equipment requirements. Researchers have experimented with various combinations of visual features and machine learning algorithms in different construction scenarios to effectively extract safety information from the site. For instance, histogram of oriented gradients (HOG) and SVM have been used to identify and classify the behaviors of excavators and dump trucks in earthworks [7]; similarly, speeded-up robust features (SURF) have been applied to radar scanning data to recognize workers near machinery and provide feedback to drivers, enhancing awareness of scene risks [8].

Technological advancements have also led to improvements in the speed and accuracy of behavior recognition. Researchers have developed bilinear interpolation algorithms and continuous frame optical flow extraction techniques that not only increase model execution efficiency but also optimize responses in dynamic scenes. Furthermore, deep learning and neural network approaches, such as convolutional neural networks (CNN), long short-term memory (LSTM) networks, and enhanced YOLO models, have been employed to automatically learn and extract behavioral features, significantly boosting recognition accuracy.

These advancements not only improve the quality of monitoring on construction sites but also effectively predict and prevent high-risk behaviors, significantly reducing the rate of site accidents. The development and application of these behavior recognition technologies provide robust technological support for construction safety management, driving the process toward automation and intelligent oversight.

6.1.3 Real-Time Positioning Technology

The application of real-time positioning technology in construction safety continues to demonstrate its importance and potential, particularly in managing and preventing safety risks at construction sites. By employing various positioning techniques such as Ultra-Wideband (UWB) [9], Global Positioning System (GPS) [10], Bluetooth, Radio Frequency Identification (RFID), and magnetic field-based distance sensing [11], researchers can monitor the positions of workers and machinery in real time, effectively identifying and preventing potential human–machine collision risks.

Studies indicate that relying solely on positional information may be insufficient to address all safety challenges, as it does not allow for the adjustment of risk identification rules based on the specific construction scenario. To address this limitation, researchers have attempted to fuse position data with other kinematic signals, such as acceleration and angular velocity, thus providing more comprehensive safety information and enabling more precise safety management decisions [12]. Subsequent research comparing various positioning technologies and corresponding data processing methods in different construction environments has validated the effectiveness of real-time location systems (RTLS) based on UWB, RFID, and GPS, particularly for safety and progress management [13]. Moreover, by integrating real-time positional data with additional data sources, such as worker posture data collected from wearable devices, researchers can more accurately determine if workers are engaging in unsafe behaviors [14, 15]. Further, analysis of accident reports has enabled the identification of typical unsafe scenarios, leading to the development of targeted safety warning systems that further enhance the effectiveness of construction-site safety management [16, 17].

In summary, real-time positioning technology, especially when integrated with other data sources, shows enormous potential in enhancing the accuracy and efficiency of safety management on construction sites, thereby contributing to better accident prevention and risk management.

6.1.4 Spatial Planning Technology

In the field of construction safety, spatial planning technology, especially the application of building information modeling (BIM), is emerging as a key tool for enhancing safety management and reducing human–machine collision risks [18]. BIM integrates three-dimensional (3D) digital models with on-site databases, not only optimizing the layout and work paths of heavy machinery such as tower cranes but also enhancing the visual management of construction sites. Research has shown that BIM can effectively combine the positional data of physical entities surrounding construction machinery with the environmental data in 3D models to provide an intuitive display of spatial relationships [19]. This integration helps operators clearly understand their surroundings to prevent potential human–machine collisions, while its rich data supports optimization algorithms that balance construction efficiency with safety, thereby mitigating risks caused by spatial conflicts [20]. The comprehensive design and process information provided by BIM allows for the optimization of path planning and space utilization even before construction commences. Automated safety inspections using BIM further help to identify and correct risk configurations that might lead to collisions or falls from heights [21]. Moreover, as a powerful data visualization platform, BIM can be combined with real-time positioning systems (such as GPS and RFID) [22, 23] and modern sensor technologies like CV [24] to integrate real-time sensor data. This data fusion not only improves the accuracy of position tracking but also enables safety management personnel to more precisely evaluate potential risks and make informed decisions.

The application of BIM technology has significantly reduced the accident rate on construction sites by ensuring safety and efficiency throughout the construction process through precise safety planning and risk management. With its integrated data and visualization functions, BIM serves as a decision support tool, providing a reliable foundation that enables managers to allocate resources and manage schedules more effectively while maintaining safety. When combined with real-time positioning technology and other sensor data, BIM not only enhances positioning accuracy but also visualizes this information, thereby improving the timeliness and effectiveness of safety warnings.

In summary, BIM technology offers an efficient solution for risk identification and prevention in construction safety management, making proactive safety control more scientific and systematic. The integration of these technologies has markedly improved safety management on construction sites, effectively reducing the occurrence of accidents and demonstrating their critical value and broad potential in modern construction projects.

6.1.5 Wearable Devices and EEG Technology

In the field of construction safety, the use of wearable technology has significantly enhanced the monitoring of workers' safety status. By integrating devices such as IMUs (inertial measurement units), EEG (electroencephalogram) headsets, and heart rate monitors, these technologies can collect real-time data on workers' postures and physiological parameters, providing comprehensive monitoring of their working conditions. Wearable devices based on IMU sensors collect joint motion data, which, when analyzed with machine learning algorithms such as SVM [25], can determine whether a worker's posture is correct, thereby providing timely warnings of potential falls or unsafe working postures [26]. In addition, by collecting EEG and heart rate data, these devices can assess a worker's stress levels [27], fatigue state [28, 29], and

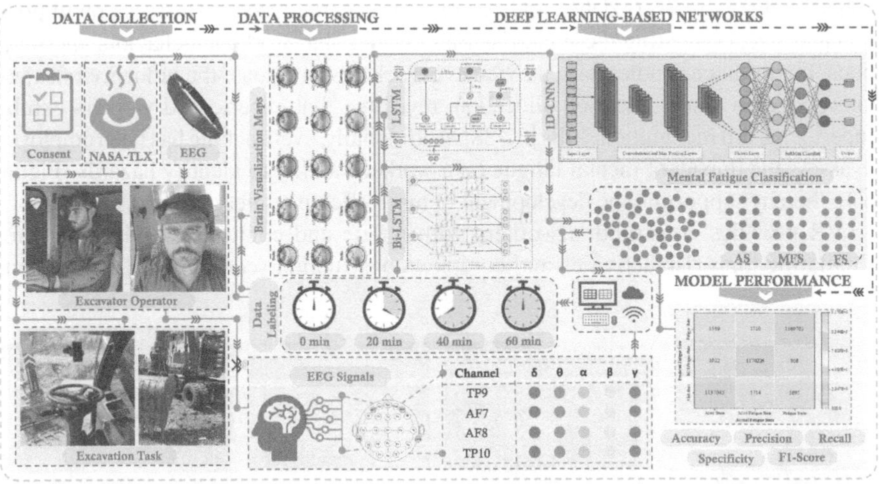

FIGURE 6.2 Deep learning-based construction equipment operators' mental fatigue classification using wearable EEG sensor data. (Source: Reference [34], Copyright 2023 by Elsevier.)

alertness to construction hazards [30, 31]. This allows for the identification of unsafe behaviors that may be induced by physiological factors.

Real-time monitoring of both physiological and postural states via wearable devices enables a more precise and proactive approach to on-site safety management, facilitating preventative measures before accidents occur [32]. Long-term monitoring of working posture can also help prevent musculoskeletal disorders resulting from prolonged unhealthy postures. Moreover, tracking physiological indicators such as heart rate and stress aids management in understanding workers' physical and mental strain, thus enabling improvements in working conditions and a reduction in overall fatigue [33]. The classification method for mental fatigue of construction equipment operators based on deep learning is shown in Figure 6.2.

In summary, the application of wearable devices in construction safety monitoring not only raises on-site safety standards but also drives a shift toward a health-oriented, preventive safety management model. This marks a move toward a more technologically advanced and systematic approach in construction safety management, providing effective technological support for reducing workplace injuries.

6.1.6 Eye-Tracking and Virtual/Augmented Reality Technologies

In construction safety, eye-tracking, virtual reality (VR), and augmented reality (AR) technologies have emerged as key tools for enhancing safety training and improving risk awareness. These technologies render safety training more interactive and immersive [35, 36], significantly boosting training effectiveness while reducing risks in real-world environments.

Eye-tracking technology works by analyzing the distribution of visual attention among workers at construction sites, thus helping researchers and safety trainers understand how workers detect and respond to potential hazards. It tracks pupil position and size, converting these metrics into insights into workers' behaviors and reactions. For instance, studies have utilized eye-tracking to evaluate the influence of safety knowledge on visual attention allocation, revealing that experienced workers exhibit superior hazard recognition skills [37, 38]. Additionally, eye-tracking has been employed to assess the impact of mental fatigue on a worker's ability to identify hazards (Figure 6.3), thereby supporting the development of more effective safety training protocols.

VR provides construction workers with a safe and controlled environment for simulating high-risk activities and complex construction scenarios. Its applications

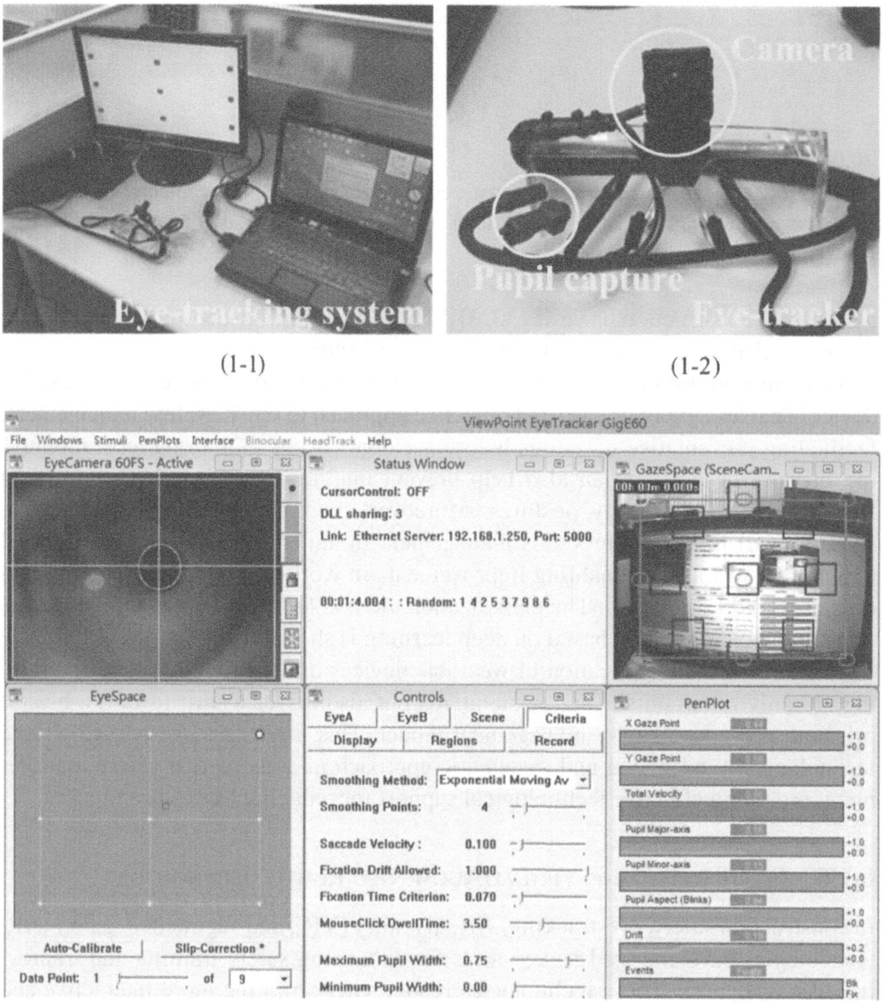

(1-1) (1-2)

(1-3)

FIGURE 6.3 Eye-tracking system: (1-1) Eye-tracking system, (1-2) Eye-tracker, (1-3) Tracker calibration process. (Source: Reference [37], Copyright 2016 by Elsevier.)

extend beyond basic training to include the rehearsal of intricate operations and high-risk tasks. Research has demonstrated that multi-user VR systems, employing region-based CNN (R-CNN) algorithms to evaluate worker behavior and personal protective equipment (PPE) compliance, are effective in assessing safety behavior under simulated hazardous conditions [39]. These VR applications have proven particularly valuable in offering real-time feedback and enhancing workers' risk perception.

Integrating eye-tracking with VR creates highly efficient safety training modules by accurately simulating and analyzing workers' visual and behavioral responses in specific work environments. This integration helps elucidate how workers process information and make decisions under actual conditions [40, 41]. Moreover, such a combined approach circumvents the dangers associated with on-site training and even facilitates remote instruction, while virtual environments can simplify the training of complex operations [42]. Customized, personalized safety training programs can also be developed based on individual needs and behavioral characteristics using this integrated methodology.

Overall, eye-tracking and VR/AR technologies introduce new perspectives and methods in construction safety training and risk assessment. By simulating real environments and monitoring worker responses, these technologies help drive the industry toward more efficient, safer operational standards.

6.2 CASE STUDIES

6.2.1 UAV Aerial Survey and Information Collection Technology

---

CASE 1: RAPID SAFETY MONITORING AND ANALYSIS OF FOUNDATION PIT CONSTRUCTION [43]

With the large-scale development of urban underground spaces, the safety monitoring of foundation pit construction has attracted widespread attention. The construction industry is considered one of the most hazardous sectors. Despite the strict rules and regulations enforced by the government, accidents still occur frequently. Traditional foundation pit safety monitoring relies on manual on-site measurements and complex equipment (labor-intensive and time-consuming) and may overlook the risk of unexpected collapse caused by severe local deformations.

This case study employs UAV imagery for rapid safety monitoring and analysis of foundation pit construction. It proposes the safety inclination of the foundation pit slopes as a safety monitoring indicator and utilizes point cloud reconstruction and surface fitting based on UAV-captured images of the pit. By introducing the local deformation distribution, the method assesses the localized safety status of the foundation pit. The approach aims to overcome the inefficiencies of traditional manual monitoring by capturing detailed UAV images of the pit and then analyzing the local deformation distribution, thereby providing a more comprehensive and rapid safety risk analysis.

Experimental results validate the potential advantages of this method. It rapidly identifies potential high-risk local areas through UAV imagery and presents a case study employing this approach. By collecting data via UAVs, this method is simpler, lower in cost, and more convenient compared to traditional techniques. The data collection and preprocessing time is reduced from several hours to no more than 10 minutes, facilitating regular and rapid measurements.

CASE 2: REAL-TIME MONITORING OF HIGH-ALTITUDE CONSTRUCTION SAFETY RISKS ON CONSTRUCTION SITES [44]

Falls from height are the most common cause of fatalities in the construction industry. This case explores the integration of UAVs into the safety planning and monitoring processes of high-rise construction. A system that integrates deep learning and UAV technology has been developed to monitor the safety of workers performing high-altitude tasks in real time, particularly addressing high-altitude work risks. Pre-trained deep learning models are employed to detect components of personal fall protection systems (such as safety harnesses, lifelines, and helmets). UAVs are used to capture images and videos of the construction site, and the data is transmitted to the model to detect safety violations. The advantages of UAVs over traditional methods are highlighted, including improved spatial resolution, the ability to capture real-time data, and enhanced worker safety through remote monitoring. The study discusses UAV operating protocols, data processing, and practical applications for continuously monitoring construction sites to identify and mitigate safety hazards. The system demonstrates high accuracy and efficiency in detecting safety violations, significantly enhancing the quality of on-site safety management.

In a case study during the construction of a high-rise building in Abu Dhabi, UAVs were employed to continuously monitor structural integrity and safety compliance. The UAVs performed daily flight missions to capture high-resolution images, which were then analyzed to track construction progress and detect any deviations from safety protocols. The overall system accuracy reached 90%, with the model's precision and recall at 97.2% and 90.2%, respectively. The average time to detect a safety violation was approximately 12 seconds. This system is fast, reliable, and cost-effective, demonstrating the effectiveness of UAVs in enhancing safety supervision. With proper implementation and usage, it has the potential to save many lives.

6.2.2 Behavior Recognition Technology

CASE 1: REAL-TIME RECOGNITION OF UNSAFE BEHAVIORS BY CONSTRUCTION WORKERS [45]

Unsafe behaviors on construction sites are a major cause of frequent accidents. Traditional safety management methods, such as Behavior-Based Safety (BBS), mainly rely on manual observation and recording. These methods are costly, time-consuming, and unable to comprehensively monitor the site and its workers. In this case study, the objective was to develop an image-based skeleton parameterization method that employs motion capture technology to identify unsafe behaviors of construction workers in real time and to validate its effectiveness in improving the efficiency and effectiveness of safety management.

Three common unsafe behaviors that often lead to accidents, such as leaning on handrails, dumping materials from height, and climbing ladders, were selected and parameterized. A mathematical model of the human skeleton was used to describe these behaviors via key joint parameters. Kinect for Windows 2.0 was used to collect body posture data from test subjects, and experiments were conducted to determine the parameter value ranges associated with unsafe behaviors. By comparing the real-time captured parameter values with predefined safe ranges, unsafe behaviors could be identified.

Experimental results indicate that the proposed method can identify and distinguish different types of unsafe behaviors in real time. Through statistical processing and analysis, the key parameters of unsafe behaviors and their standard value ranges were determined. The use of Kinect technology reduces data redundancy and circumvents the training phase required by traditional RGB video methods. This study demonstrates the feasibility and efficiency of the proposed method, providing technical support for construction safety management practices and enhancing the timeliness and accuracy of safety supervision. This approach offers an effective technical means for on-site safety monitoring, which helps reduce the occurrence of construction accidents. Future research directions include improving data stability and incorporating additional feature parameters to identify a broader range of unsafe behaviors.

CASE 2: IFC DATA EXTENSION FOR REAL-TIME SAFETY MONITORING IN AUTOMATED CONSTRUCTION OF HIGH-RISE BUILDINGS [46]

As the number of high-rise buildings continues to increase, Automated Construction Systems (ACS) have been widely adopted worldwide. ACS improve construction efficiency and safety by integrating automation and robotics technologies. However, ACS operations, especially during lifting phases, require real-time safety monitoring to prevent accidents, making real-time data access critical for effective safety management.

This study aims to extend the Industry Foundation Classes (IFC) standard to seamlessly integrate real-time safety information with digital models, thereby enhancing both the safety and efficiency of ACS. The research identifies the key information necessary to describe ACS components and safety monitoring projects and then extends the IFC schema to create a comprehensive description of safety-related data. Additionally, an innovative parameter mapping technique was developed to incorporate dynamic sensor data into the corresponding ACS components within the BIM model.

Initially, through literature review and on-site investigations, the essential information required for ACS components and safety monitoring projects was identified. Then, based on the IFC4 ADD3 standard, an IFC-based extension model was developed to describe ACS components and safety monitoring data. Utilizing the xBIM™ toolkit, a program was created to map sensor data in real time onto the IFC model. The extended IFC schema accommodates both static and dynamic information in ACS operations, ensuring seamless integration and interoperability with various systems. The integration of real-time sensor data enables stakeholders to rapidly identify potential safety issues, monitor changes, and respond promptly to evolving construction site conditions. Finally, the case study demonstrates the effectiveness of the proposed IFC extension model in real-time ACS safety monitoring.

The IFC extension model developed in this study is critical for improving the safety and efficiency of ACS operations in high-rise building construction. The application of this method in high-rise projects is feasible and lays a solid foundation for future data-driven efforts. Using this approach, safety monitoring data can be obtained more intuitively and promptly, providing strong support for on-site emergency response. Through automated real-time data mapping technology, sensor data is seamlessly integrated with the BIM model, bringing innovation to the field of construction safety monitoring.

CASE 3: GENERATION OF COMPUTER VISION SYNTHETIC DATA FOR CONSTRUCTION WORKER SAFETY MONITORING BASED ON A GAME ENGINE [47]

On construction sites, monitoring worker safety is of utmost importance. Traditional safety management methods that rely on manual supervision are inefficient and unreliable. To improve safety, researchers have proposed an automated safety monitoring approach based on CV. However, training CV models requires a large, diverse, and accurately labeled dataset. Data that is difficult or costly to collect in real-world environments. In this case study, the researchers propose a synthetic data generation method driven by a game engine to simulate different construction scenarios and train CV models, with a particular focus on improving the detection performance for small personal protective equipment used by construction workers.

Research using the Unity game engine has built a 3D virtual environment to generate customized images of safety incident scenarios. An automated data annotation module is employed within this game engine-based 3D virtual environment to produce synthetic images. This environment offers various backgrounds, lighting conditions, and viewpoints by adjusting the lighting, camera angles, and distances to simulate different accident scenarios. A CNN-based object detection model was trained using real-world data, synthetic data, and a mixed dataset to validate the performance of the proposed synthetic data generation process.

Experimental results indicate that incorporating synthetic data into training can boost detection performance by up to 30.4%. Using synthetic data reduces the risks associated with collecting real-world data. Synthetic data significantly cuts down the cost and time needed to train CV models, as it enables rapid generation of large amounts of labeled data and allows for the creation of a more comprehensive training dataset, thereby enhancing the generalization ability of CV models for real-world scenarios. This method provides an effective safety monitoring technology for the construction industry, especially under conditions of limited resources or challenging data acquisition.

The study demonstrates the potential of a game engine-based synthetic data generation method in construction worker safety monitoring, particularly for improving the detection performance of small-scale personal protective equipment. By leveraging a virtual environment to generate customized images and automate annotation, the data preparation workload is greatly reduced, thereby increasing both the training efficiency and accuracy of the safety monitoring system. The combination of synthetic data and precise annotations effectively improves the performance of CV models in detecting small objects, which is critical for construction safety management. Moreover, synthetic data generation can serve as a cost-effective approach, reducing the manual labor and expense involved in building datasets for training CV models. Future research will explore methods to narrow the reality gap between virtual and real environments and to apply synthetic data for safety monitoring across multiple construction sites. Additionally, this approach helps overcome privacy and safety concerns associated with on-site data collection, offering a new technological pathway for construction safety management.

6.2.3 Real-Time Positioning Technology

CASE 1: REALIZING POSITIONING IN CONSTRUCTION USING RFID AND IOT INTEGRATION [48]

In the construction industry, real-time and precise positioning of construction materials and personnel is crucial for ensuring project progress and safety. In construction supply chain management systems, assessments of labor, materials, and equipment are mostly based on on-site supervision or manual observations by engineers. Due to the complexity of the construction environment, tracking multiple construction resources is both challenging and time-consuming. Poor management of inventory, equipment, and workforce safety can hinder the efficient operation of supply chain systems.

RFID technology has been extensively explored across various stages of a building's lifecycle, including planning, design, construction, operation, and maintenance, to enhance and automate supply chain systems within the construction industry. In this case study, a positioning system integrating RFID and IoT (Internet of Things) technologies was developed for resource positioning and safety monitoring. RFID technology (Figure 6.4) shows significant potential in improving the visibility of supply chain management and ensuring real-time safety measures.

The case introduces the three key components of an RFID system: RFID tags, readers, and backend databases. It explores methods such as received signal strength indicator (RSSI), phase, time of arrival (ToA), and time difference of arrival (TDoA) to estimate the distance between the tag and the reader. The latest

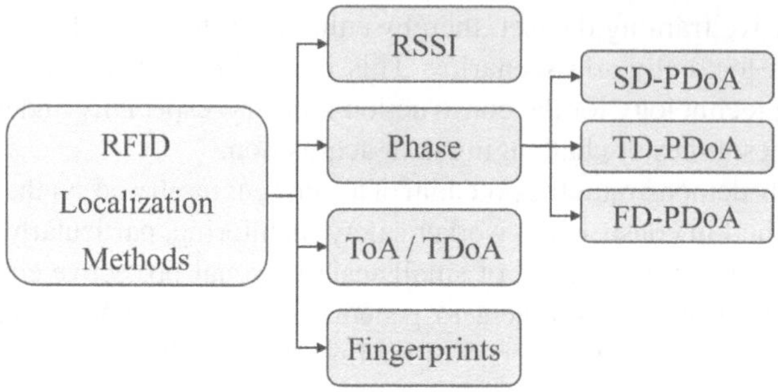

FIGURE 6.4 Localization methods of RFID system.

advancements in RFID positioning systems using RSSI, phase, ToA, TDoA, and fingerprinting are analyzed. The application of RFID technology on construction sites is discussed, particularly the challenges in metal-rich environments. A case study is presented that integrates RFID with IoT and other technologies (such as infrared, GPS, GIS, BIM, and ZigBee), proposing a chipless RFID system tailored for the construction industry, offering cost-effectiveness and security.

The case finds that RFID positioning faces numerous challenges in the construction industry, including the impact of metal environments, cost issues, reading accuracy, and workforce attitudes toward RFID technology. It demonstrates various applications of IoT and RFID integration in the construction industry, such as logistics management, progress monitoring, and safety improvements. The proposed chipless RFID system has advantages in cost-effectiveness, resilience to metal environments, long service life, and adaptability to harsh conditions.

RFID technology offers many benefits to the construction industry, including increased visibility of inventory, equipment, and labor, which helps projects stay on time and within budget. However, the study also reveals specific challenges in the construction industry, including positioning and security issues. By proposing solutions that integrate IoT technology and advanced security measures with chipless RFID, existing obstacles can be overcome, significantly improving the efficiency and effectiveness of supply chains in the construction industry. The case hopes this research will inspire further innovation and contribute to the continuous growth and success of the construction industry.

CASE 2: UWB POSITIONING PERFORMANCE DEGRADATION IN CONSTRUCTION PROJECTS [49]

In construction projects, real-time material tracking and progress estimation are critical for project success. UWB positioning systems, a promising indoor positioning technology, are affected by obstructions and environmental changes during construction. This study explores UWB system performance degradation under full blockage conditions using wood and steel as blocking materials across three stages, proposing effective system implementation recommendations.

Three experimental stages were designed, including two laboratory studies and one field implementation study. By analyzing UWB system performance in different environmental conditions, the position dependency of system errors was evaluated, and the possibility of position-based error modeling was explored. UWB system performance was evaluated under different blockage conditions, and its performance changes during construction were analyzed.

The UWB system performed well in unblocked conditions but showed performance degradation when obstructed by materials like metal and wood. The UWB system's performance degraded significantly in active construction environments. The study found that even under full metal blockage, the UWB system's average error remained below 50 cm, and location could be estimated within a 1.0-meter radius with 95% confidence, which is accurate enough for many construction-related applications. These findings provide important guidance for effective UWB system implementation in indoor construction projects. The application of UWB technology in construction must consider environmental factors and dynamic changes. Future research will focus on improving UWB system stability and accuracy in complex construction environments.

CASE 3: AUTOMATING CONSTRUCTION-SITE EQUIPMENT OPERATION ANALYSIS USING GPS DATA [50]

In construction-site management, continuous tracking of equipment operational efficiency and safety is crucial for improving productivity and ensuring site safety. The lack of methods for analyzing equipment operations using real-time positioning data limits a deeper understanding of equipment performance and safety monitoring. This study automated the spatio-temporal analysis of construction equipment operations using low-cost GPS devices.

In this case, commercially available low-cost GPS devices were used to continuously record the location and timestamps of equipment to enhance the automation level of construction equipment operation analysis. First, the error rate of the GPS devices was evaluated, followed by defining methods and rules for analyzing the operational data of construction equipment. The study also developed a detailed software interface that allows users to set up, analyze, and visualize equipment parameters, providing support for more realistic equipment operation analysis in simulation models.

The experimental results show that the developed technology can automatically identify and track equipment activities and safety-related information, providing support for performance and layout decisions on construction sites. This work will assist project managers in better planning, managing, and controlling equipment-related tasks on construction sites, particularly in dynamic and complex environments. GPS technology provides detailed and real-time information about equipment operations, helping to improve the performance and safety of construction equipment. The user interface makes the data easy to understand and use, aiding construction managers in making more informed decisions.

6.2.4 Spatial Planning Technology

CASE 1: AUTOMATED SAFETY RULE CHECKING IN CONSTRUCTION BASED ON BIM [51]

Construction safety is a global issue, with a large number of workers losing their lives each year due to safety accidents. Traditional safety planning methods rely on manual observation and paper records, which are not only inefficient but also difficult to accurately predict and prevent potential safety risks. In this case, the integration of automated safety rule checking with BIM addresses safety issues in building design and construction planning. A rule-based engine was implemented on a commercial BIM platform, demonstrating the feasibility of this method.

The research team proposed an automated safety rule checking system based on BIM (Figure 6.5), which uses a rule engine to automatically analyze building models and detect safety risks. This allows potential safety hazards, especially fall-related dangers, to be identified early in the building planning and design stages. The implementation case includes its application to real-world construction projects, demonstrating how these hazards can be automatically identified and corrected. Through this platform, construction engineers and project managers can know when, where, why, and what safety measures are needed to prevent fall-related accidents before the project begins. This automation tool significantly improves the efficiency of safety management and reduces accidents and safety hazards on construction sites.

The system in this case is divided into the following four steps:

1. *Safety Rule Interpretation*: Convert safety rules from natural language into machine-readable formats for computer processing.
2. *Building Model Preparation*: Ensure the BIM model contains all objects, attributes, and relationships required for rule checking.
3. *Rule Execution*: Automatically execute rule checks, identify unsafe conditions, and provide solutions.
4. *Rule Check Report*: Generate reports that include visualizations of applied safety equipment and detailed textual information.

By implementing the automated safety rule checking system on the BIM platform, the efficiency and accuracy of construction safety planning can be

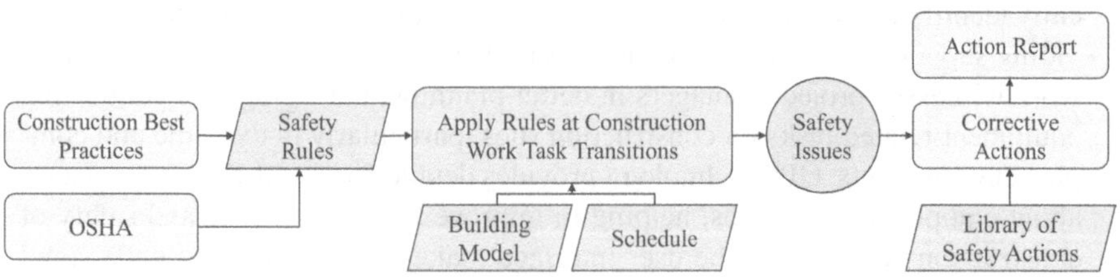

FIGURE 6.5 Framework for implementing an automated rule-based safety checking in BIM.

significantly improved. This system helps engineers and managers understand the required safety measures before construction begins, thereby preventing related safety incidents.

CASE 2: BIM-BASED FALL HAZARD IDENTIFICATION AND PREVENTION [52]

This case explores how to identify and eliminate potential fall hazards early in the planning phase of a construction project through BIM. The research demonstrates how to integrate automated safety rule checks into BIM to improve construction safety.

A framework was developed, including a BIM-based automated safety rule checking algorithm. By testing the model in office buildings and residential construction projects in Finland, the study compared manual and automated safety models and simulated the detection and prevention of fall hazards. Using existing safety rules, guidelines, and best practices combined with 3D design and scheduling information, an automated safety rule checking system was created. The developed prototype was tested using models of office buildings and residential projects. It demonstrates how the framework was applied to project schedules and simulates fall hazard detection and prevention.

The research provides an effective integrated safety framework for automated rule checking. The developed tool can automatically detect unprotected slab edges and successfully install the required guardrail system. BIM software's built-in features make it easy to calculate the number of guardrails. Additionally, the automatic installation of guardrails for slab edges and window openings can later be manually modified by the user. The contribution lies in providing a method for construction practitioners to detect and prevent fall-related hazards. Furthermore, it discusses the commercialization of the developed prototype and its potential impact on traditional safety management practices.

CASE 3: BIM-BASED SAFETY RISK ASSESSMENT FOR CONSTRUCTION PROJECTS DURING THE DESIGN PHASE [53]

This case study explores how BIM integration can be used to conduct safety risk assessments during the design phase to identify and manage potential risks before project implementation. A new method for conducting quantitative construction safety risk assessments during the design phase of construction projects is proposed. The method includes three indicators: probability, consequences, and exposure, which are calculated using occupational injury, death, and specific construction planning data.

By combining BIM with safety assessment tools, a system was developed to automatically detect and assess safety risks in construction designs. The case analyzed specific construction projects, demonstrating how BIM tools can be used for risk identification and management. A database for injuries and deaths, as well as a construction planning database, was created to collect the data required for risk assessment. The collected data was then linked to the BIM platform to quantify safety risks. Finally, safety risk indices for different occupations were calculated, and concepts of relative fatality risk and relative injury risk were proposed.

This method allows architects and engineers to identify potential risks early in the design phase, enabling them to take preventive measures and reduce safety incidents during construction. This integrated approach promotes the shift from reactive to preventive safety management, enhancing the overall safety of the project.

6.2.5 Wearable Devices and EEG Technology

CASE 1: PRE-WORK FATIGUE SCREENING FOR CONSTRUCTION WORKERS [54]

Construction workers may already feel fatigued before starting their work, and this fatigue could be caused by insufficient sleep, alcohol consumption, or emotional changes. Such pre-work fatigue may lead to decreased job performance and increased accident risks. Therefore, identifying whether workers are fit for high-intensity labor is a crucial step in ensuring construction-site safety. This study developed a pre-service fatigue screening method based on EEG signal analysis to assess the work suitability of construction workers.

To ensure workers get sufficient rest before starting work, this case presents a novel screening method based on wearable EEG devices. The method assesses workers' mental fatigue levels by analyzing EEG signal spectral parameters, such as gravity frequency and power spectral entropy. The researchers conducted laboratory-based fatigue tests, where participants performed a Stroop test (a classic cognitive interference assessment) while EEG data were recorded to validate the proposed method. Finally, a simulated excavation task was used as a verification experiment.

The experimental results showed that by comparing participants' test performance and reaction times, fatigued individuals could be successfully identified, thus preventing safety accidents caused by fatigue. The quantitative method developed by the researchers effectively assessed workers' mental fatigue levels and provided a basis for identifying workers unfit for work.

CASE 2: MONITORING WORKERS' ATTENTION AND ALERTNESS IN CONSTRUCTION ACTIVITIES [55]

Maintaining workers' attention and alertness in high-risk construction environments is key to accident prevention. The high-risk nature of the construction industry requires workers to remain highly alert when performing tasks.

However, due to prolonged exposure to risks, workers may become less sensitive to potential dangers, increasing the likelihood of accidents.

Researchers developed a wireless, wearable EEG system to monitor construction workers' attention and alertness in real time. The system quantifies workers' perceived risk levels by processing brain signals, particularly focusing on frequency, power spectral density, and signal spatial distribution. Experiments were conducted on construction sites, including monitoring workers' EEG while performing routine construction tasks. By analyzing EEG data, brain activity patterns associated with reduced attention and alertness were identified. The researchers analyzed EEG signal patterns of workers when avoiding different obstacles and how these patterns reflected their alertness.

Through real-time data monitoring, the study successfully identified moments when workers' attention was not focused, allowing for intervention measures before potential safety risks occurred. The study found that specific EEG signal attributes, particularly activity in the low gamma frequency band and the left frontal EEG clusters, directly reflected workers' alertness. These findings contribute to the future implementation of wearable EEG devices through data filtering and channel optimization. This monitoring approach is expected to significantly enhance the efficiency and effectiveness of on-site safety management, especially in high-risk operational environments.

CASE 3: PSYCHOLOGICAL FATIGUE CLASSIFICATION OF CONSTRUCTION EQUIPMENT OPERATORS USING WEARABLE EEG SENSOR DATA [34]

In the construction industry, workers' fatigue states are critical to work performance and safety. Equipment operators experience mental fatigue due to prolonged equipment operation, which may lead to decreased attention, operational errors, and an increased risk of accidents. This case presents a method based on wearable EEG sensor data and deep learning networks for real-time monitoring and classification of operators' psychological fatigue states. The researchers used LSTM, bidirectional LSTM (Bi-LSTM), and 1D convolutional neural networks (1D-CNN) models to process and analyze EEG data.

On the construction site, the researchers conducted a 1-hour excavation operation experiment with 15 operators while recording brain activity patterns using EEG sensors. In the experiment, the operators' psychological fatigue states were subjectively assessed using the NASA-TLX score. The deep learning model's performance was evaluated based on accuracy, precision, recall, specificity, and F1 score. The results showed that the Bi-LSTM model performed the best in classifying psychological fatigue states, achieving an accuracy of 99.941%, indicating that the method effectively identified and classified the psychological fatigue state of construction equipment operators, providing a new tool for improving health and safety management on construction sites.

6.2.6 Eye-Tracking and Virtual/Augmented Reality Technologies

CASE 1: IDENTIFYING AND CLASSIFYING CONSTRUCTION EQUIPMENT OPERATORS' MENTAL FATIGUE WITH WEARABLE EYE-TRACKING TECHNOLOGY [56]

In construction, operator mental fatigue is a leading cause of equipment-related accidents. Their mental state directly impacts operational safety, as fatigue can lead to decreased performance, increased errors, and even accidents. This study uses eye-tracking technology to identify and classify mental fatigue, aiming to enhance work safety, as shown in Figure 6.6.

In this case, a new method was developed. It uses wearable eye-tracking technology. This technology can record operators' eye-movement data in real time in a simulated operating environment. This helps to identify and classify operators' multi-level mental fatigue. Data was collected through simulated excavator

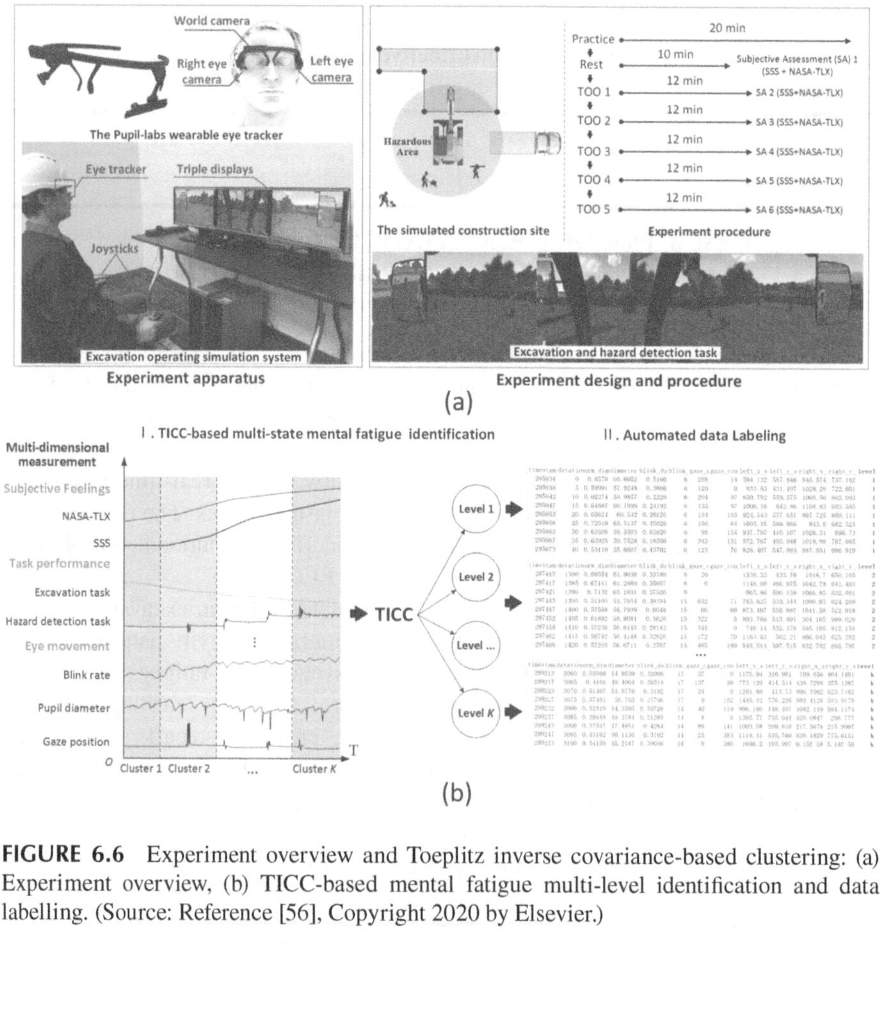

FIGURE 6.6 Experiment overview and Toeplitz inverse covariance-based clustering: (a) Experiment overview, (b) TICC-based mental fatigue multi-level identification and data labelling. (Source: Reference [56], Copyright 2020 by Elsevier.)

operation experiments. Then, the Toeplitz inverse covariance-based clustering (TICC) method was applied. This method analyzed eye-movement data such as pupil changes and fixation durations. Thus, it could determine mental fatigue levels. Six participants were recruited to collect relevant data. The study first used the TICC method to identify three levels of mental fatigue. Then, four sets of eye-movement features suitable for different construction scenarios were extracted. A supervised learning algorithm was used to classify operators' multi-level mental fatigue.

Classification performance analysis shows that eye-movement features of fatigued operators are significantly different from those of non-fatigued ones. Fatigued operators have more unstable fixations and slower pupil responses. These changes can monitor operators' fatigue levels in real time. This can provide a scientific basis for construction-site safety management. Thus, work plans can be adjusted timely to prevent accidents. Support vector machine is the most suitable algorithm for classifying mental fatigue when facing different construction scenarios and individual deviations (with an accuracy rate between 79.5% and 85.0%). The study generally proves the feasibility of applying wearable eye-tracking technology to identify and classify mental fatigue in construction equipment operators.

CASE 2: EVALUATING THE IMPACT OF MENTAL FATIGUE ON CONSTRUCTION EQUIPMENT OPERATORS' HAZARD-DETECTION CAPABILITIES [40]

Equipment-related accidents, such as collisions between construction equipment and pedestrians, are a major challenge to construction-site safety. Operators are prone to attention decline after long-term task execution, a key cause of accidents. This case uses wearable eye-tracking technology to assess the impact of mental fatigue on operators' hazard-detection capabilities and the corresponding patterns of visual attention allocation.

Twelve healthy participants performed simulated excavator-operation tasks in a lab. Operators were tested for extended periods in a simulated operating environment, with their eye-movement behavior recorded via eye-tracking technology. Pupil size, fixation duration, and their changes were analyzed to assess operators' attention distribution and hazard-identification responses. Subjective mental-fatigue assessments, hazard-detection task performance, and eye-tracking indices were recorded and analyzed.

Experimental results show that as subjective mental-fatigue levels increase, operators' hazard-detection capabilities decline. This is shown by a significant increase in response time to hazards and the number of false alarms. Eye-tracking indices, such as unstable fixations and dull pupil responses, can effectively predict mental-fatigue states. This finding is crucial for the construction industry. It offers a way to assess operator fatigue and take safety measures by monitoring eye-movement behavior. The study provides insights into the impact of mental fatigue on operators' hazard-detection capabilities from a visual attention perspective and grounds for developing effective safety interventions and attention-based safety management training methods.

CASE 3: COMPARING THE SEARCH PATTERNS OF EXPERIENCED AND INEXPERIENCED WORKERS IN CONSTRUCTION-SITE HAZARD IDENTIFICATION USING AN EYE TRACKER [37]

The high accident rate in construction is mainly due to safety inspectors' lack of sufficient safety knowledge and experience. Effective hazard identification is essential for preventing construction-site accidents. This case explores the differences in visual-search behavior between experienced and inexperienced workers when identifying potential hazards. Integrating knowledge and experience into training can help improve safety awareness and hazard-identification skills.

In this study, a construction-site hazard identification experiment based on a virtual environment was set up. It included images of workplaces with obvious and non-obvious hazards. Eye-tracking technology recorded workers' visual-search patterns, including fixations, saccade paths, and response times. Participants' eye-tracking records during image inspection were analyzed. Twenty-five volunteers participated, including experienced construction managers and graduate students without construction work experience. The experiment included four images: two with obvious hazards and two with non-obvious ones.

The results show that experienced workers identify hazards faster than inexperienced ones, with more purposeful and consistent saccade paths to quickly locate key hazard areas. However, there was no significant difference in hazard-identification accuracy. Experienced workers spent significantly less time identifying obvious and non-obvious hazards than inexperienced workers. Additionally, fixation frequency analysis shows that inexperienced workers are less confident in determining whether attention points are hazards. They had more fixations on almost all attention points than experienced workers. The finding highlights experience's importance in improving hazard identification efficiency and offers improvement directions for training new workers, such as simulating experienced workers' saccade strategies. The research results can provide valuable information for safety trainers and educators.

6.3 PRACTICAL APPLICATIONS

Monitoring construction safety occurs in three stages: pre-construction worker behavior assessment, pre-construction training for safety enhancement, and real-time monitoring of unsafe behaviors during construction. Table 6.1 summarizes the key technologies for these stages.

TABLE 6.1
Key Points of Construction Safety Behavior Technology

Technology Type	Application Stage	Advantages	Disadvantages	Future Research Direction
Virtual reality	Diagnosis	Provide immersive safety behavior simulation	Limited accuracy, needs wider hazard scenario integration	Dynamic navigation and speed control interface integration
Eye-tracking	Diagnosis	Assessing workers' visual attention to safety risks	Reliance on static images and a lack of dynamic environment testing	Mobile eye trackers in action
Predictive modeling	Diagnosis	Automated security behavior prediction	Data collection is costly and requires integration with video surveillance systems	Behavior-based security and inter-organizational security collaboration
Basic computer training	Prepare	Simulate real dangerous work scenarios to improve safety training effectiveness	Lack of emphasis on written/oral communication skills	Long-term impact evaluation and personalized learning resources
Drone/ sensor monitoring	Monitor	On-site data capture for security monitoring	Difficulty in using the equipment, need to adapt to working conditions	Instant alarm application for smartphone sensors
Vision-based behavior monitoring	Monitor	Detect unsafe behavior using live images/videos	A large number of samples is required for deep learning training	Innovation of 3D body key point estimation model
Real-time positioning	Monitor	Locate workers and warn of exposure to dangerous areas	GPS positioning accuracy issues, ineffectiveness in indoor environments	3D proximity model to prevent collision accidents

6.4 CONCLUSIONS AND FUTURE DIRECTIONS

Based on the above case studies, we can make an initial outlook on the current mainstream construction safety behavior monitoring technologies.

6.4.1 Virtual Reality Technology

Despite its accuracy limitations in simulating actual movements, VR technology can simulate a wide range of hazard scenarios. If it can cover extensive hazard scenarios, it can become a viable tool. It can assess workers' ability to identify safety risks by tracking their visual attention, but it usually relies on static images rather than dynamic real-world situations. If integrated with advanced 3D head-mounted displays, it may become more feasible. Future research will introduce more precise motion trackers in simulated environments to achieve more realistic results in predicting construction workers' safety behaviors. Future digital wayfinders can enhance the experience of workers and site visitors and help supervisors analyze construction-site safety accessibility through effective use of blue-dot navigation, web-based maps, route planning, and location alerts.

6.4.2 Eye-Tracking Technology

Eye-tracking is an increasingly growing field in safety behavior technology research. It has been used to evaluate workers' responses to simulated work scenarios. However, laboratory experiments using static photos instead of dynamic real-world environments have been questioned. Some argue these photos are just static screen images and cannot reflect actual construction-site conditions. Thus, construction workers' real-world responses may differ. To address these limitations, future research is encouraged to conduct eye-tracking tests on real construction-site workers using head-mounted eye trackers. With this wearable tool, more reliable results might be obtained.

6.4.3 Real-Time Positioning Technology

Currently, real-time positioning technologies like GPS trackers locate individuals who may enter hazardous areas and trigger alarms. For high-density construction sites of high-rise and multi-story buildings, this is an area needing further research. In crowded sites, workers' unsafe behaviors and human error risks are high. Future research can focus on proximity estimation to prevent collision accidents. This may involve combining 3D spatial reconstruction of building materials and construction machinery with worker density estimation to improve proximity estimation and enhance site safety.

6.4.4 Precise 3D Body Estimation Models Using Wearable Smart Cameras

Capturing and analyzing on-site photos and videos helps construction managers identify unsafe behaviors and personal protective equipment violations. Regarding CV technology, new breakthrough image-processing techniques may enable the accurate estimation of workers' unsafe behaviors. Improving camera accuracy and placement can effectively avoid occlusions when capturing on-site photos.

REFERENCES

1. H. Kim, K. Kim, H. Kim, Vision-based object-centric safety assessment using fuzzy inference: Monitoring struck-by accidents with moving objects, Journal of Computing in Civil Engineering 30(4) (2016) 04015075.
2. S. Gupta, S. Nair, A review of the emerging role of UAVs in construction site safety monitoring, Materials Today: Proceedings (2023).
3. D. Kim, M. Liu, S. Lee, V.R. Kamat, Remote proximity monitoring between mobile construction resources using camera-mounted UAVs, Automation in Construction 99 (2019) 168–182.
4. S. Bang, H. Kim, H. Kim, UAV-based automatic generation of high-resolution panorama at a construction site with a focus on preprocessing for image stitching, Automation in Construction 84 (2017) 70–80.

5. J. Park, Y. Kim, Obstacle detection and collision avoidance of quadrotor UAV using depth map of stereo vision, AIAA Guidance, Navigation, and Control (GNC) Conference, 2013, p. 4994.
6. M. Itani, A. Haroun, W. Fahs, Obstacle avoidance for ultrasonic unmanned aerial vehicle monitoring using android application, 2018 International Arab Conference on Information Technology (ACIT), IEEE, 2018, pp. 1–4.
7. M. Golparvar-Fard, A. Heydarian, J.C. Niebles, Vision-based action recognition of earthmoving equipment using spatio-temporal features and support vector machine classifiers, Advanced Engineering Informatics 27(4) (2013) 652–663.
8. Y.K. Cho, M. Gai, Projection-recognition-projection method for automatic object recognition and registration for dynamic heavy equipment operations, Journal of Computing in Civil Engineering 28(5) (2014) A4014002.
9. F. Vahdatikhaki, A. Hammad, H. Siddiqui, Optimization-based excavator pose estimation using real-time location systems, Automation in Construction 56 (2015) 76–92.
10. J. Wang, S.N. Razavi, Low false alarm rate model for unsafe-proximity detection in construction, Journal of Computing in Civil Engineering 30(2) (2016) 04015005.
11. J. Park, E. Marks, Y.K. Cho, W. Suryanto, Performance test of wireless technologies for personnel and equipment proximity sensing in work zones, Journal of Construction Engineering and Management 142(1) (2016) 04015049.
12. R. Akhavian, A.H. Behzadan, Construction equipment activity recognition for simulation input modeling using mobile sensors and machine learning classifiers, Advanced Engineering Informatics 29(4) (2015) 867–877.
13. H. Li, G. Chan, J.K.W. Wong, M. Skitmore, Real-time locating systems applications in construction, Automation in Construction 63 (2016) 37–47.
14. T. Cheng, G.C. Migliaccio, J. Teizer, U.C. Gatti, Data fusion of real-time location sensing and physiological status monitoring for ergonomics analysis of construction workers, Journal of Computing in Civil Engineering 27(3) (2013) 320–335.
15. T. Cheng, J. Teizer, G.C. Migliaccio, U.C. Gatti, Automated task-level activity analysis through fusion of real time location sensors and worker's thoracic posture data, Automation in Construction 29 (2013) 24–39.
16. S. Chae, T. Yoshida, Application of RFID technology to prevention of collision accident with heavy equipment, Automation in Construction 19(3) (2010) 368–374.
17. W. Wu, H. Yang, D.A. Chew, S.-h Yang, A.G. Gibb, Q. Li, Towards an autonomous real-time tracking system of near-miss accidents on construction sites, Automation in Construction 19(2) (2010) 134–141.
18. A. Marefat, H. Toosi, R.M. Hasankhanlo, A BIM approach for construction safety: Applications, barriers and solutions, engineering, Construction and Architectural Management 26(9) (2018) 1855–1877.
19. G. Lee, J. Cho, S. Ham, T. Lee, G. Lee, S.-H. Yun, H.-J. Yang, A BIM-and sensor-based tower crane navigation system for blind lifts, Automation in Construction 26 (2012) 1–10.
20. H. Son, H. Seong, H. Choi, C. Kim, Real-time vision-based warning system for prevention of collisions between workers and heavy equipment, Journal of Computing in Civil Engineering 33(5) (2019) 04019029.
21. L. Heng, D. Shuang, M. Skitmore, H. Qinghua, Y. Qin, Intrusion warning and assessment method for site safety enhancement, Safety Science 84 (2016) 97–107.
22. H. Kim, H.-S. Lee, M. Park, B. Chung, S. Hwang, Automated hazardous area identification using laborers' actual and optimal routes, Automation in Construction 65 (2016) 21–32.
23. O. Golovina, J. Teizer, N. Pradhananga, Heat map generation for predictive safety planning: Preventing struck-by and near miss interactions between workers-on-foot and construction equipment, Automation in Construction 71 (2016) 99–115.

24. A.S. Kulinan, M. Park, P.P.W. Aung, G. Cha, S. Park, Advancing construction site workforce safety monitoring through BIM and computer vision integration, Automation in Construction 158 (2024) 105227.
25. K. Yang, C.R. Ahn, M.C. Vuran, S.S. Aria, Semi-supervised near-miss fall detection for ironworkers with a wearable inertial measurement unit, Automation in Construction 68 (2016) 194–202.
26. H. Kim, C.R. Ahn, K. Yang, Identifying safety hazards using collective bodily responses of workers, Journal of Construction Engineering and Management 143(2) (2017) 04016090.
27. X. Xing, H. Li, J. Li, B. Zhong, H. Luo, M. Skitmore, A multicomponent and neurophysiological intervention for the emotional and mental states of high-altitude construction workers, Automation in Construction 105 (2019) 102836.
28. X. Xing, B. Zhong, H. Luo, T. Rose, J. Li, M.F. Antwi-Afari, Effects of physical fatigue on the induction of mental fatigue of construction workers: A pilot study based on a neurophysiological approach, Automation in Construction 120 (2020) 103381.
29. Y. Wang, Y. Huang, B. Gu, S. Cao, D. Fang, Identifying mental fatigue of construction workers using EEG and deep learning, Automation in Construction 151 (2023) 104887.
30. J. Chen, Z. Lin, X. Guo, Developing construction workers' mental vigilance indicators through wavelet packet decomposition on EEG signals, Construction Research Congress 2018, 2018, pp. 51–61.
31. D. Wang, H. Li, J. Chen, Detecting and measuring construction workers' vigilance through hybrid kinematic-EEG signals, Automation in Construction 100 (2019) 11–23.
32. X. Yan, H. Li, A.R. Li, H. Zhang, Wearable IMU-based real-time motion warning system for construction workers' musculoskeletal disorders prevention, Automation in Construction 74 (2017) 2–11.
33. H. Jebelli, S. Hwang, S. Lee, EEG-based workers' stress recognition at construction sites, Automation in Construction 93 (2018) 315–324.
34. I. Mehmood, H. Li, Y. Qarout, W. Umer, S. Anwer, H. Wu, M. Hussain, M.F. Antwi-Afari, Deep learning-based construction equipment operators' mental fatigue classification using wearable EEG sensor data, Advanced Engineering Informatics 56 (2023) 101978.
35. R. Sacks, A. Perlman, R. Barak, Construction safety training using immersive virtual reality, Construction Management and Economics 31(9) (2013) 1005–1017.
36. A. Albert, M.R. Hallowell, B. Kleiner, A. Chen, M. Golparvar-Fard, Enhancing construction hazard recognition with high-fidelity augmented virtuality, Journal of Construction Engineering and Management 140(7) (2014) 04014024.
37. R.-J. Dzeng, C.-T. Lin, Y.-C. Fang, Using eye-tracker to compare search patterns between experienced and novice workers for site hazard identification, Safety Science 82 (2016) 56–67.
38. S. Hasanzadeh, B. Esmaeili, M.D. Dodd, Measuring the impacts of safety knowledge on construction workers' attentional allocation and hazard detection using remote eye-tracking technology, Journal of Management in Engineering 33(5) (2017) 04017024.
39. M. Mohd Saudi, A.H. Ma'arof, A. Ahmad, A.S. Mohd Saudi, M.H. Ali, A. Narzullaev, M.I. Mohd Ghazali, Image detection model for construction worker safety conditions using faster R-CNN, International Journal of Advanced Computer Science and Applications 11(6) (2020).
40. J. Li, H. Li, H. Wang, W. Umer, H. Fu, X. Xing, Evaluating the impact of mental fatigue on construction equipment operators' ability to detect hazards using wearable eye-tracking technology, Automation in Construction 105 (2019) 102835.
41. S. Park, C.Y. Park, C. Lee, S.H. Han, S. Yun, D.-E. Lee, Exploring inattentional blindness in failure of safety risk perception: Focusing on safety knowledge in construction industry, Safety Science 145 (2022) 105518.

42. X. Li, W. Yi, H.-L. Chi, X. Wang, A.P. Chan, A critical review of virtual and augmented reality (VR/AR) applications in construction safety, Automation in Construction 86 (2018) 150–162.
43. J. Wu, L. Peng, J. Li, X. Zhou, J. Zhong, C. Wang, J. Sun, Rapid safety monitoring and analysis of foundation pit construction using unmanned aerial vehicle images, Automation in Construction 128 (2021) 103706.
44. M.Z. Shanti, C.-S. Cho, B.G. de Soto, Y.-J. Byon, C.Y. Yeun, T.Y. Kim, Real-time monitoring of work-at-height safety hazards in construction sites using drones and deep learning, Journal of Safety Research 83 (2022) 364–370.
45. Y. Yu, H. Guo, Q. Ding, H. Li, M. Skitmore, An experimental study of real-time identification of construction workers' unsafe behaviors, Automation in Construction 82 (2017) 193–206.
46. R. Hu, K. Chen, W. Jiang, H. Luo, IFC data extension for real-time safety monitoring of automated construction in high-rise building projects, Automation in Construction 162 (2024) 105408.
47. H. Lee, J. Jeon, D. Lee, C. Park, J. Kim, D. Lee, Game engine-driven synthetic data generation for computer vision-based safety monitoring of construction workers, Automation in Construction 155 (2023) 105060.
48. S.I. Khan, B.R. Ray, N.C. Karmakar, RFID localization in construction with IoT and security integration, Automation in Construction 159 (2024) 105249.
49. A. Shahi, A. Aryan, J.S. West, C.T. Haas, R.C. Haas, Deterioration of UWB positioning during construction, Automation in Construction 24 (2012) 72–80.
50. N. Pradhananga, J. Teizer, Automatic spatio-temporal analysis of construction site equipment operations using GPS data, Automation in Construction 29 (2013) 107–122.
51. S. Zhang, J. Teizer, J.-K. Lee, C.M. Eastman, M. Venugopal, Building information modeling (BIM) and safety: Automatic safety checking of construction models and schedules, Automation in Construction 29 (2013) 183–195.
52. S. Zhang, K. Sulankivi, M. Kiviniemi, I. Romo, C.M. Eastman, J. Teizer, BIM-based fall hazard identification and prevention in construction safety planning, Safety Science 72 (2015) 31–45.
53. Y. Lu, P. Gong, Y. Tang, S. Sun, Q. Li, BIM-integrated construction safety risk assessment at the design stage of building projects, Automation in Construction 124 (2021) 103553.
54. H. Li, D. Wang, J. Chen, X. Luo, J. Li, X. Xing, Pre-service fatigue screening for construction workers through wearable EEG-based signal spectral analysis, Automation in Construction 106 (2019) 102851.
55. D. Wang, J. Chen, D. Zhao, F. Dai, C. Zheng, X. Wu, Monitoring workers' attention and vigilance in construction activities through a wireless and wearable electroencephalography system, Automation in Construction 82 (2017) 122–137.
56. J. Li, H. Li, W. Umer, H. Wang, X. Xing, S. Zhao, J. Hou, Identification and classification of construction equipment operators' mental fatigue using wearable eye-tracking technology, Automation in Construction 109 (2020) 103000.

7 HVAC Equipment Monitoring

Zeyu Zhao, Xinrou Han, Shuang Yang, and Bin Yang

7.1 INTRODUCTION

Since the late 1980s, with the rapid advancement of automation technologies, fault detection and diagnosis (FDD) systems have been widely adopted in heating, ventilation, and air conditioning (HVAC) systems and building engineering. The core objective of FDD is to ensure system stability and optimize energy efficiency through real-time monitoring and analysis, enabling the timely identification and resolution of system anomalies. This technology plays a pivotal role in enhancing system reliability, reducing energy consumption, and improving occupant comfort.

During this period, research efforts focused primarily on core refrigeration equipment and critical piping systems. Key equipment includes chillers, chilled water pumps, cooling water pumps, header pumps, heat pumps, and heat exchangers, whose performance directly impacts the cooling efficiency and energy usage of HVAC systems. Piping systems, particularly chilled water loops and cooling water loops, were also prioritized, as they act as the "circulatory network" for thermal energy transfer, directly influencing energy distribution efficiency and operational reliability.

FDD systems typically deploy temperature, pressure, and other sensors at critical nodes to measure and record operational parameters. Through continuous monitoring and correlation analysis of these parameters, FDD detects anomalies such as temperature deviations, pressure fluctuations, and thermodynamic imbalances. These signals, processed by intelligent algorithms, are translated into diagnostic insights for potential issues, including blockages, leaks, and equipment wears. Furthermore, FDD systems leverage historical data for trend analysis, enabling predictive maintenance to preempt failures.

The application of FDD not only extends equipment lifespan and reduces maintenance costs but also enhances building energy efficiency and occupant comfort through stabilized indoor temperatures and optimized energy use. As the technology evolves, modern FDD systems integrate diverse sensors, such as humidity sensors, airflow sensors, and water quality sensors, and employ advanced data processing techniques like machine learning and artificial intelligence (AI), significantly improving fault detection accuracy and adaptive capabilities. This integration marks a shift toward intelligent, high-efficiency HVAC systems, offering innovative solutions for building energy management and environmental sustainability.

DOI: 10.1201/9781003611455-7

7.2 HVAC EQUIPMENT MONITORING METHODS

Traditional FDD methods in the HVAC field can be categorized into three main classes: quantitative model-based methods, qualitative model-based methods, and process history-based methods. Quantitative models are built on well-established physical or engineering principles but involve complex computations. While qualitative models are easier to develop and apply, they heavily rely on developers' expertise, and certain rules may become ineffective in complex systems. Process history-based models are black-box approaches constructed when the system's physical characteristics are not fully understood. Although their development logic is straightforward, they require extensive operational data from HVAC systems for training.

Traditionally, HVAC operational data are structured datasets collected through building automation systems (BASs). Contact-based measurement methods, such as temperature and humidity sensors, are used to acquire operational data (e.g., temperature, humidity) from equipment or pipelines. While simple to implement, these methods may interfere with normal system operations, introducing measurement errors that compromise fault diagnosis accuracy. Therefore, the development of non-contact measurement methods based on image signals, capable of rapid and accurate fault diagnosis, holds groundbreaking significance for FDD in HVAC systems.

Current approaches primarily utilize thermal imaging cameras and standard optical cameras to capture thermal and visible-light images, respectively. Infrared (IR) imaging and optical techniques are then employed to analyze these images, enabling non-contact equipment monitoring. This section reviews the commonly used monitoring methods and IR imaging applications in detail.

7.2.1 Common Monitoring Methods

The cornerstone of ensuring efficient operation of HVAC systems lies in precise monitoring technologies, which form the foundation for advancing system intelligence and automation. As building environments demand higher energy efficiency, comfort, and sustainability, the application of advanced sensor-based monitoring technologies in HVAC systems has become increasingly critical. These sensors employ contact and non-contact measurement methods to collect and analyze key environmental parameters (such as temperature, humidity, and air quality) in real time, providing comprehensive feedback to enable dynamic adjustments based on actual conditions.

Real-time monitoring technologies empower HVAC systems with rapid responsiveness, allowing them to adapt flexibly to indoor and outdoor environmental fluctuations. For instance, when outdoor temperatures shift abruptly or indoor humidity levels rise abnormally, the system can automatically adjust airflow, temperature settings, and humidity control parameters by analyzing these data, ensuring a consistently comfortable and healthy indoor environment. Such intelligent responsiveness not only enhances occupant well-being but also improves system stability and reliability.

Precise monitoring is equally vital for optimizing energy efficiency in HVAC systems. By continuously tracking and analyzing operational data, inefficiencies and potential energy waste can be identified. For example, the system may detect uneven airflow distribution in specific zones or mismatched fan speeds for current load demands, prompting adjustments to damper positions or fan optimization. Such intelligent regulation minimizes unnecessary energy consumption, significantly boosting overall efficiency.

Since the mid-20th century, sensor-based monitoring technologies have undergone remarkable evolution. From early mechanical thermostats to electronic sensors and the integration of microprocessor technology, each advancement has enhanced the granularity of monitoring and control. Modern sensors not only achieve higher accuracy and stability but also integrate seamlessly with smart control systems, enabling sophisticated data analytics.

Despite their simplicity, contact-based measurement methods risk interfering with system operations and introducing data errors, which can compromise fault diagnosis accuracy. To address these limitations, non-contact measurement methods have emerged as transformative solutions in the FDD of HVAC systems. Vision-based approaches, such as infrared thermography (IRT) for detecting equipment hotspots or assessing human thermal comfort, minimize operational disruptions while delivering highly accurate data. By capturing thermal radiation emissions, these methods enable rapid and precise fault identification, enhancing diagnostic reliability. Non-contact technologies also provide intuitive insights into indoor airflow patterns, temperature distribution, and humidity levels, further strengthening system oversight.

As technology advances, future HVAC monitoring systems will grow more intelligent and integrated. By combining AI, Internet of Things (IoT), and big data analytics, HVAC systems will achieve finer predictive control, driving unprecedented improvements in energy efficiency and indoor environmental quality.

7.2.2 Infrared Thermography

IRT originated in 1800 when astronomer Sir William Herschel discovered IR radiation while studying the thermal effects of sunlight through colored filters, laying the groundwork for subsequent IR technologies [1]. In 1840, his son Sir John Herschel recorded images termed "thermograms" using carbon and alcohol [2], pioneering a new era in temperature measurement. Following these early advancements, IR thermographic nondestructive testing (NDT) remained largely confined to military applications, such as inspecting engine nozzle bonding quality or detecting pipeline leaks, due to cost and precision limitations [3]. During World War II, Germany first deployed IR communication and night-vision equipment in its military. At this stage, IR technology was primarily exploratory, with its advantages over other methods not yet fully realized.

A milestone occurred in 1964 when Sweden's AGA Corporation developed the world's first industrial IR thermal imager, marking the formal entry of IR technology into NDT applications. Developed nations such as the United States, the United Kingdom, and Sweden accelerated IR advancements, particularly in aerospace, power electronics, and petrochemical industries [4–6]. The US currently leads in IR technology, hosting numerous equipment manufacturers, advancing thermographic applications, and establishing industry standards (e.g., ASNT, ASTM) alongside operational manuals and training systems.

IR thermographic techniques are categorized into active and passive methods [7]. The concept of active thermography was proposed in the 1930s, with Green and Alzofon systematically outlining its theoretical foundations and applications in the 1960s [8]. This spurred global research into IR thermographic NDT and evaluation. By the 1970s, the integration of heat transfer theory deepened thermographic research. Carlslaw's work on homogeneous heat conduction [9] and contributions by Maldague [10], Vavilov [11], Almond [12], and Favro [13], who modeled one- and two-dimensional heat transfer under thermal/cooling stimuli, established the theoretical basis for modern IRT. In the late 1980s, advancements in IR imager performance and digital signal processing elevated detection precision. Novel thermographic evaluation techniques emerged, propelling military and civilian applications to new heights. The 1990s saw IRT evolve toward multi-spectral and high-resolution capabilities with improved detector performance and reduced costs [14]. The advent of cooled and uncooled focal plane array (FPA) IR thermographic products enabled mass industrialization, revolutionizing applications.

In the 21st century, IRT has achieved widespread adoption in medical diagnostics, industrial inspection, security surveillance, and aerospace, driven by maturation and accessibility [14]. Micro-electromechanical systems (MEMS) have further miniaturized and reduced the cost of IR detectors. Modern thermographic devices remain vital in military contexts while expanding into civilian domains such as building energy efficiency, fire rescue, medical imaging, and environmental monitoring.

Overall, IRT holds vast potential as it transitions from military to civilian markets. With technological advancements and cost reductions, its applications in security, healthcare, industrial inspection, and smart transportation continue to grow. Notably, during the COVID-19 pandemic, IRT demonstrated significant value in rapid body temperature screening. Future integration with AI will further expand its role in automated and intelligent systems, with each technological leap driving profound societal impacts.

7.2.2.1 Infrared Thermal Imaging Technology and Infrared Thermal Imaging Camera

IR is a common form of radiation in the electromagnetic spectrum, and any object above absolute zero will produce IR radiation energy due to the irregular motion of its internal molecules and atoms. This form of energy occupies an important position in the electromagnetic spectrum, especially in the application of thermal imaging technology, which has a unique advantage. IRT is a technology that converts optical signals into electrical signals by absorbing IR light radiated by a target object, i.e., converting IR radiation invisible to the naked eye into a visible image [15]. This technique converts invisible IR radiation into an image visible to the naked eye, providing researchers and engineers with an intuitive and effective tool for detecting and analyzing the temperature distribution of a target object.

In principle, IRT converts the IR radiation of the object under test into electrical signals, which are then presented as thermal images reflecting the temperature distribution of the object's surface through a complex process of signal processing and image generation. The key to this process is the use of the IR radiation properties of the object, that is, the intensity of the object's thermal radiation as a function of its

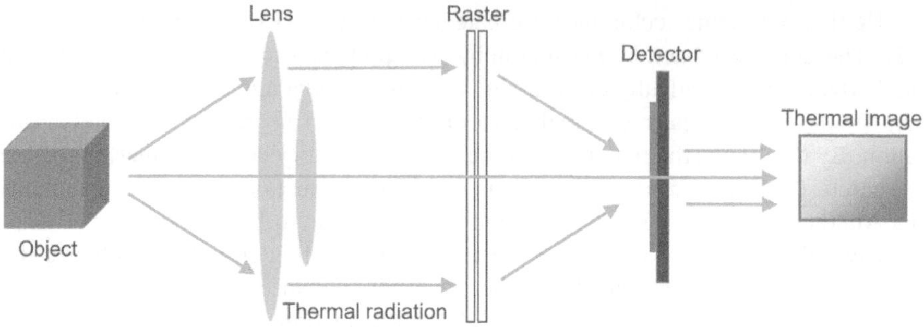

FIGURE 7.1 Infrared thermal imaging camera schematic.

surface temperature. The technology not only monitors the temperature of the object but also obtains a map of its thermal energy distribution, thus providing a more accurate means of measuring and analyzing temperature for scientific research, engineering applications, and other fields.

The use of IR thermal imaging technology, according to different needs, has developed and designed a variety of temperature measurement equipment; the most widely used should be the IR thermal imaging camera, which is based on the relationship between IR thermal radiation and temperature, and uses IR radiation to detect the target object. Through the photoelectric conversion, electrical signal processing, and other means, the temperature distribution of the target object is converted into a high-tech video image, as shown in Figure 7.1. It involves the synthesis and intersection of optics, mechanics, microelectronics, physics, computers, image processing, and other disciplines, and the product mainly consists of an IR optical system, a focal plane detector, follow-up circuits, and embedded image processing software. Due to its advantages of flexible shooting, accurate scanning, rich data, various measurable states, and comprehensive data processing methods [16], it has been widely used in various fields of society and is now showing an extremely broad market prospect.

IRT is used in such a wide range of fields, mainly because of its ability to measure temperature non-contact and image in real time. In the military field, IRT is widely used in night vision equipment and target reconnaissance, which enables the army to effectively identify targets in total darkness or under complex meteorological conditions, thus enhancing the covertness and safety of military operations. In the medical field, IRT is used in the early diagnosis and monitoring of diseases, such as inflammation, tumors, and other abnormal temperature changes in the lesion area due to its non-invasive and non-radiation characteristics. In the industrial field, IRT is used for equipment detection and troubleshooting, especially in electrical and mechanical equipment, to quickly locate potential overheating areas and prevent major equipment failures or fire hazards. In addition, the security field has achieved personnel tracking and intrusion detection in complex environments through thermal imaging, greatly enhancing the reliability of security systems. Finally, the construction field uses IRT for building energy efficiency assessment and quality inspection, which can effectively identify heat loss and water leakage problems in buildings, thus improving the energy efficiency of buildings.

IRT has become an indispensable monitoring tool in various fields by virtue of its multifunctionality and comprehensive interdisciplinary application. With the advancement of technology and the application of new materials and techniques, the accuracy, resolution, and the wide range of application scenarios of IRT equipment will certainly be further enhanced.

7.2.2.2 Thermal Infrared Image Enhancement Algorithm and Image Processing

Computer image processing technology takes digital images as input and processes them through a series of algorithms and processing methods to obtain clearer and more useful information. It originated in the early 20th century and was first applied to the newspaper industry. Subsequently, with the development of computing hardware and the advancement of processing units, digital image processing techniques began to take shape and were applied in several fields [17]. In particular, in 1964, the Jet Propulsion Laboratory (USA) successfully applied digital image processing techniques for the first time to improve the visual quality of digital images transmitted by Ranger [18]. In addition, image processing techniques have made breakthroughs in the field of medical imaging, and in the early 1970s, the development of computerized tomography (CT) technology, which generates two-dimensional images of the interior of objects and three-dimensional volumetric images from X-rays, provided an important tool for medical diagnosis.

Image processing technology is characterized by its high level of accuracy and precision, which enables a detailed analysis of the image content, resulting in a richer and clearer representation of the details in the image reconstruction and enhancement process. This technology applies operations to image data through algorithms that not only improve image quality but also enhance the readability and visualization of images. Image processing techniques also have the advantage of reproducibility, which means that they can provide stable and predictable results, even in different contexts, provided that the same algorithms and parameters are used. Image processing techniques are also very versatile and flexible, as both visible and non-visible images, such as X-rays, thermal IR images, or ultrasound images, can be analyzed using the same digital image processing techniques.

With the continuous development of computer technology, image processing techniques are also advancing. Image processing techniques are not limited to traditional tasks such as image enhancement, alignment, and edge detection, but also cover complex image analyses based on deep learning (DL), such as target detection, recognition, and segmentation. DL algorithms, especially convolutional neural networks (CNNs), have demonstrated excellent performance in areas such as image classification, target detection, segmentation, and medical imaging. For example, in medical imaging, DL techniques are used for tumor detection, pathology image analysis, etc., the core of which lies in the ability of its convolutional layer to automatically learn the features in the image, thus improving the accuracy of classification.

DL has been widely used in image processing, including but not limited to image recognition, image segmentation, and image enhancement. In the field of image recognition, CNNs can be used to identify objects in an image, such as face recognition and license plate recognition. Image segmentation techniques, on the other hand, can be used to divide images into different regions, such as semantic segmentation, instance segmentation, etc. Image enhancement techniques can also be used to achieve super-resolution reconstruction of images, denoising, defogging, and other effects by learning the features of a large number of high-quality images and automatically adjusting the parameters (such as brightness, contrast, sharpness, etc.) of low-quality images. With the continuous progress of DL technology, the performance and application scope of image processing technology will be further expanded and deepened.

With the rapid development of IR technology, IR thermal cameras are widely used in military, medical, industrial, and other fields. IR images are obtained based on the thermal radiation energy of the object itself, which can highlight thermal targets hidden in the background [19]. Compared with visible images, IR images are less affected by nighttime, glare, and bad weather, but they have disadvantages such as a large dynamic range, low contrast, high noise, and blurred details [20], and it is often difficult to obtain satisfactory results by directly using these raw images for analysis. Therefore, thermal IR image enhancement algorithms are widely used to enhance image quality and make IR images more suitable for further analysis and applications. Thermal IR image enhancement algorithms can significantly improve the visualization quality and analysis accuracy of thermal imaging data. By applying these algorithms, the contrast of the image can be enhanced and details can be highlighted, thus improving human visual recognition and machine analysis. Thermal IR image enhancement techniques also play an important role in improving the decision-making capabilities of automated systems. For example, in fault detection and preventive maintenance, enhanced images can identify potential problems more quickly and accurately, thereby improving the safety and reliability of the system. As technology continues to advance, thermal IR image enhancement algorithms have become more sophisticated and efficient, providing strong support for their application in a variety of fields.

The core objective of thermal IR image enhancement techniques is to optimize the quality and visualization of the original image through pre-processing and post-processing methods. The pre-processing steps mainly include noise removal, image smoothing, and denoising, while the subsequent processing focuses on enhancing the contrast, sharpness, and edge features of the image. Various enhancement methods have been developed in this field, including histogram-based methods, filters, transform domain, morphology, saliency extraction, fuzzy logic, machine learning, optimization algorithms, and so on [21].

Lv et al. [20] proposed two innovative algorithms for high gray level IR image enhancement. As shown in Figure 7.2, the MDDC_HDRII algorithm introduces the concept of dehazing to solve the problem of image overexposure and improves the detailed information of the image through the multi-scale detail enhancement technique. Meanwhile, the PAEA_HDRII algorithm takes into account the sensitivity of the human eye's vision and concentrates the image information mainly in the blue and red channels to enhance the image detail performance. The algorithm combines the ideas of debate fusion and exposure fusion, and proposes a large grayscale IR image enhancement method based on multi-scale fusion. Compared with the existing

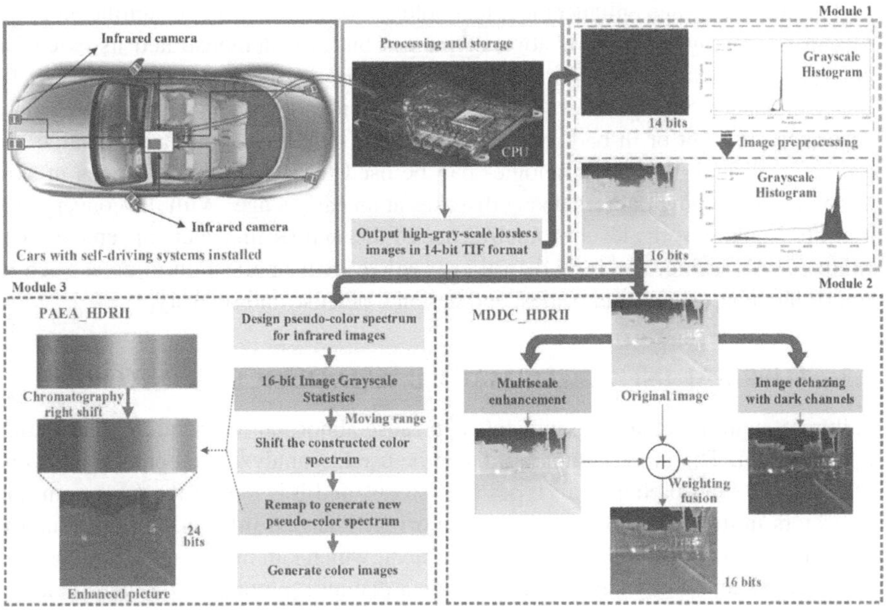

FIGURE 7.2 Schematic diagram of the framework of two high gray level infrared image enhancement algorithms. (Source: Reference [20], Copyright 2023 by Elsevier.)

techniques, these algorithms show significant improvement in several evaluation indexes in the enhancement application of vehicle-mounted IR images and synthetic aperture radar images, effectively enhancing the visual effect of the images.

Gao et al. [22] further proposed an algorithm for contrast enhancement of IR images based on an atmospheric scattering model. The algorithm enhances the overall contrast of the image while highlighting the texture details, providing excellent visual effects. Although existing primitive image enhancement methods perform well in compressing images and improving contrast, they have some limitations. For example, the enhanced image may be overexposed, the noise level may rise as the contrast increases, making it difficult to obtain an image that is both high contrast and low noise, and the prominence of targets in images enhanced using this method is sometimes insufficient.

In order to overcome these limitations, Wang et al. [23] designed an inverse enhancement framework aiming at simultaneously solving the problems of overexposure, noise amplification, and target unremarkability. They improved the traditional image enhancement framework by introducing an inverted framework based on the fundamental prior of IR images and designed an inverted nonlinear grayscale mapping curve. This innovative approach effectively avoids the drawbacks of overexposure and over-amplification of noise, and furthermore, utilizing the laminar decomposition model and the gamma correction technique, it improves the quality and accuracy of the computational results. In addition, extended applications of the framework demonstrate its potential in stimulating the effectiveness of algorithms in related fields.

Thermal IR image enhancement technology has not only made significant progress in improving the visualization of images but also demonstrated its extensive potential in practical applications. For example, in the driverless field, thermal IR image enhancement algorithms can help vehicles identify roads and obstacles more accurately at night or in bad weather conditions, thus improving driving safety. In the medical field, these technologies can be used to detect subtle changes in skin temperature, helping doctors detect diseases at an early stage. With the convergence of AI technologies, the level of intelligence of thermal IR image enhancement algorithms is increasing, which will further promote their application in areas such as smart manufacturing, smart cities, and national defense and security.

7.3 THERMAL INFRARED-IMAGE-BASED DEVICE MONITORING

Failure of equipment and components can cause abnormal temperature distribution in the system. Therefore, temperature can be used to analyze the operational status of equipment and piping in all types of systems, and it is one of the most common indicators in the field of equipment monitoring. Thermal infrared (TIR) camera is a non-contact condition monitoring instrument, which can measure the temperature of the object and its dynamic changes from a distance, obtain the thermal image of the equipment or components, and then monitor and analyze the temperature anomalies to remind the personnel to maintain in time to prevent failures. As shown in Figure 7.3, IR imaging technology is becoming more and more mature and has been widely used in condition monitoring in a variety of fields such as machinery monitoring, electrical equipment monitoring, civil structure monitoring, nuclear industry monitoring, etc. [24–29].

7.3.1 Generalized Mechanical Monitoring Field

Mechanical monitoring is a key component in ensuring stable operation and maintaining the productivity of industrial equipment. Its main purpose is to identify potential equipment failures and performance degradation promptly through real-time or regular inspections, so that preventive maintenance measures can be taken to avoid the high costs associated with unplanned downtime and emergency repairs. However, the field of machinery monitoring currently faces a number of challenges, including ensuring the accuracy and reliability of data, the complexity of processing and analyzing large amounts of data, and achieving real-time monitoring and rapid response. These challenges require monitoring systems that are not only highly accurate and efficient but also adaptable to rapidly changing technical environments and escalating regulatory requirements. With the development of technology, especially the application of AI and machine learning, it is expected to improve data processing capabilities, optimize the monitoring process, reduce costs, and enhance the intelligence of monitoring systems.

The process of monitoring the condition of machinery and processes is called condition monitoring. The purpose of condition monitoring is to prevent unexpected failures, maximize equipment, and reduce associated hazards. It detects problems before major failures occur in a machine or component. Several non-destructive

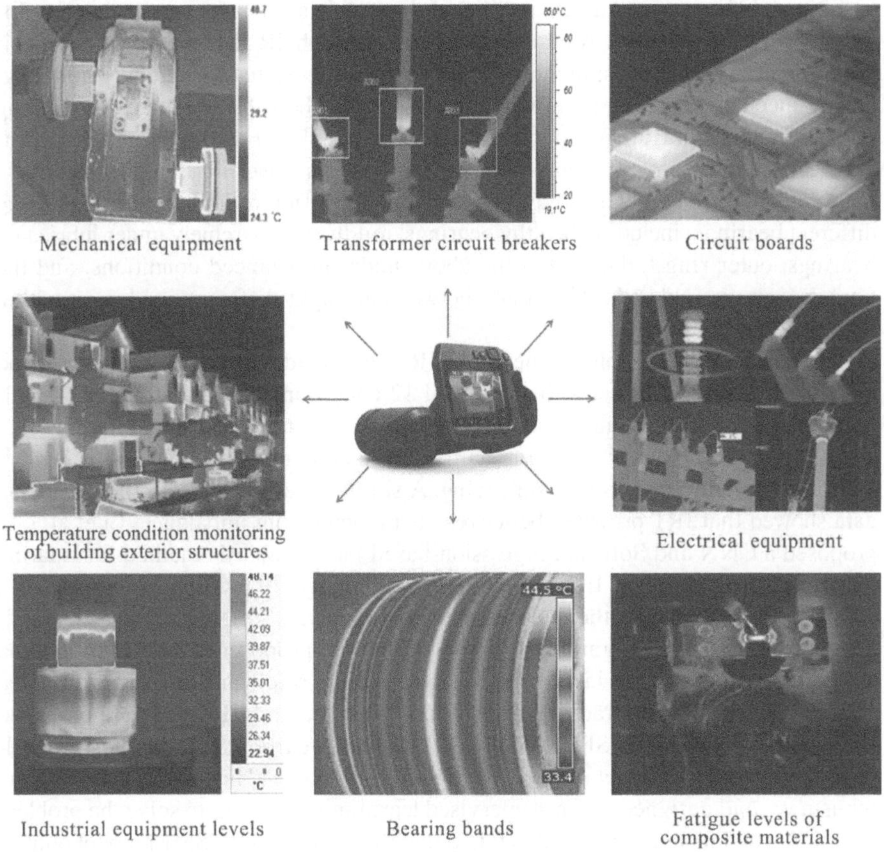

FIGURE 7.3 Multiple condition monitoring applications for TIR.

testing techniques, such as radiography, eddy current testing, ultrasonic testing, acoustic emission, and vibration analysis, are commonly used for condition monitoring [30]. It is well known that temperature is one of the most useful parameters to indicate the structural health of an object [31]. Whereas thermal IR imagers can remotely measure the temperature of an object, fault diagnosis methods based on thermal IR images have been successfully applied in many fields. It also plays an important role in meeting the challenges in the field of machinery monitoring by providing high-quality images of temperature distribution to help improve data quality and enable real-time monitoring and warning. This technology enables rapid identification of equipment anomalies and reduces the costs associated with technology updates, while enhancing the management of risks associated with new processes and materials. The stability of thermal IR imaging allows it to remain effective under variable operating conditions, providing strong support for intelligent operations and maintenance. It also offers advantages in terms of data security and privacy protection, while reducing the risk of manual inspections, providing a safer and more reliable means of monitoring industrial production.

Li et al. [32] proposed a new intelligent diagnostic system for classifying the different operating states of rotating machinery through IR thermal imaging. Their proposed system can assist in monitoring the operating state of machinery and diagnosing mechanical faults by effectively classifying the different states of rotating machinery through IR thermal images. Janssens et al. [33] took the rotating machinery bearings as their research object and proposed a novel IR imaging automatic fault detection system. The study introduced eight different faults/conditions using different bearings, including healthy bearings, mildly and extremely under-lubricated bearings, outer ring failures, and the above under unbalanced conditions, and the final system was able to differentiate between all eight different conditions with a high degree of accuracy (~90%).

Janssens et al. [34] explored the use of IRT combined with image processing and machine learning techniques that detected 12 conditions and faults such as unbalance and hard particle contamination in rotating machinery, and found that the IRT system performed well in detecting hard particle contamination, but had some difficulty in detecting faults in the outer ring. A similar analysis through thermocouple data showed that IRT provided better results in identifying imbalance. Li et al. [32] proposed a CNN and SoftMax regression-based fault diagnosis method for rotating machinery, which utilizes IRT to capture thermal images of rotating machinery, then applies a CNN to extract the fault features, and finally, a SoftMax regression classifier for fault pattern recognition. It is shown that the proposed CNN-based method has better performance in identifying various faults in rotors and bearings compared to other DL models and traditional vibration-based methods. Li et al. [35] proposed a new IR imaging-based SPSMM model for helical gearbox fault diagnosis, developed a probabilistic output strategy to enhance the model's diagnostic accuracy and robustness, and designed a semi-supervised learning framework to solve the problem of insufficient labeled samples. It is found that their proposed SPSMM model outperforms other existing methods in terms of fault diagnosis accuracy.

Nowadays, the application of IRT technology in the field of broad machinery monitoring is rapidly evolving toward intelligence and integration. We can foresee that in the future, IRT technology will be deeply integrated with AI and machine learning algorithms to form highly automated intelligent diagnostic systems, which are capable of realizing rapid identification and classification of mechanical faults, as well as real-time monitoring and predictive maintenance of equipment health status. Through the integration of IoT, IRT monitoring will support remote data collection and analysis, making equipment maintenance no longer limited by geographic location. Meanwhile, combined with augmented reality (AR) and virtual reality (VR) technologies, the visualization of IRT data will be more intuitive, providing on-site technicians with an immersive troubleshooting experience.

In addition, with the application of big data analytics tools, IRT technology will be able to process and analyze vast amounts of equipment monitoring data, revealing potential patterns and trends in the occurrence of faults, thereby optimizing maintenance strategies and reducing downtime. The development of adaptive learning algorithms will further enable the system to automatically optimize diagnostic models based on new data, improving accuracy and efficiency. Customized solutions will be developed to meet the unique needs of specific industries or equipment, and as the cost of the technology decreases, more small and medium-sized enterprises will be able to adopt this highly effective monitoring technology.

7.3.2 Electrical Equipment Field

In recent years, IRT technology has become an indispensable tool, especially for condition monitoring of electrical equipment [36]. The device is capable of revealing a wide range of potential problems within electrical equipment by sensing the IR energy (i.e., temperature) emitted by the target device. As electrical equipment and its constituent parts age, the equipment may experience overheating, which not only increases the risk of equipment failure but can also lead to serious consequences such as unexpected power outages, personal injuries, and even fires. Not only that, the efficiency of the electrical system may drop significantly before a failure occurs, resulting in a significant amount of energy being wasted in the form of heat. Therefore, it is crucial to implement early preventive measures to avoid these undesirable consequences. Thermal anomalies can be quickly and accurately identified and assessed by automated diagnostic systems; however, their accuracy is highly dependent on the ability to correctly identify and select the detection area in the thermographic image. If the region is not selected correctly, it may lead to inaccurate thermal condition analysis [37].

The thermal effect of electrical equipment due to the passage of current under normal operating conditions is a phenomenon that must be closely monitored. Current-induced heat that exceeds the normal range may indicate the existence of faults or defects in the equipment, especially the abnormal hot spot phenomenon, which, if not recognized and dealt with promptly, will likely trigger serious safety hazards or even lead to equipment damage or failure.

Failures of electrical equipment can be broadly categorized into external and internal thermal failures. External thermal faults usually originate from overheating of some localized area of the equipment, which results in a continuous emission of IR rays from that area, which may lead to the failure of exposed joints and connectors. One of the distinguishing features of such failures is the uniformity of the distribution of hot spots in the IR thermogram. Using this feature, IR thermograms are able to visually identify the presence of localized heating faults and pinpoint the exact location of the faults and their degree of damage by means of a temperature distribution map. This provides a reliable basis for taking rapid and effective measures to prevent potential accidents.

Compared to external heat faults, internal heat faults tend to develop more slowly. In the normal operation of electrical equipment, a certain level of heat generation is acceptable. This steady state of heat generation is difficult to detect effectively by IR detection techniques, because the contact between fault points may trigger phenomena such as heat conduction, convection, and radiation, which result in the heat inside the equipment being continuously emitted to the outside, altering the temperature and heat field distribution outside the equipment. Therefore, the diagnosis of internal heat generation faults requires more in-depth analysis and evaluation to ensure that potential failure points can be effectively identified and appropriate preventive measures taken.

Faults in power systems can be categorized into problems such as poor electrical connections, short circuits or circuit breaks, overloads, load imbalances, and improper installation of equipment [38–41], with poor electrical connections being a prevalent problem in power system transmission and distribution lines [42]. According to a survey between 1999 and 2005 [43], thermal problems on conductor connection attachments and bolted connections accounted for almost half of the problems, which are usually caused by loose connections, corrosion, rust, or inadequate use of inhibited grease. These problems can be identified by detecting thermal patterns with the IRT camera, where the highest temperature point indicates the location of the fault, which shows an example of an oxidized connection of a miniature circuit breaker and its corresponding IR image. During the inspection, the oxidized cable connection resulted in a hot spot temperature of over 76.6°C and signs of burning. Ideally, the electrical system should be designed and sized to meet the load demands of the facility. However, over time, additional loads due to building expansions have caused electrical system equipment to exceed its original design capacity, and improperly sized or undersized electrical equipment and components can also lead to overload conditions. Overload conditions typically result in uniform thermal modes throughout the circuit, and the total heat generated by the equipment depends on the load and the ambient operating temperature of the equipment. The temperature of a certain circuit breaker rises above 67°C, while the rest of the circuit breakers operate below 40°C.

In IR thermal images, heating components are usually detected as red or brighter areas. A thermal image of a circuit breaker is shown in Figure 7.4. In Figure 7.4(a), the brighter area (R phase) indicates a high resistance, resulting in overheating of that phase. The probable cause of this high resistance is a poor or dirty connection. Figure 7.4(b) shows the temperature distribution along line L_0, reflecting temperature variations on the surface of the device. Based on the information fed back from the thermal IR image, defects in the component are identified and recorded, and maintenance procedures are planned and executed before equipment failure occurs.

IRT plays an important role in electrical equipment inspection and maintenance with its unique capabilities. This technology captures the invisible IR radiation emitted

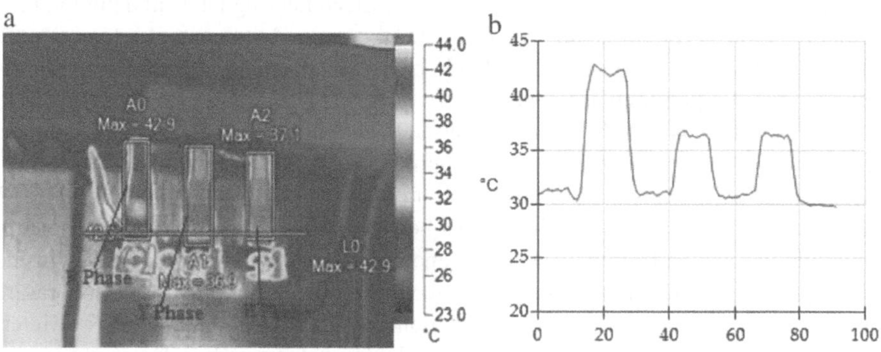

FIGURE 7.4 (a) Thermal diagram of R-phase overheating in a circuit breaker; (b) L_0 line temperature distribution. (Source: Reference [44], Copyright 2013 by Elsevier.)

by the object itself, but it can be visualized as an image through IRT. The color change in the image intuitively reflects the temperature change when the equipment fails, enabling the inspector to quickly identify the area of temperature change and perform a comparative temperature analysis to efficiently locate the problem point [45].

IRT not only reduces inspection and repair time but also significantly improves inspection efficiency and provides strong predictive maintenance capabilities. The technology measures and collects thermal imaging data from equipment during operation, and combines it with advanced data analysis methods to provide an in-depth assessment of key performance indicators. This assessment not only identifies potential equipment failure risks but also allows for advanced planning of maintenance strategies to minimize the impact of failures on production. IR thermal imaging technology can classify equipment failures in detail, and systematically record and share the cause of failure and maintenance information. This information-sharing mechanism provides a valuable reference for subsequent maintenance and reduces the probability of repeated equipment failures, thereby effectively reducing the maintenance costs of electrical equipment and minimizing production stoppages due to failures.

Compared with traditional detection methods, IR thermal imaging technology has obvious advantages. First, its detection speed is faster, and it can complete complex monitoring tasks without the need for a large number of human resources. Secondly, unlike traditional equipment that requires frequent power outages, IRT usually only requires short outages when a problem is detected and repairs are needed. This means that repairs can be more flexibly scheduled during planned outages without additional disruption to the production process. In addition, IRT is particularly good at locating hotspot areas in equipment and can quickly identify potential faults. However, in some complex scenarios, the interpretation of IRT images still relies on experienced and specialized technicians who can take into account a number of influencing factors and accurately assess the condition of the equipment. Therefore, while IRT offers significant advantages in terms of time and cost efficiency, its effectiveness is still partly dependent on the level of expertise of the operator.

7.3.3 Applications of Thermal Infrared Imaging in the HVAC Field

The HVAC system is complex, encompassing multiple components such as the refrigeration system, electrical system, and air system, and integrating a variety of equipment. Failures in the entire system are interrelated and have complex causes, making them difficult to diagnose. Failures caused by damage to induction motors, mechanical failures, and thermal failures are common in HVAC systems [46–48].

Among them, heat exchangers play a pivotal role in the HVAC field, serving as the core components for effective heat transfer. They control the conversion and distribution of heat and cold sources through the heat exchange process, thereby maintaining indoor comfort. Efficiently operating heat exchangers has a decisive impact on enhancing the energy efficiency of the entire HVAC system. Their flexibility and stability enable them to adjust the amount of heat exchange according to the temperature requirements of different situations, meeting diverse heating and cooling needs. In HVAC systems, heat exchangers account for a relatively high proportion of costs (20%–40%), and their performance directly affects the overall performance and economy of the system. The selection and configuration of heat exchangers must be carefully considered in system design and operation to ensure their optimal performance. The heat exchanger industry not only serves the HVAC sector but also involves multiple industries, such as pressure vessels, reclaimed water treatment equipment, chemical and petroleum industries, forming a vast industrial chain. In industrial production, heat exchangers, as general-purpose thermal equipment, are widely used in industrial processes. They are also the main equipment for waste heat recovery and utilization, as well as the development of industrial secondary energy [49]. Therefore, it is crucial to conduct fault diagnosis and performance testing on heat exchangers.

IRT, as a highly specialized non-contact and non-destructive detection method, is widely used for the condition monitoring of industrial equipment [50]. Its core advantage lies in the ability to monitor the surface temperature of equipment in real-time, helping engineers to promptly detect temperature anomalies and thereby predict potential equipment failure risks. Since the detection process does not require direct contact with the equipment, IRT is particularly suitable for hazardous or hard-to-reach environments such as high-temperature and high-pressure conditions, like heat exchangers and other industrial equipment, greatly ensuring the safety of operators. The technology can perform comprehensive inspections without interfering with the normal operation of the equipment, eliminating the need for shutdowns or disassembly, which significantly reduces maintenance time and costs. With its high-resolution imaging capabilities, IRT can also detect minute temperature changes, helping to identify problems early on and prevent minor equipment faults from developing into more serious failures, thereby further enhancing the precision of equipment monitoring and the effectiveness of preventive maintenance.

As heat exchangers operate over the long term, various problems can arise in the fluid flow and heat transfer processes within them, severely affecting their performance and operating efficiency. Fouling is one of the common issues in heat exchangers [51], where impurities and minerals in the fluid gradually accumulate on the surface of the heat exchanger, forming a thermal resistance layer. This significantly reduces heat transfer efficiency and consequently increases energy consumption. Figure 7.5 shows a typical fouling resistance curve. Moreover, the chemical nature of the heat transfer medium or external environmental factors can lead to corrosion of the metallic components of the heat exchanger [52]. Corrosion not only weakens the structural strength but, in severe cases, can also cause leakage of the medium, posing safety hazards. Blockage of the heat exchanger's flow passages is another frequent problem, usually caused by fouling, corrosion by-products, or other foreign matter. This blockage hinders the normal flow of fluid, leading to overheating or insufficient cooling in localized areas.

If the fluid distribution inside the heat exchanger is not properly designed or if the operating conditions are unstable, it will lead to uneven heat exchange, which is manifested as reduced local heat transfer efficiency or abnormal temperatures. Scaling on the inner walls of the heat exchange tubes is particularly significant. This not only reduces the heat exchange efficiency but also increases the fluid flow resistance, thereby affecting the overall performance of the system.

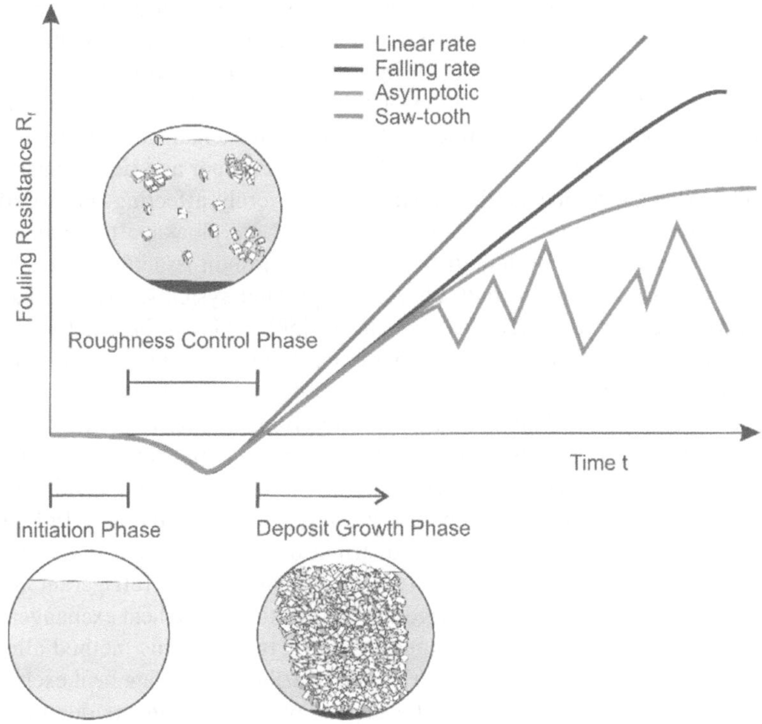

FIGURE 7.5 Typical fouling resistance curve. (Source: Reference [51], CC BY 4.0, https://creativecommons.org/licenses/by/4.0/.)

As a core component in industrial systems, the performance of heat exchangers has a direct impact on overall energy efficiency and safety [53]. Therefore, regularly monitoring the temperature distribution on the surface of heat exchangers is crucial for ensuring their long-term efficient operation. First, abnormal temperature distribution is usually an early sign of equipment failure. IRT can promptly detect issues such as fouling [54], corrosion [55], and blockages [56] inside heat exchangers. This allows for maintenance or cleaning measures to be taken before problems escalate, thereby avoiding equipment downtime or major failures. Second, regular inspection using IRT helps to identify potential problems in the equipment at an early stage, reducing excessive wear on critical components and thus extending the service life of heat exchangers. Meanwhile, by monitoring the temperature distribution, engineers can timely adjust operating parameters to optimize the heat exchange process, reduce energy consumption, and improve the overall system efficiency. The non-contact and real-time monitoring advantages of IRT enable inspection to be carried out while the equipment is in normal operation, thereby reducing the frequency of regular shutdown inspections and further enhancing production efficiency. This series of measures not only ensures the safe operation of the equipment but also improves its economy and reliability.

Plate heat exchangers have a development history of over 100 years and are characterized by their high efficiency and compactness. They account for approximately

28% of the market share and are the second most widely used heat exchange devices after shell-and-tube heat exchangers [57]. The uniformity of fluid distribution in plate heat exchangers for refrigeration is an important factor that affects heat exchange performance and the stable operation of refrigeration systems. When used as an evaporator, the refrigerant at the inlet of the plate heat exchanger is mostly in a two-phase state, with significant velocity slip between the vapor and liquid phases. This can lead to noticeable uneven flow distribution, severely affecting the overall heat transfer performance and potentially causing local ice blockage that can damage the heat exchanger. Uneven fluid distribution is a common issue in heat exchangers and is particularly severe in multi-channel or parallel systems, such as plate heat exchangers and microchannel heat exchangers [58]. IRT can be used to estimate the refrigerant distribution in plate evaporators. Zhang et al. [59] captured thermal images of the side and end surfaces of the evaporator using an IR camera, performed on-site calibration, and then established a simplified evaporator model based on the corrected thermal images. This model represents the division between the two-phase region and the superheated region of the refrigerant, thereby estimating the uneven distribution of the refrigerant between and within the channels. IRT can also quantify the two-phase refrigerant flow distribution in brazed plate heat exchangers (BPHEs). Li et al. [60] estimated the two-phase length of the refrigerant channels from IR images and adjusted the refrigerant distribution in the heat exchanger model to match the two-phase length identified in the IR images. This method allows for the quantification of the two-phase refrigerant distribution in plate heat exchangers.

The accumulation of deposits on the surface of heat exchangers due to impurities in the fluid is a common problem in industry that significantly degrades the performance of heat exchangers. Berce et al. [54] used IRT to observe the thermal images of heat exchangers and non-invasively assess the thermal performance of heat exchangers. This method provides spatial information on flow distribution and the location of severe deposit areas. The experimental results show that after the formation of deposits, the thermal images reveal a clear difference between the clean and deposit-formed areas, which is measurable even under different operating conditions. IRT can serve as a basis for the study of deposits in heat exchangers, offering new approaches for research on dynamic operation, transient response, and computational models.

Bobič et al. [61] proposed a combined theoretical and experimental approach to analyze the dynamic response of plate heat exchangers to inlet temperature disturbances under different fluid flow configurations. By establishing a 1D discrete semi-empirical model based on the finite volume method and validating it experimentally with IRT, the study found that the mass flow rate of the fluid significantly affects the transient temperature response of the heat exchanger. This provides important guidance for the optimal design of plate heat exchangers and the development of control algorithms, emphasizing the importance of understanding dynamic characteristics for the development of efficient district heating and cooling systems.

Dudzik et al. [62] employed IRT to study heat exchangers and proposed a new algorithm for calculating thermal power consumption based on the results obtained. The algorithm is based on an artificial neural network (ANN) and uses measurement data from IRT to train the network. The combination of IRT and ANN can effectively calculate the thermal power consumption of heat exchangers in a steady heat transfer state. By integrating advanced imaging techniques with machine learning algorithms, it is possible to non-invasively monitor and calculate the thermal energy efficiency of heat exchangers, which holds potential application value for improving energy utilization efficiency and reducing costs.

Some researchers have monitored the heat transfer conditions on the surfaces of heat exchangers or condenser fins and tube walls based on IR images [63, 64]. Ge et al. [65] used IR imaging technology to obtain overall and local surface temperature distribution maps of the condenser to monitor the heat transfer performance of air-cooled condensers. They verified that ambient air temperature, natural airflow, and surface defects can affect the performance of the unit. Sarraf et al. [66] investigated the impact of steam superheating on the performance of plate heat exchangers used as condensers based on IR images. Superheated steam exists at the condenser inlet, and under conditions where the wall temperature is lower than the saturation wall temperature, it can enhance local heat transfer. Based on IR images, researchers have also explored the effects of factors such as fin count, shape, height, width, and Reynolds number on heat exchanger performance [67–69]. For air-cooled condensers in power generation units, using shorter continuous wavy fins can improve heat transfer efficiency [70]. At low Reynolds numbers, selecting an appropriate number of fins can enhance the heat transfer performance of steam chamber heat sink fins. At high Reynolds numbers, choosing a larger number of fins can improve heat transfer performance [71]. Equipment condition monitoring based on TIR ensures the normal and healthy operation of equipment, improves system heat transfer efficiency, and reduces maintenance costs.

IRT can promptly detect issues such as uneven heat transfer or fouling accumulation in heat exchangers. Based on this information, engineers can carry out timely cleaning and maintenance, which significantly enhances the thermal efficiency of the equipment, reduces energy consumption, and consequently lowers the operational costs for businesses. It also effectively decreases the emission of greenhouse gases, contributing to the achievement of energy-saving and emission-reduction goals. In addition, IRT can monitor temperature anomalies in heat exchangers in real-time, which is crucial for preventing safety accidents caused by overheating or leakage, thereby ensuring the safety of production processes. More importantly, the precise temperature data generated by IR imaging technology provides comprehensive information on the operational status of the equipment. This helps engineers to develop scientific and rational maintenance strategies, plan equipment maintenance and cleaning schedules more effectively, ensure that the equipment operates in its optimal condition, and improve the accuracy and efficiency of decision-making.

He et al. [72] developed a non-contact method for troubleshooting and detecting faults in refrigeration machine rooms based on inspection robots using video and audio signals. As shown in Figure 7.6, the inspection robot collects video and audio information using an IR camera, a regular camera, and a microphone. First, the regular camera captures images of the equipment in the machine room at preset locations and classifies the images using the AlexNet CNN. If the image is classified as a dial, the dial reading is obtained using image morphology methods to determine whether the pipeline is operating normally. If the image is classified as a pump, the

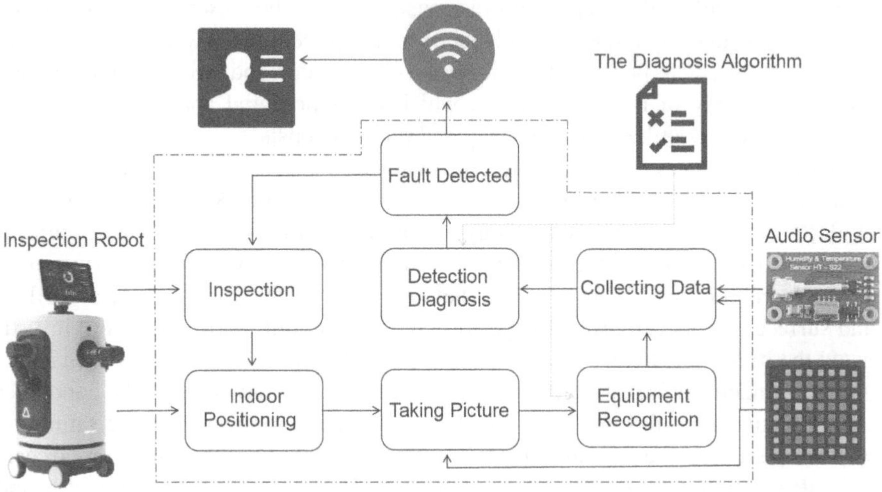

FIGURE 7.6 Detection and diagnosis framework based on robotic inspections.

audio sensor is used to check if the sound is normal, and the IR camera captures the thermal image of the pump. The highest temperature of the pump on the IR image is identified using optical character recognition (OCR) to determine whether the pump is overheated. The study was experimentally validated in an air-conditioning machine room in Shanghai, China. The results show that the proposed method can achieve mechanical fault detection of equipment and thermal fault detection of pipelines. However, the current method is mainly designed and applied for detecting faults in pumps and dials in refrigeration machine rooms. Non-contact fault diagnosis methods applicable to other equipment in refrigeration machine rooms still need further development.

7.4 EQUIPMENT MONITORING BASED ON VISIBLE IMAGES

Compared with thermal IR images, visible images combined with video image processing technology can provide more visual information. Equipment health monitoring based on visible images has been a research direction that has attracted much attention in recent years. There have been new developments, especially in the problems of frosting on heat pump surfaces and condensation on heat exchanger surfaces.

This section will first summarize the research progress in the field of frost and dew meteorological observation based on Computer vision (CV) technology, and analyze the potential directions for the future development of automated frost and dew weather observation. It will then review the application of CV technology and AI algorithms in addressing the issues of frosting and condensation on heat exchanger surfaces. The section will elaborate on the defrosting control strategies for heat exchanger surfaces and the measurement techniques for condensation heat transfer, and provide an outlook and suggestions for future development directions. This will help to broaden the research ideas of CV technology and AI algorithms in the fields of frost and dew meteorological observation, and heat and mass transfer to a certain extent.

7.4.1 Research Background of Frost and Dew Phenomena Observation

7.4.1.1 Frost and Dew Meteorological Observation

Frost and dew weather are closely related to human production and life, as well as the natural environment. Accurate monitoring of frost and dew phenomena has important practical guiding significance for fields such as agricultural production, traffic safety, and ecological protection. In the 1960s, the World Meteorological Organization (WMO) officially included frost and dew phenomena in the category of ground condensation phenomena in meteorology and recommended systematic observations at standard meteorological stations to ensure the scientific nature and consistency of the data.

In the early observations of frost and dew weather, reliance was primarily on manual methods. Observers, drawing on years of experience, used various sensory means such as sight and touch to make qualitative judgments about frost and dew. Although this method could provide relatively accurate results in the short term, its shortcomings were also quite evident. First, manual observation required a significant amount of labor and was inefficient. Second, human subjectivity and fatigue could affect the objectivity of the observation results. Moreover, manual methods made it difficult to continuously track and finely analyze the changes in frost and dew weather over the long term, limiting the precision and reliability of the data.

With the advancement of science and technology, technical equipment such as leaf wetness sensors, water balance method, microwave radiometers, paper lysimeters, plate lysimeters, and container lysimeters has gradually been introduced into the observation of frost and dew phenomena. These instruments are capable of automatically recording the appearance and dissipation processes of frost and dew. However, although these devices can improve the efficiency of observation to some extent, there are often significant differences in the measurement results between different types of instruments, and manual observation is still required for assistance and verification.

To overcome these limitations, automated observation technologies in the field of meteorology have become an important direction for development. By integrating advanced sensing technologies, data processing algorithms, and meteorological models, automated observation systems are able to monitor frost and dew phenomena in real-time and with high precision. This provides a more scientific basis for decision-making in human production and daily life, and offers more continuous and comprehensive monitoring data for the dynamic changes in nature.

7.4.1.2 Frost Formation on Heat Exchanger Surfaces

Frost and dew formation are common condensation phenomena in nature and have significant impacts on the natural ecological balance as well as human production and daily life. As an important natural phenomenon, frost and dew are closely related to fields such as agriculture and construction, and also frequently occur in engineering fields such as astrophysics, cryogenic refrigeration, and aerospace, directly affecting

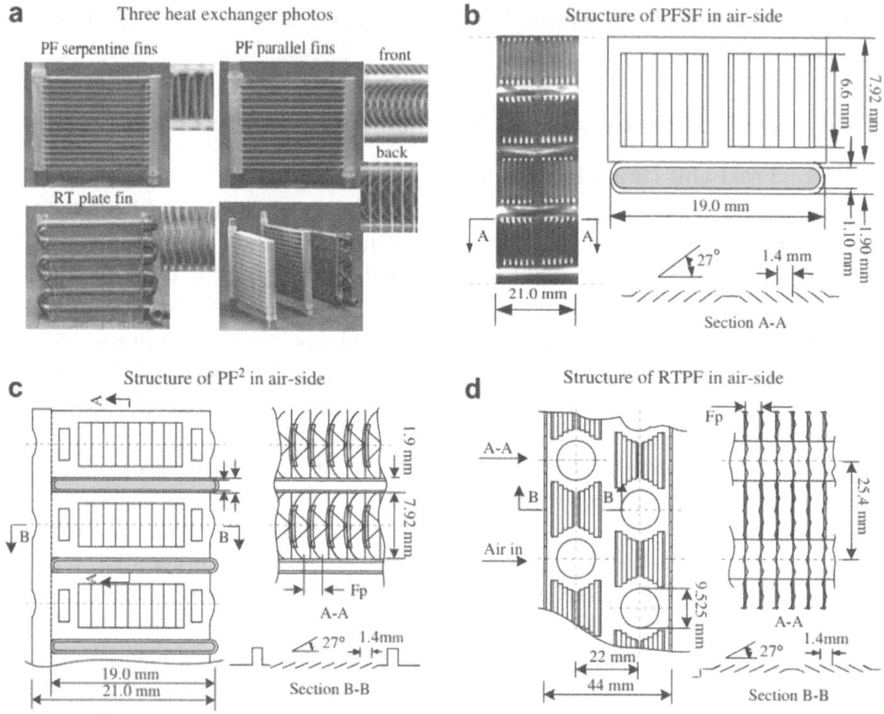

FIGURE 7.7 Three heat exchanger configurations: (a) Three heat exchanger photos, (b) Structure of parallel flow serpentine fin (PFSF) in air-side, (c) Structure of parallel flow parallel fin (PF^2) in air-side, (d) Structure of round tube plate fin (RTPF) in air-side. (Source: Reference [73], Copyright 2009 by Elsevier.)

the heat transfer efficiency and operational stability of systems. Heat exchangers, as key energy conversion devices in these fields, have work efficiencies that are directly related to the energy utilization rate and stability of the entire system.

Heat exchangers are widely used in industrial and civilian systems. As the core components for heat transfer, energy conversion, and temperature control, their design and performance are crucial to system operation. Three types of heat exchanger structures are shown in Figure 7.7. They are key devices for energy conversion and temperature control. Through efficient heat transfer processes, they not only improve the energy utilization efficiency of the system but also ensure the stability of the process and the quality of the products. In refrigeration and air-conditioning systems, heat exchangers regulate the temperature of the medium to provide a comfortable environment for people. Moreover, heat exchangers also play an important role in environmental protection. Through waste heat recovery and reduction of harmful emissions, they promote the green development of industry. With the advancement of technology, the design and materials of heat exchangers are also constantly innovated to meet higher performance requirements and more complex application scenarios. However, frosting on heat exchanger surfaces is particularly prominent in refrigeration and air-conditioning systems. This phenomenon can significantly affect the performance and operating efficiency of the equipment, especially in low-temperature and humid environments, where frost layers can easily form on the surfaces of outdoor heat exchangers. As the frost layer accumulates, the heat transfer resistance increases, and the heat transfer efficiency drops significantly, which may lead to unstable operation of the equipment and even sudden shutdown of the unit due to excessive frost thickness. This issue not only affects system performance but also has a negative impact on the thermal comfort of people indoors.

The widespread use of heat pump systems in heating and air conditioning has made defrosting an important issue. The defrosting process of heat pumps consumes a large amount of energy, which can account for as much as 10%–30% of the total energy consumption, and during defrosting, the heat pump cannot provide heating, which affects temperature control. Precise defrost control, which starts defrosting when necessary, avoids unnecessary defrosting, and stops defrosting promptly after completion, can significantly improve the overall energy performance of heat pumps and extend the service life of components. Therefore, exploring the mechanism of frosting, researching methods to inhibit frosting and defrosting, and further achieving precise defrosting have important practical significance and have attracted widespread attention from researchers.

In the initial stage of frosting, delaying and suppressing frosting on the surface of heat exchangers is commonly approached by modifying the shape and structure of the heat exchangers themselves, or by applying hydrophobic coatings to the surface of the heat exchangers for surface modification to facilitate the drainage of condensate. However, these methods of delaying frosting have limited effectiveness. As the system continues to operate, frosting is ultimately inevitable, and more practical defrosting methods need to be considered.

There are many traditional defrosting methods, but in terms of defrosting principles, they can be summarized into three main categories: thermal defrosting, mechanical defrosting, and hybrid defrosting. Thermal defrosting methods include electric heating defrosting, reverse cycle defrosting, four-way reversing valve defrosting, hot gas bypass defrosting, water spraying defrosting, phase change heat storage defrosting, etc., which melt the frost layer on the surface of the heat exchanger by heating. Mechanical defrosting methods include electric field defrosting, ultrasonic defrosting, high-pressure airflow defrosting, etc., which remove the frost layer by using mechanical external forces to prevent frost accumulation. Hybrid defrosting methods include combinations such as electric heating and bypass hybrid defrosting, electric heating, and four-way reversing valve hybrid, etc., which use two or more defrosting methods together to overcome the drawbacks of a single defrosting method, leveraging their strengths to improve the defrosting effect.

With the continuous advancement and wide application of heat pump technology, finding precise and efficient defrosting control strategies in practical applications has become a hot issue in both academia and industry. Researching the mechanism of frosting and exploring new defrosting technologies, especially improving the precision and efficiency of the defrosting process, remain key issues that need to be urgently addressed. In the future, the optimization of defrosting technology will provide important technical support for improving the overall energy efficiency of heat pumps and air conditioning systems and reducing their operating costs.

7.4.1.3 Dew Formation on Heat Exchanger Surfaces

The phenomenon of dew formation on the surface of heat exchangers is very common in air conditioning and refrigeration systems, usually occurring when the surface temperature of the heat exchanger is below the dew point temperature of the surrounding air. When water vapor in the air condenses into liquid water droplets on the cold surface, condensate is formed. To some extent, this phenomenon is a normal system response, but if left uncontrolled, it can cause a series of problems. For example, the accumulation of condensate can lead to damp walls, mold growth, and even corrosion of equipment components, thereby shortening the service life of the equipment. Therefore, it is crucial to control and manage dew formation when designing and operating air conditioning and refrigeration systems to avoid potential damage and related maintenance costs.

Condensation is an important mode of energy transfer and is widely used in fields such as energy, chemical engineering, aerospace, building heating and cooling, and seawater desalination. Condensation can be divided into two types: dropwise condensation (DWC) and filmwise condensation (FWC). FWC forms a liquid film on the wall, which increases thermal resistance. Under the same condensation conditions, the heat transfer efficiency of DWC is five to seven times that of FWC [74]. Therefore, condensation phenomena, especially DWC, have become a research area of common interest to many scholars over the past century [75–77].

In condensation heat transfer measurement techniques, traditional methods mainly rely on the coolant enthalpy change method. This method uses temperature sensors such as thermocouples to measure the inlet and outlet temperatures of the coolant in the pipe, and then calculates the heat flux density and heat transfer coefficient based on the temperature difference. However, due to the non-uniform temperature distribution of the coolant in the pipe, as well as the limitations of sensor accuracy and installation location, this method has a large measurement error. In addition, different flow conditions and complex pipe geometries further increase the limitations of this measurement method, resulting in uncertainties in heat transfer coefficients and heat flux density that usually exceed 20%.

To address these issues, researchers have proposed a condensation heat transfer measurement technique based on enthalpy change. This technique involves installing a condensate collector at a suitable location inside the condensation chamber to directly measure the mass of the condensate, thereby estimating the condensation heat transfer efficiency. This method no longer relies on temperature sensors, avoiding the temperature error issues in traditional methods. However, accurately collecting condensate from different condensation surfaces and ensuring the accuracy of the measurement results remains a challenge for this technology. Therefore, despite its potential, it has not been widely applied in practice.

In condensation experiments, tests are usually conducted in a vacuum chamber to reduce the adverse effects of non-condensable gases on heat transfer performance [78], so as to more accurately investigate the heat and mass transfer mechanisms of condensation phenomena on heat exchangers. In recent years, with the rapid development of CV technology and AI algorithms, researchers have gradually introduced these advanced technologies into the field of condensation heat transfer measurement. By combining image processing and intelligent algorithms, it is possible to monitor and analyze the condensation process in real time, thereby more accurately enhancing the DWC on the surface of heat exchangers. The application of this technology can significantly improve the overall heat transfer efficiency of heat exchangers and has significant practical significance in the fields of air conditioning, refrigeration, and other industries.

In summary, dew formation and condensation heat transfer on heat exchanger surfaces have significant impacts in various industrial applications. By thoroughly investigating the mechanisms of dew formation and condensation and employing advanced heat transfer measurement techniques, it is possible to effectively enhance system performance and energy efficiency, thereby promoting technological progress and development in related industries.

7.4.2 Frost and Dew Meteorological Observation Based on CV Technology

CV technology plays a crucial role in frost and dew meteorological observation. It provides continuous and objective observational data through automated image analysis, significantly enhancing the accuracy and efficiency of monitoring. CV technology reduces the subjective errors associated with human observation and can identify and record frost and dew phenomena in real time. This is vital for improving the reliability of weather forecasting, guiding agricultural production, ensuring aviation safety, and conducting in-depth research on climate change. Moreover, the application of CV technology reduces the reliance on human labor, increases the frequency of data collection, and maintains the continuity of observation. This provides meteorologists with richer and more precise data, making the understanding and prediction of these complex meteorological events more accurate.

7.4.2.1 Mechanism of Frost and Dew Formation

7.4.2.1.1 Mechanism of Surface Frost Formation

The specific process of frost formation is quite complex and involves an unsteady heat and mass transfer process. Each stage of the frost growth process has distinct characteristics that significantly influence subsequent growth. Hayashi et al. [79] divided the frost growth process into three stages based on its characteristics: the initial growth period of frost crystals, the growth period of the frost layer, and the full growth period of the frost layer. The formation process of the frost layer is shown in Figure 7.8. In the initial growth stage of frost crystals, water vapor in the air condenses into small droplets when it comes into contact with a cold surface. These small droplets gradually aggregate and grow. When the temperature of the cold surface is lower than the dew point temperature of the air, the droplets freeze into frost nuclei. Frost crystals form and continue to grow in different directions with the frost nuclei as the center, and this stage is very short. As the frost layer enters the growth period, the frost crystals grow in a dendritic manner and continuously expand to

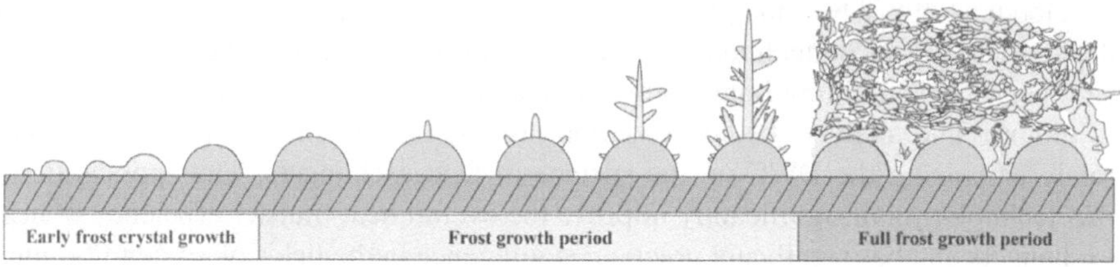

FIGURE 7.8 The process of frost formation.

form a network, gradually evolving into a frost layer. The density and thickness of the frost layer gradually increase. As the frost layer grows, its thickness increases at a gradually slower rate. In the full growth period of the frost layer, condensation causes the surface temperature of the frost layer to rise continuously. Some frost crystals are melted into water, which infiltrates into the interior of the frost layer and freezes again, increasing the density and thickness of the frost layer.

7.4.2.1.2 Mechanism of Surface Dew Formation

DWC was first discovered by Schmidt et al. in the 1930s. Researchers in the field of heat transfer, such as Tanasawa, Rose, Westwater, and Umer, have conducted a series of theoretical and experimental studies on the mechanism of DWC. The initial droplet formation mechanism is the basis of the heat transfer mechanism of DWC, and there are currently two main theories: the liquid film rupture hypothesis and the fixed nucleation center hypothesis. In 1936, Jakob [80] proposed the liquid film rupture hypothesis, which suggests that vapor first forms an initial condensation liquid film on a supercooled surface. The liquid film continues to thicken until it reaches a critical thickness, at which point it ruptures. Under the action of surface tension, the fragments of the condensation liquid film contract into small droplets. Tammann et al. [81], on the other hand, believed that there are randomly distributed condensation nuclei on the supercooled surface. These nuclei are covered by small droplets, which grow continuously as the condensation process continues and begin to merge with each other. During the merging process, the position of the droplets shifts. When the droplet size reaches the sliding size, the droplets begin to slide off the condensation surface. In the dry trace left after the droplet slides off, new small droplets continue to nucleate. This theory, which considers DWC as a nucleation phenomenon, is known as the fixed nucleation center hypothesis.

The fixed nucleation center hypothesis has been accepted by most scholars in explaining the mechanism of droplet formation. This hypothesis suggests that droplets nucleate at randomly distributed positions on a solid surface, typically at surface defects such as pits, grooves, and scratches. The size of these defects, which can form liquid nuclei, depends on the degree of supercooling. Although this theory has been widely accepted, there has long been a debate in the scientific community about whether there is an extremely thin liquid film connecting droplets.

In 1991, Song et al. [82] proposed the liquid film coexistence hypothesis. This theory suggests that the formation of droplets during condensation must meet two conditions: first, the surface energy of the cooling wall must be low, and second, there must be nucleation centers on the wall. They pointed out that there are no macroscopic-scale "clean" surfaces on the condensation surface. The liquid film existing between droplets will have an important impact on condensation heat transfer. There are a large number of droplets of different sizes on the condensation surface, which are distributed according to certain rules. There is a continuous thin liquid film of a certain thickness between droplets. The liquid generated by the condensation of steam on the surface of the liquid film continuously flows along the film surface to the surrounding droplets.

The liquid film coexistence hypothesis further explains the detailed process of heat transfer between condensation droplets and the condensation wall. It suggests that the distribution of droplets on the condensation wall spans several orders of magnitude, from initial droplets to the largest droplets that are about to detach. Surfaces covered by smaller droplets generate higher local heat flux densities, while surfaces covered by larger droplets have lower local heat flux densities due to the higher thermal resistance of the larger droplets. This non-uniformity of surface heat flux limits the heat flow near the condensation surface and increases the heat transfer resistance. The additional thermal resistance caused by the non-uniformity of heat flux is called the limiting thermal resistance.

As observation and research continue to advance, the liquid film coexistence hypothesis has gradually been accepted, providing a new perspective for understanding and predicting complex phenomena during the condensation process. The development of this theory not only enriches our understanding of the condensation process but also offers an important theoretical foundation for the experimental design and industrial applications of condensation heat transfer.

7.4.2.2 Research Progress in Frost and Dew Meteorological Observation Based on CV Technology

7.4.2.2.1 Instrument-Based Automated Observation

At present, the equipment and methods for automated observation of frost and dew phenomena are mainly divided into two categories. The first category is indirect observation methods. These methods establish mathematical models to link meteorological elements such as temperature and humidity with frost and dew phenomena, thereby inferring the formation of frost and dew. The advantage of this method is that it can utilize existing meteorological station data. However, the downside is that it may produce significant errors in a single observation, as it relies on the accuracy of the model and the precision of the input data. The second category is direct observation, which continuously identifies frost and dew phenomena using instruments and equipment such as sensors, meteorographs, satellites, and automated weather observation systems. The observation process of this method is more efficient and resource-saving, and the observation results are more objective and accurate. Direct observation methods can provide continuous and real-time data streams, which are crucial for understanding the formation and development of frost and dew phenomena. For example, automated image recognition technology can monitor and record the formation process of frost and dew in real time, while sensors can provide precise temperature and humidity data. These data are very helpful for verifying and improving mathematical models. The following is a brief discussion of direct observation methods.

Capacitive sensors map the changes in capacitance during the condensation process of frost and dew to changes in frequency values, thereby determining whether frost and dew phenomena have occurred and the extent of their growth [83]. Weremczuk et al. [84] found that compared to the classic impedance-based frost and dew detector, the main advantage of the impedance matrix-based detector is that it can identify the distribution of water and ice and display the morphology of the condensation layer by the difference in dielectric constant values. Fiber optic measurement technology can simultaneously measure air pressure, humidity, and temperature, and minimize electromagnetic interference. Photoelectric detection technology combines optical and electronic techniques. Based on the photoelectric effect, photoelectric sensors convert optical signals into electrical signals, achieving high-precision and high-speed measurements. When connected to a computer, it facilitates information control and storage, advancing the automation of frost and dew observation. Nie et al. [85] proposed a dew point temperature measurement method based on tunable diode laser absorption spectroscopy (TDLAS). This method can use a single near-IR laser source to simultaneously measure the temperature, pressure, and concentration of water vapor, with low cost and high accuracy. Since the propagation of ultrasound in liquids can cause effects such as acoustic cavitation, acoustic streaming, heating, and atomization, Kozak et al. [86] developed a patch antenna sensor operating at a frequency of 2.378 GHz. By monitoring the resonant amplitude and frequency of the transmission coefficient between the antenna sensor and a broadband receiver, it can detect frost and ice. This sensor can distinguish between ice, frost, and water, and also measure the thickness and surface coverage of ice. It is low-cost, real-time, and high-precision, and does not require additional equipment such as long transmission lines, simplifying the frost and ice detection device. The TSVS series visibility sensors are a new generation of meteorological visibility detection equipment. Using a downward scattering angle, the light emitter sends out pulsed light, while the light receiver detects the intensity of the forward-scattered pulsed light from aerosol particles in the atmosphere. All measurement information is collected by a microprocessor controller and converted into meteorological optical range through a mathematical model algorithm. The instrument sends meteorological visibility values and status information to the monitoring center via a communication interface. It offers higher measurement accuracy and stronger anti-interference capability, providing strong data support for meteorological observation and analysis.

7.4.2.2.2 Automated Observation Based on CV Technology and Artificial Intelligence Algorithms

With the continuous innovation of new technologies such as big data, AI, 5G, and digital twins, automated observation technologies based on CV technology and AI algorithms have been further proposed. CV technology combines image acquisition methods and image processing techniques to simulate the human visual system, as shown in Figures 7.9 and 7.10. Image acquisition devices such as cameras capture images and use computers to recognize image features, analyze and process image signals, and implement corresponding control strategies. This process can be

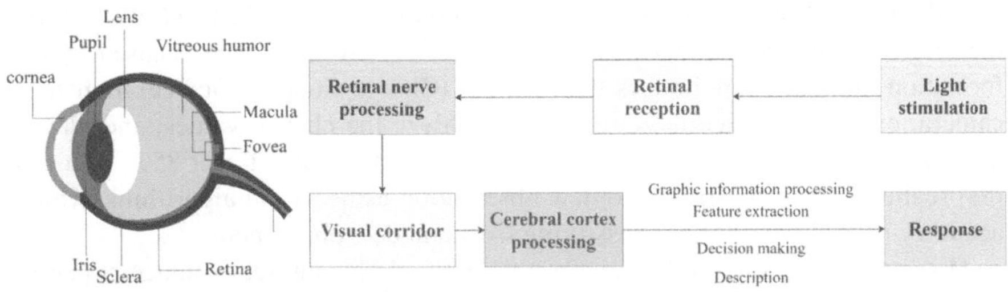

FIGURE 7.9 Human visual system.

analogized to the human eye seeing an image and then analyzing the image information through the brain to give a response. The key to the entire process lies in image digitization and image processing. Image digitization converts optical signals, such as pixel distribution, brightness, and grayscale, into digital signals. The image processing system then performs calculations on the digital signals, combining them with AI algorithms such as machine learning and DL to extract target features and output results based on set thresholds. This enables the identification of the occurrence time, duration, and condensation amount of frost and dew phenomena.

Frost and dew, as meteorological condensation phenomena close to the ground, require specific carriers for observation. However, due to the unevenness of natural vegetation surfaces and their susceptibility to environmental factors such as wind speed, automated frost and dew observation is difficult to implement on these surfaces, making them unsuitable as basic carriers for image detection. To improve the accuracy and reliability of observation, CV technology often uses smooth or frosted glass slides as contact surfaces, creating a flat observation carrier. Such carriers provide a stable and predictable environment, which helps to collect more accurate data. Moreover, by analyzing changes in image features using machine learning algorithms, frost and dew phenomena can be effectively identified. This method not only improves the precision of observation but also enhances the credibility of the data,

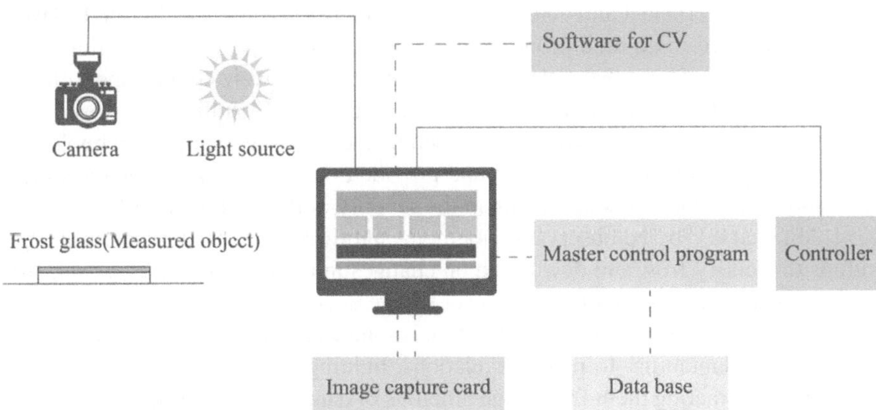

FIGURE 7.10 Observation systems based on CV techniques.

providing strong technical support for meteorological research. Li et al. [87] used the brightness difference between smooth and frosted glass slides caused by dew formation to obtain dew images with a CCD (Charge Coupled Device) high-speed camera, extract the brightness difference, analyze the change values, and identify dew formation, achieving an automatic dew recognition rate of over 85%. Zhu et al. [88] realized automatic frost and dew observation using visual algorithms, with an accuracy rate of over 90% in classifying frost and dew phenomena.

However, this method relies on training with a large dataset of images to ensure the accuracy of the classifier. For ground-level frosting, Zhu et al. [89] designed a carrier suitable for image detection and proposed a frosting detection algorithm based on change detection, change modeling, and saliency detection. The algorithm identifies potential frosting images based on texture features, filters out similar optical property changes caused by accidental interference, uses manifold learning to perform saliency detection on each superpixel to obtain possible frosting areas, and trains a Support vector machine (SVM) classifier online with frosting and non-frosting image samples to make a secondary judgment on candidate frosting areas. The results showed that this method achieved an accuracy of 87% in observations at WMO standardized meteorological observation stations. Focusing on the characteristics of frost and dew images, Zhu et al. [90] extracted three explicit visual feature information: texture, color, and structure, and proposed a method that uses an SVM classifier to classify and detect images with only a small number of samples. Compared with image classification based on neural networks, this method achieved higher accuracy and required less training time. Wang et al. [91] performed image preprocessing using CLANE, bilateral filtering, and morphological processing based on the optical differences and shape features between dew areas and the surrounding environment. They then used edge detection, circle detection, and area measurement to identify the shape of dew as a recognition feature to determine whether dew has formed. The ratio of the pixel area of the observed dew area to the carrier glass slide was defined as the dew coverage rate, which reflects the amount of dew condensation. Compared with the manual method as the standard reference, this method achieved a dew detection consistency of 95.9% and a dew coverage rate judgment error of 10.8%, enabling efficient and objective detection of ground dew. The Manitoba Hydro system is located in an area prone to ice disasters. Since the 1980s, to mitigate ice disasters, the province's hydroelectric company has selectively created short circuits on overhead transmission lines to melt the ice on them. In addition, a warning ice detection system has been further developed to use visual recognition technology to automatically obtain ice information and provide timely warnings, allowing staff to take appropriate measures according to the severity of the ice disaster [92].

Traditional CV-based observation methods mainly rely on machine learning algorithms to identify frost and dew based on changes in image features. However, frost and dew features are easily affected by objective environmental factors such as temperature and lighting, and feature extraction is somewhat challenging. DL algorithms, due to their adaptability to pixel correlations, lighting, and deformations, can learn features by extracting them from large amounts of data and have stronger generalization capabilities. Therefore, combining DL algorithms with frost and dew observation to further improve recognition accuracy has become a new research direction.

Zhou et al. [93] proposed a classification method based on multi-path feature fusion and DL. On the basis of the ResNet-50 model, they added multiple paths from shallow to deep network layers, combining shallow and deep features to enhance the frost and dew feature information in subsequent convolutional operations. Test results showed that the classification accuracy of this method reached 99.2%.

In summary, there is still room for improvement in the automated observation of frost and dew weather. On the one hand, the research on the mechanism of frost and dew formation is not yet complete. Through machine learning algorithms and DL frameworks, it may be possible to extract new features by learning from a large number of images, reflecting the textural features and spatial structural changes of the frosting and dewing process, which will facilitate pattern recognition and in-depth analysis of the formation mechanism of frost and dew. On the other hand, further exploration is needed for the automated quantitative observation of frost and dew amounts. At present, the condensation amount is still indirectly reflected through the dew coverage rate.

7.4.3 Monitoring of Frost Formation on Heat Pump Surfaces

Frosting significantly affects the heat transfer performance of heat exchangers. Research on frosting both domestically and internationally mainly focuses on the formation mechanism and characteristic analysis of frost, simulation of heat exchanger frosting characteristics, and defrosting control. Air source heat pump systems are highly efficient, energy-saving, and environmentally friendly, and are widely used in residential and commercial buildings around the world, becoming a typical alternative to traditional coal-based space heating technologies. To maintain the safe operating performance of air source heat pumps, regular defrosting is necessary. Over the past few decades, various methods for suppressing frost, delaying frosting, and defrosting have been continuously developed, but defrosting operations that reduce energy efficiency still cannot be avoided. To improve system energy efficiency and reduce energy consumption during the defrosting process, it is necessary to monitor more intuitive frosting factors, determine more precise defrosting start and stop control points, build more reasonable defrosting logic, and implement more intelligent defrosting control strategies.

Based on the type of monitored parameters, defrosting control strategies can be divided into indirect defrosting and direct defrosting. Intelligent defrosting control strategies introduce image processing technology and new AI technologies, such as machine learning and DL algorithms, into the defrosting control logic. This is a new type of defrosting control strategy developed on the basis of direct defrosting.

7.4.3.1 Indirect Defrosting Control Strategies

Indirect defrost control strategies involve measuring relevant parameters of the air-source heat pump system to assess the frost growth condition, setting defrost logic, and thus carrying out defrost control. The monitored parameters mainly consist of two categories. The first category reflects the conditions for frost formation. Common control strategies include time-based defrost control, which starts defrosting according to a preset schedule; temperature difference control, which determines whether to initiate defrosting based on the temperature difference between the air and the evaporator surface; Time–Temperature (T–T) control, which triggers defrosting when the heating operation time, the surface temperature of the outdoor heat exchanger, and the ambient temperature all reach their set values; and Time–Humidity–Temperature (T–H–T) control, which activates defrosting when the heating operation time, the surface temperature of the outdoor heat exchanger, and the air humidity all reach their set values. The second category of parameters reflects the impact caused by frosting. Common control strategies include air pressure difference control, which is based on the principle that frosting increases the air resistance of the outdoor heat exchanger. By establishing a mathematical relationship between the thickness of the frost layer on the heat exchanger surface and the air pressure difference across the heat exchanger, the air pressure difference is continuously monitored. When it reaches a set value, the control system starts defrosting, and vice versa. There is also the maximum average heat supply control, which calculates the actual average heat supply of the system based on the obtained data. The optimal defrost start and end points are determined according to the ratio between the actual measured value and the theoretical value. In addition, there are refrigerant superheat control and fan input power control methods. For the evaporator of a multi-room refrigeration system, Maldonado et al. [94] proposed the thermal variation easiness (TVE) value as an indirect parameter to indicate its frosting degree.

Indirect defrost control methods are simple to operate and cost-effective, which makes them widely used in air-source heat pump units. Although indirect parameters can reflect the frost growth stage to some extent, the complexity of frosting as a non-steady-state heat and mass transfer process often leads to errors in the assessment of frosting conditions. This affects the defrost control logic and can easily result in incorrect defrost operations. That is, defrosting is initiated when it is not needed, and it is not started in time when it is required. This not only wastes energy but also affects the thermal comfort of the occupants indoors.

7.4.3.2 Direct Defrost Control Strategy

Direct defrost control strategies utilize various new types of instruments and equipment to directly monitor the thickness of frost layers, overcoming the drawbacks of indirect defrost control strategies. They can more intuitively determine the state of frost layers and carry out relatively accurate defrost control. Contact measurement devices include digital micrometers, vernier calipers, and scanning probe microscopes, while non-contact measurement methods include laser displacement meters, photoelectric conversion technology, and capacitive detection technology.

7.4.3.2.1 Contact Direct Measurement

Lee et al. [95] used a digital micrometer to measure the frost layer on a horizontal cylinder in cross-flow. Due to the non-uniform distribution of frost layer surface temperature and ambient air humidity, the measurement of the frost layer was uneven. It could only measure one point of the frost layer and was not able to measure the entire frost layer. Kandula et al. [96] proposed using a vernier caliper to measure the thickness of the frost layer on an aluminum plate. The average value of five measurement points was used as the frost layer thickness. Although the accuracy was improved, it was still not sufficient to characterize the overall frost layer thickness. Matsumoto et al. [97] used a scanning probe microscope to measure the frost layer thickness at the nanoscale. This method can collect multi-line data of frost layer thickness, making the measured thickness value closer to the average thickness of the entire frost layer. However, the initial cost of the probe is high, the measurement time is longer, and it is difficult to avoid the possible damage to the frost layer caused by the scanning probe, which may affect the accuracy of the measurement.

7.4.3.2.2 Non-Contact Direct Measurement

Qu et al. [98] measured the local changes in frost thickness and morphology using a laser displacement meter, with a thickness measurement uncertainty of 0.01 mm. The laser displacement meter is highly real-time and easy to use. However, the heat carried by the laser itself may disrupt the temperature field of the cold surface, causing measurement errors. Moreover, its high initial cost is not conducive to its widespread application. Photoelectric measurement technology directly detects frost thickness through the conversion of photoelectric signals. Changes in frost thickness directly affect the light source of the photoelectric sensor, which in turn causes changes in the output voltage and current of the conversion circuit. The fluctuation of the electrical signal is used to determine whether to initiate defrosting. Capacitive detection technology works on a similar principle, monitoring frost thickness by detecting changes in capacitance caused by the frost layer. Xiao et al. [99] conducted 21 sets of experiments and verified the relationship between the output voltage of the photoelectric sensor and the frost height with 600 sets of data, and proposed using this relationship for defrost control. Wang et al. [100] developed a tube encircled photoelectric sensor (TEPS) and conducted experimental studies under standard frosting conditions, finding that the sensor can effectively extend the defrost interval, reduce the number of defrost cycles, and improve the heat exchange efficiency and COP of the unit. Ge et al. [101] conducted on-site tests in Beijing for two heating seasons, proving the feasibility and correctness of defrost control based on TEPS. Bai et al. [102] carried out on-site tests and found that in mild and humid areas, compared with the T–T control strategy, the defrost control strategy based on TEPS is more conducive to accurate defrosting, and the operating performance of the air-source heat pump unit is better, with the COP and heat exchange capacity increased by 11.3% and 12.1%, respectively.

Direct defrost control strategies, on the one hand, have a high dependence on various instruments and equipment, are easily restricted by the location and number of measurement points, have low accuracy, and the related technologies are prone to affecting the target frost layer, causing inaccurate monitoring results. On the other hand, the system is more complex, with higher initial and maintenance costs. Therefore, it has not been widely put into production and use at present.

7.4.3.3 Intelligent Defrost Control Strategy

Given the many shortcomings of existing indirect and direct defrost control methods, it is an inevitable trend to develop a convenient and low-cost high-precision frosting detection technology. This will promote the rational defrosting of air-source heat pump units, improve the energy efficiency of heat pump systems, and boost energy conservation and emission reduction. Thus, intelligent defrost control strategies based on digital image processing technology and AI algorithms have come into the public view. They can obtain the growth status of frost layers in real time through videos or images and develop prediction models for frost growth, which has become an effective means of exploring the precise timing for defrosting.

In the early visual studies, the characteristic parameters of frost layers included frost thickness, frost coverage, and fractal dimension, which quantified the degree of frosting from three dimensions: thickness, area, and density.

7.4.3.3.1 Introduction of the Frost Characteristic Parameter – Fractal Dimension

During the formation of frost, the shape and structure of frost crystals change continuously over time. Although experimental studies on the initial growth stage have described the state of ice crystals under different environmental conditions, these descriptions lack specific indicators to quantitatively distinguish different morphologies. The modeling of frost layer structure is overly simplified and does not match the actual frost layer. The fractal dimension is a structural parameter that can describe the complexity of frost crystals and their ability to fill space. Introducing the fractal dimension allows for the quantification of the degree of frosting from the perspective of frost crystal structure, which has significant guiding significance for more precise defrosting control [103].

Researchers first conducted a series of studies to determine whether frost crystals have fractal characteristics. Hao et al. [104] were the first to point out that at the critical state of frosting, the distribution of frost crystals on a cold surface has fractal characteristics. Hou et al. [105] used the box-counting dimension method to conduct a fractal analysis of the morphology of frost crystals in the early stage of frost layer growth and calculated the fractal dimension, finding that the fractal dimensions of frost crystals under different conditions vary significantly. Ice crystals that have not fully grown do not have fractal characteristics, while fully grown dendritic frost crystal structures become more complex, occupy more space, and therefore have a larger fractal dimension. Liu et al. [106] established a two-dimensional model for frost crystal growth based on the DLA model of fractal theory and conducted computer simulations. By changing the number of iterations in the program and the probability of generating random particles, they simulated the growth time and density changes during the growth process of frost crystals. They integrated incomplete boundary grids in the image processing process and optimized the box-counting dimension calculation method for binary images, improving the accuracy and stability of fractal dimension calculations. They found that the fractal dimension and volume fraction have different meanings in characterizing frost layer properties. The fractal dimension mainly reflects the shape and structural characteristics of frost crystals, while the volume fraction represents the density of frost crystals [107].

The introduction of fractal dimension parameters into the heat conduction model has promoted the mechanistic study of the heat transfer characteristics of frost layers and the heat exchange performance of heat exchangers, further proving the rationality of using fractal dimension as a characteristic parameter of frost layers. Early researchers established a structural model for frost growth based on fractal percolation structures and calculated the thermal conductivity of this structural model using the corresponding heat conduction model. They found that in the early stages of frosting, the thermal conductivity of the frost layer increased slowly. When the density of the frost layer reached a certain value, that is, when the frost crystals formed a connected state, the thermal conductivity suddenly increased. Compared with other models, the thermal conductivity calculated based on the fractal theory-established frost layer structural model is more accurate. Chen et al. [108] established a DLA model based on fractal theory and numerically simulated the frosting process on the surface of a finned-tube vaporizer. They also experimentally observed the morphology of frost growth and obtained images of frost growth at different times. The comparison between the simulated images and the experimental images showed good consistency, verifying the rationality of the numerical simulation. On this basis, a fractal model for frost layer heat conduction was established, and the thermal conductivity expression of the frost layer was given using the thermal resistance method. The thermal conductivity of the frost layer determined by this method is consistent with the range of measured effective thermal conductivity values of the frost layer. Yao et al. [109], based on the application of fractal theory to obtain the effective thermal conductivity of actual frost layers, further calculated the heat flux density through the cryogenic surface during frosting. They found that the heat flux density first increased and then gradually decreased in the early stages of frosting. The fluctuation of the heat flux density made the finned-tube vaporizer unstable in cryogenic conditions. Negrelli et al. [110] established a frost morphology model for heterogeneous porous media based on fractal dimension using the diffusion-limited aggregation method, which can predict the thermal conductivity of frost layers formed under different working conditions. Tahavvor et al. [111] further combined fractal theory with ANNs. The fractal geometric model based on the Von-Koch curve can accurately simulate the thickness and density of frost and reproduce the frost growth process. The backpropagation (BP) algorithm was used to predict frost growth characteristics. This model has strong generalization ability and can be applied to the prediction of frost growth on plates under various conditions.

Using the fractal dimension to describe the growth state of frost layers can make the established frost layer structural model more consistent with the actual growth state of frost layers. Intelligent defrost control strategies can use frost thickness, frost coverage, and fractal dimension as direct control information to characterize the degree of frosting from different dimensions. The defrost system can be activated or deactivated based on whether the frost characteristic parameters reach the set threshold, achieving more precise control of the defrost system's start and stop.

7.4.3.3.2 Defrost Control Strategy Based on Image Processing Technology

Obtaining videos or images through high-speed cameras and analyzing the growth state of frost layers using image processing technology can explore more precise start and stop control points for defrosting systems, effectively reducing defrosting energy consumption and achieving intelligent defrosting.

Digital image processing technology involves analyzing, processing, and manipulating images using computers. It mainly includes methods such as noise removal, enhancement, restoration, segmentation, and feature extraction. Over the past few decades, scholars both domestically and internationally have used CCD high-speed

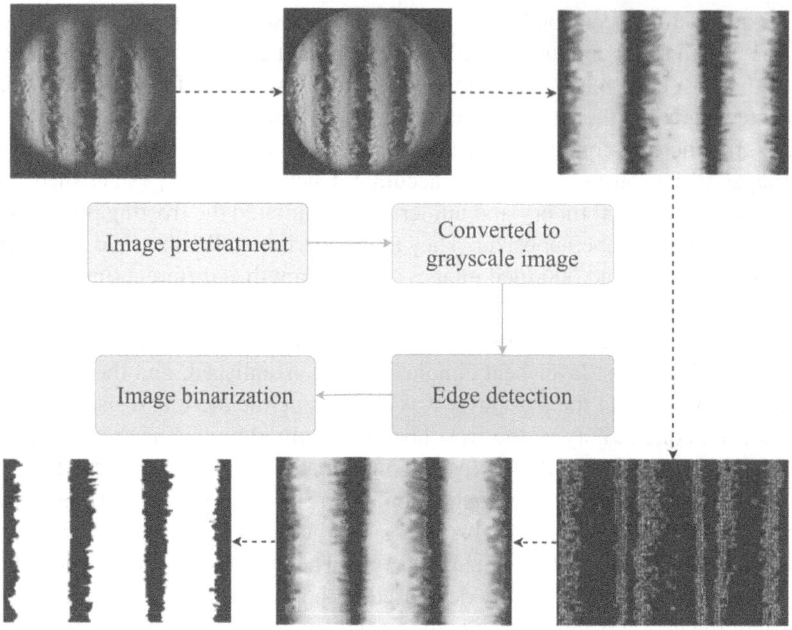

FIGURE 7.11 Image processing flow of frost.

cameras to capture images and quantitatively measure the growth state of frost layers through digital image processing technology, conducting a series of studies on the specific frosting process. As shown in Figure 7.11, image processing technology typically includes the following processes:

1. *Image Preprocessing*: Segment the region of interest (ROI) from the original frost image.
2. *Grayscale Transformation*: Convert the three-channel color image into a grayscale image with only the luminance channel.
3. *Edge Detection*: Identify the edges of the frost layer based on the differences in grayscale values among different pixels.
4. *Image Binarization*: Distinguish between the frost foreground and the background of the plate in the image. Select a reasonable grayscale threshold for threshold segmentation.

The growth characteristics of frost layers on cold surfaces are influenced by the physical properties of air and the nature of the cold surface, including air temperature, humidity, velocity, as well as surface temperature, geometry, and surface wettability. Based on binary images and according to the corresponding reference system, frost thickness can be calculated, and fractal dimension and frost coverage, among other frost characteristic parameters, can be calculated based on the grayscale values of grayscale images. This allows for the investigation of the shape, structure, and growth characteristics of frost crystals under different operating conditions, quantification of the degree of frosting, and analysis of the frosting process. Cheng et al. [112] introduced a CCD high-speed camera and a microscopic imaging system to observe the changes in frost thickness and growth on a circular tube and a flat plate under different environmental variables. Based on image processing technology, the frost thickness was calculated, and the results showed that the error between the calculated and measured values was only 7%. Wu et al. [113] observed the frosting process of a parallel-flow evaporator under three different operating conditions using a CCD camera and studied the relationship between crystal distribution in the frost layer and frost thickness. It was found that during the frost growth and full growth periods, the closer to the cold surface of the fin, the higher the proportion of frost crystals. Malik et al. [114] proposed a hybrid system with both monitoring and defrosting functions, which monitored the frost thickness on the evaporator in real time and found that defrosting when the frost thickness reached the optimal threshold of 6 mm could save 10% of the energy consumption of a household refrigerator. In addition, some researchers have started from the perspective of surface wettability and compared the effects of surfaces with different wettability on frost thickness, density, mass, and other growth characteristics through frosting images [115, 116]. In terms of calculating frost thickness, the accuracy of side-view digital image processing measurement still needs to be improved in cases of uneven frost thickness.

Based on image processing technology, scholars have further explored defrost control methods for air-source heat pumps. Yoo et al. [117] used measured data from an air source heat pump system and, based on image processing technology, estimated the amount of frosting per time step and further estimated the total amount of frosting to reasonably determine the optimal defrost start time when system performance declines. Zheng et al. [118] proposed a novel defrost control method based on image processing technology (Temperature–Humidity–Image (T–H–I) method). Using image preprocessing, grayscale transformation, and multi-threshold segmentation techniques, the visual features of frosting were quantified, and non-frosting, moderate frosting, and severe frosting areas were distinguished. A frosting coefficient was introduced to evaluate the degree of frosting, and the optimal criteria for defrost start (P1=0.3) and end (P2=0.05) were determined and verified. The T–H–I method can directly detect the formation of frost layers, thus making more rational defrost decisions and avoiding incorrect defrosting. Miao et al. [119] improved the T–H–I method by using fractal theory and image processing technology to realize frost thickness and fractal dimension as direct control information to characterize the degree of frosting from two aspects: thickness and structure. Through experimental research on frosting characteristics, the T–H–I defrost control strategy defrosts when the frost thickness is 0.726 mm, and the fractal dimension is 2.839, and stops when the fractal dimension drops to 2.324. The optimized T–H–I defrost control strategy has greater improvements in accuracy and energy efficiency. Li et al. [120] proposed a frosting detection method based on image recognition and a characteristic parameter F that can be evaluated through image recognition to quantify the degree of frosting on the outdoor heat exchanger surface. A series of experiments was conducted from the aspects of shooting angle, imaging pixels, illumination, and outdoor heat exchanger surface temperature to verify the applicability of the method. It was found in practical applications that only ambient illumination affects the detection accuracy of the method. Considering the impact of illumination changes on image recognition accuracy, Wang et al. [121] proposed a method of surface source compensation illumination to improve the novel frosting detection method. They selected a reference illumination surface source and used a frosting threshold correction factor to compensate for the influence of the illumination environment, developing an air-source heat pump image recognition frosting detection technology based on light adaptation. This overcame the impact of outdoor illumination environment changes on the accuracy of image recognition frosting detection and ensured the accuracy of image recognition frosting detection.

7.4.3.3.3 Defrost Control Strategies Based on Artificial Intelligence Algorithms

Frost formation on heat exchanger surfaces is a complex heat and mass transfer process, influenced by various factors such as environmental conditions, refrigerant types, and the geometric structure of heat exchangers. Current experimental data and models have significant limitations. In recent years, with the development of the Internet, big data, and AI technologies, the field of defrost control has further explored data-driven models such as machine learning and DL. AI algorithms are highly flexible and can be used to address many complex nonlinear problems. Different algorithms and models can be utilized to predict frost growth under various operating conditions, and defrost control strategies can be adjusted based on the prediction results.

Initially, some researchers attempted to use parameters such as the roughness and temperature of cold surfaces, air temperature, and relative humidity as input parameters. They trained neural network models based on DL and other AI algorithms to predict frost growth on simple structures like pipes and plates. Defrosting is initiated when the frost thickness reaches a set threshold. Tahavvor et al. [122] used air temperature, surface temperature, relative humidity, cylinder diameter, and time as inputs for an ANN to predict frost thickness on a horizontal cylinder under natural convection conditions. Experimental results showed that this method was highly effective. Subsequently, Tahavvor utilized an ANN once again to predict frost thickness on a plate by inputting parameters such as ambient temperature, cold surface temperature, air humidity, Reynolds number, and frosting time [111]. Zendehboudi et al. [123] applied AI algorithms in the field of cryogenic engineering, using the temperature of the cold surface, air temperature, relative humidity, and time as input parameters for the model. They trained and tested four intelligent models: multiple linear regression (MLR), ANN, least squares support vector machine (LSSVM), and adaptive network-based fuzzy inference system (ANFIS) to predict frost thickness on vertical cryogenic surfaces. The performance of the four models was evaluated and compared using statistical error analysis. The results indicated that the ANFIS model was superior in both speed and accuracy for predicting frost thickness, promoting the development of research on frost growth on cryogenic surfaces. Additionally, surface roughness has a significant impact on the initial performance of frosting on cold surfaces, and it may be possible to include roughness as an input parameter in the future to improve the ANFIS model. These research findings have demonstrated the feasibility of using data-driven models to predict frost growth conditions. However, since these models are applied to simple structures, they are not suitable for predicting the entire system.

Scholars have further established more complex frost growth models for heat pump systems based on AI algorithms. In addition to traditional physical parameters, they have also taken into account various factors that may affect frosting, such as the geometric structure of the heat exchanger and the properties of the refrigerant, to optimize defrost control strategies. Datta et al. [124] creatively proposed a defrost control strategy for retail display cases based on a shallow neural network (SNN). The SNN model was trained using wind speed. When the predicted wind speed reached a set value, defrosting was initiated. Zhao et al. [125] combined ridge wave neural networks with the stochastic finite element method, introducing fuzzy stochastic parameters to describe the uncertainty of the temperature field, and conducted a more accurate heat transfer analysis of finned evaporators. They established a prediction model for the frosting weight and cooling capacity of finned evaporators. Comparisons with simulation results showed that the model could correctly predict the frosting condition of the evaporator, providing a basis for taking reasonable defrosting measures. Tang et al. [126] comprehensively considered factors affecting the frosting process, such as environmental conditions, refrigerant properties, and heat exchanger geometric parameters, and established a dimensionless artificial neural network (DANN) frosting model. They also proposed a new adaptive control strategy, which effectively enhanced the model's generalization ability and could predict frosting amounts under different operating conditions. The relative error between the predicted results and experimental data was approximately 7.55%. This frost prediction model is highly beneficial for analyzing the frosting process, delaying frosting in air source heat pumps during winter, and optimizing the system design of heat pumps. Eom et al. [127] trained a multi-output regression model through DL using data from sensors and manufacturers to quantitatively predict the performance degradation of air source heat pumps caused by frosting, such as heating capacity, power consumption, and COP. Defrosting is initiated when the performance of air source heat pumps drops to a set threshold, realizing a new defrost start control method. By comparing the prediction accuracy of three DL frameworks (fully connected deep neural network (FCDNN), CNN, and long short-term memory (LSTM)), it was found that the FCDNN model performed the best. Rabbi et al. [128] experimentally investigated the effects of fin spacing, surface temperature, and air relative humidity on fin frosting rate under different surface wettability. They constructed a dataset based on experimental data and trained an ANN to more accurately predict the fin frosting rate. Wang et al. [129] proposed a defrost start strategy based on the internal operating parameters of the air source heat pump system. They trained a CNN defrost model using internal operating parameters, learning the defrost logic from the existing time-based control strategy. The maximum prediction error was 12%, and the minimum prediction error was 2%. The results showed that the CNN could provide accurate frosting predictions and make correct defrost start decisions, avoiding incorrect defrosting. Iranagh et al. [130] successfully predicted frost thickness on square fins by training an ANN model using the average temperature of different refrigerants, ambient temperature, and air humidity as input parameters, which also fully demonstrated the significant impact of these parameters on frost thickness. To test the frosting characteristics of microchannel heat exchangers in electric vehicle heat pump systems, Chen et al. [131] proposed a correlation considering the influence of environmental and structural parameters. They predicted the frost thickness on fins through dimensionless analysis, with a relative error analysis result of 8.44%, proving the feasibility of the method. They also proposed the effective blockage ratio (EBR) parameter and found that when the EBR reached 30%, it was the optimal defrost start point, avoiding unnecessary defrost cycles and achieving efficient operation of the electric vehicle heat pump system.

Defrost control strategies based on AI algorithms have also been widely applied to scenarios prone to frosting, such as cold storage facilities. Based on DL and image recognition technologies, actual cold storage images are collected as a dataset, and frost-free and frost-covered images are manually labeled. Neural networks are then trained on the dataset to achieve image recognition, thereby controlling the defrost system and effectively reducing the number of defrost cycles. The results showed that this method could avoid the impact of personnel entry and exit on the temperature of the cold storage facility and reduce the energy consumption losses caused by timed defrosting.

7.4.4 Monitoring of Surface Condensation on Heat Exchangers

7.4.4.1 Image Processing-Based Measurement of Condensation Heat Transfer

7.4.4.1.1 Visual Investigation of Condensation on Cold Surfaces

The dynamic behaviors between droplets and between droplets and surfaces significantly influence local and global heat and mass transfer processes. However, acquiring micron- or nanoscale condensation image data and extracting quantifiable feature parameters from these images is not straightforward. To further simplify and quantify heat transfer performance, some researchers have begun introducing CV technologies and AI techniques, such as machine learning and DL, into the field of heat and mass transfer. These approaches enable visual studies of droplet dynamics in DWC, combining droplet statistics with classical phase-change theory to quantify condensation performance.

Current research methods for studying droplet growth and distribution primarily fall into two categories: numerical simulations and experimental observations. Numerical simulations involve constructing mathematical models to replicate droplet nucleation and growth processes. Experimental observations, on the other hand, rely on instruments such as environmental scanning electron microscopy (ESEM) or high-speed cameras to directly capture and record condensation processes. While ESEM provides exceptionally high imaging quality, its prohibitive cost and inability to measure heat flux density or heat transfer coefficients limit its practicality. Consequently, using high-speed cameras to collect image datasets at appropriate frame rates, combined with image processing and CV techniques for analysis, has become a more widely adopted visual research method.

Researchers worldwide have shown growing interest in image-based visual studies of DWC. As research progresses, it has become evident that droplet morphology characteristics, such as size distribution, maximum detachment diameter, detachment frequency, and condensation coverage area, are closely linked to heat and mass transfer performance, as illustrated in Figure 7.12.

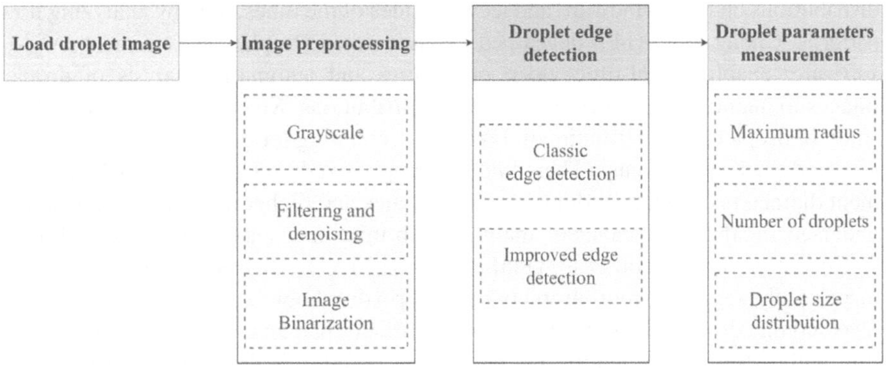

FIGURE 7.12 Framework of condensate droplet detection system.

International research in this field dates back to the 1970s, when Tower and Westwater observed that the detachment radius of condensing droplets varied across different surfaces, establishing detachment radius as a critical parameter in droplet size distribution. During the 1980s, researchers explored the maximum detachment radius in DWC. For instance, Rose [132] derived an empirical expression for the maximum droplet radius during pure steam condensation using dimensional analysis and experimental fitting. Extrand et al. [133] estimated the maximum detachment radius in centrifugal force fields. By the 2010s, advanced tools like ESEM enabled deeper insights. Dietz et al. [134] visualized droplet detachment frequency during DWC on superhydrophobic surfaces using ESEM, while Miljkovic [135] employed ESEM to study droplet morphology on superhydrophobic nanostructured surfaces and its impact on growth kinetics.

Recent decades have seen the integration of computational and experimental methods. Kulju [136] validated multiphase flow simulations using high-speed videography and CV techniques, demonstrating CV's precision in characterizing droplet dynamics. Mudugamuwa et al. [137] developed an active droplet generation system comprising three modules: droplet generation, image acquisition, and a control algorithm. By applying image processing, they quantified droplet morphology parameters (e.g., coverage area, perimeter, and average diameter in pixels) and tracked generation frequency. Their experiments revealed that droplet diameter and generation frequency gradually decreased over time due to factors such as the chemical properties of internal flow channels and the geometric design of micro-control devices. Domestic research in this field began with the work of Song et al. [82], who discovered a correlation between droplet detachment radius and surface properties: surfaces with larger contact angles exhibited smaller droplet detachment radii. This marked the inception of DWC visualization studies in China.

Subsequent advancements integrated computational tools. Hu et al. [138] utilized AutoCAD software to process droplet images and perform statistical analyses. Guan et al. [139] employed CCD cameras to observe convective condensation heat transfer of wet flue gas on horizontal finned tubes. They recorded droplet growth, coalescence, and detachment processes, applying image processing to measure droplet diameter distributions on both windward and leeward sides of the tubes, thereby analyzing flow patterns. Chong et al. [140] developed an Automated Droplet Measurement (ADM) software, enabling real-time video processing and automatic analysis of droplet images to quantify droplet counts and size distributions. Xu et al. [141] expanded the study of droplet size distribution in DWC by accounting for overlapping droplets and repeated nucleation events. Their findings emphasized that smaller droplet detachment diameters correlate with faster renewal rates and higher heat transfer efficiency, underscoring the significance of analyzing sub-micron droplet distributions [142].

Recent work by Zhou et al. [143] leveraged MATLAB-based image processing to precisely segment microdroplets on superhydrophobic surfaces. Their analysis revealed that droplets on superhydrophobic surfaces had an average diameter of 1–10 μm, with ~50% falling within this range. Compared to hydrophobic or hydrophilic surfaces, superhydrophobic surfaces minimize solid–liquid contact area, reduce droplet adhesion, and enhance coalescence-driven self-removal frequency, ultimately achieving superior condensation heat transfer efficiency.

7.4.4.1.2 Visual Investigation of Indirect Evaporative Cooling Systems

As shown in Figure 7.13, evaporative cooling technology utilizes the heat absorption from water evaporation to cool fresh air, characterized by its simple structure, cleanliness, and high efficiency. Although the indirect evaporative cooler (IEC) is also a type of heat exchanger, it differs from traditional vapor-compression systems by relying solely on water pumps and fans as power-consuming components. With

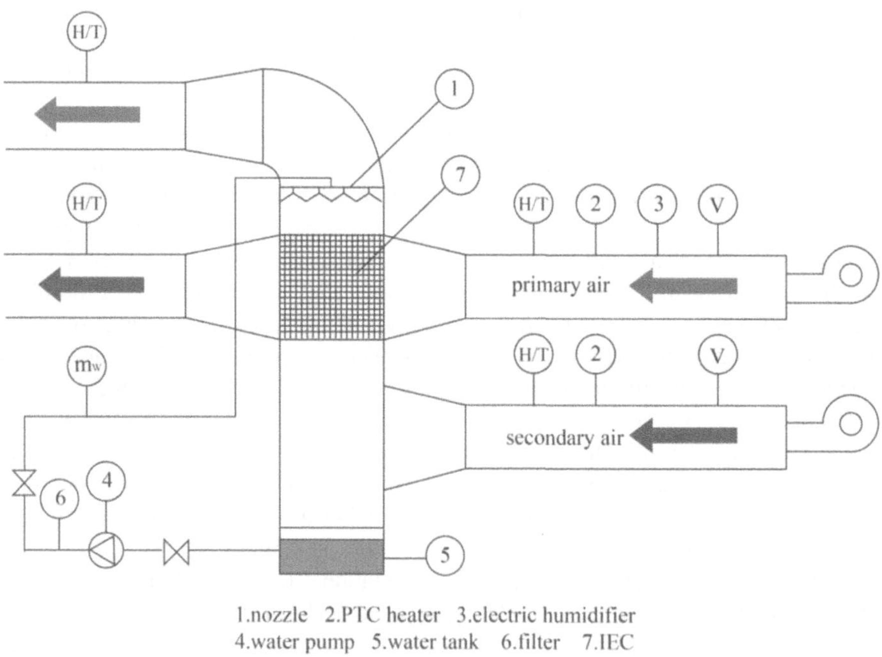

FIGURE 7.13 The schematic diagram of the indirect evaporative cooling system.

advantages such as low energy consumption, high comfort, pollution-free operation, ease of maintenance, and significant energy-saving potential, IEC systems have undergone rapid development over the past few decades [144–147].

Researchers worldwide have conducted theoretical and experimental studies on the heat and mass transfer mechanisms of IEC systems. These studies analyze factors such as internal structures, geometric shapes and dimensions, physical properties of inlet air, and airflow organization. Models including one-dimensional parallel flow, counterflow, and two-dimensional crossflow have been proposed. Subsequent refinements to these models incorporated various influencing factors in heat and mass transfer processes, such as longitudinal conductive thermal resistance of heat exchange walls, wall wettability area, spray water temperature, and thermal resistance of spray water film conduction. Continued optimizations in IEC system design (spanning structures, materials, spray system configurations, and control strategies) aim to enhance heat and mass transfer efficiency and maximize engineering practicality [148–151]. As shown in Figure 7.14, the spray water film directly influences the heat and mass transfer processes in IEC systems. Early research primarily focused on improving the surface wettability of secondary channels to enhance the accuracy of heat transfer models. Experimental studies on IEC surface modifications also emphasized optimizing secondary channel wettability for higher evaporation efficiency, with limited attention to heat and mass transfer enhancement in primary channels under condensation conditions.

However, in recent years, the potential of IEC applications in humid and hot regions has expanded. In such environments, hot and humid fresh air in primary channels can be cooled below the dew point, triggering condensation. This transforms the IEC into a heat recovery device, enabling simultaneous cooling, dehumidification, and energy savings. With the advancement of IEC-based air conditioning energy recovery technologies, the heat and mass transfer mechanisms of condensate films in primary channels have begun attracting scholarly interest.

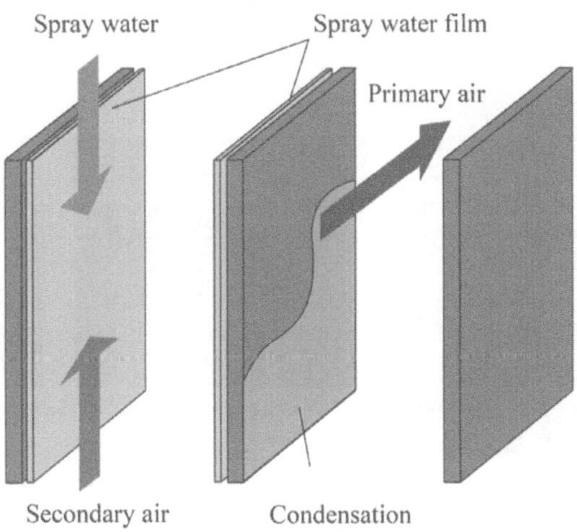

FIGURE 7.14 Heat and mass transfer mechanism diagram of indirect evaporative cooling system.

In theoretical research, Chen et al. [152] explored the analytical model of IEC, proposing criteria to classify three condensation states in the primary channel: non-condensing, partial condensation, and full condensation. Chen et al. [153] further investigated and developed a novel numerical model, which verified that although condensation reduces wet-bulb temperature, it enhances heat transfer and dehumidification in IECs. However, this model has limitations as it neglects variations in spray water film thickness within the secondary channel under condensation conditions, which do not fully align with practical scenarios. Researchers constructed a visualization experimental setup [154] to capture condensation images on the primary channel plate surface using high-speed cameras. These images were processed and analyzed, while experimental data from sensors were used to calculate IEC performance metrics, establishing a connection between condensation and IEC heat and mass transfer performance.

In experimental research, most prior studies focused on investigating the effects of primary air inlet temperature, humidity, airflow velocity, and IEC structural parameters on heat exchange performance using sensor-based data. Few explored the relationship between condensation phenomena in IEC primary channels and overall heat transfer performance through direct visualization. In recent years, researchers have begun constructing visualization setups to directly capture condensation images in primary channels via high-speed cameras. These images are processed and analyzed alongside sensor-derived experimental data to calculate IEC performance metrics, aiming to link condensation with heat and mass transfer efficiency. Meng et al. [155] established a visualization system to observe condensation in crossflow IEC primary air channels, identifying transition points from non-condensation to partial condensation and from partial to full condensation. Their experimental study on IEC performance under varying inlet air temperatures and humidity levels demonstrated that condensation increases outlet primary air temperature and water consumption, reduces wet-bulb efficiency, and enhances total heat exchange by releasing latent heat. As shown in Figure 7.15, Min et al. [154] further investigated the microscopic

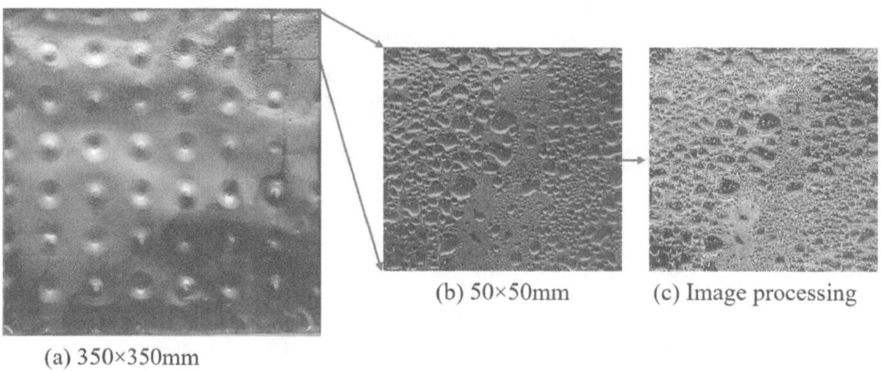

FIGURE 7.15 Condensed image processing: (a) The visualized part of the aluminum heat exchanger plate, (b) A single grid image of the visualized plate surface, (c) The visualization effect of droplet contours. (Source: Reference [154], Copyright 2020 by Elsevier.)

condensation mechanisms in IEC primary air channels through visualization and empirical data, developing analytical models and experimental frameworks for IEC condensation heat transfer under different dehumidification conditions. Using image processing, they quantified the area ratio of DWC to FWC on flat surfaces. By observing DWC and FWC coverage areas, they applied an analytical model incorporating heat flux densities from both condensation modes to theoretically analyze overall heat transfer performance. Their study quantitatively examined the effects of inlet primary air temperature, relative humidity, and velocity on IEC condensation heat transfer. Key findings include: the DWC/FWC area ratio remained stable at approximately 0.4/0.6 with increasing inlet air temperature, while total heat flux density slightly increased; primary air relative humidity significantly influenced total surface heat flux density, with higher humidity increasing FWC dominance and reducing the growth rate of heat transfer coefficients; higher airflow velocities improved condensation heat transfer by increasing the DWC area ratio.

Coating hydrophilic layers on the secondary channels of IEC systems can significantly enhance surface wettability and improve evaporation efficiency [156]. Inspired by research on secondary channel surface modifications and droplet dynamics theory, scholars have explored strengthening DWC in primary channels to boost IEC heat transfer performance. Min et al. [157] applied silicon-based nano-hydrophobic coatings to primary air channel surfaces, investigating their impact on IEC heat transfer in hot and humid climates. Through visual experiments and image processing, they compared condensation mechanisms and performance factors between hydrophobic-coated IEC and uncoated IEC. Results showed that large, irregular droplets forming water films were observed on bare aluminum plates, while hydrophobic surfaces promoted smaller droplet condensation. Minimized droplet detachment size and frequent droplet removal enhanced convective heat transfer of treated air along coated surfaces. Hydrophobic coatings on primary channel surfaces improved IEC energy efficiency by 8.5%–17.2%, demonstrating potential for dehumidification-focused air conditioning applications.

7.4.4.2 Condensation Heat Transfer Measurement Based on CV Technology and AI Algorithms

Although CV technology provides new insights for studying condensation droplet dynamics and heat transfer measurement, it still has limitations. Traditional CV techniques, while capable of rapid multi-target analysis, require strict environmental control and manual intervention for accurate feature extraction, making them unsuitable for long-term condensation experiments. As a result, many researchers have utilized measurement instruments and sensor devices to obtain experimental data, build datasets, and train frameworks or models based on AI algorithms, such as machine learning and DL, to predict heat exchanger performance. These efforts have further demonstrated the effectiveness of AI-based predictive models, advancing intelligent condensation heat transfer measurement.

Pacheco-Vega et al. [158] used datasets of wet airflow velocity, dry-bulb and wet-bulb inlet temperatures, fin spacing, and heat transfer rates to train an ANN to predict the heat transfer rate of a condensing heat exchanger. Experimental results provided upper bounds for estimation errors, confirming the feasibility of this method. Hosoz et al. [159] developed an ANN based on the BP algorithm, using dry-bulb temperature, relative humidity of air, and air/water flow rates as input parameters to predict performance metrics of a direct evaporative air cooler (DEC), including outlet air dry-bulb temperature, relative humidity, mass flow rate of water evaporated into the airflow, and heat transfer rate. Their results demonstrated the accuracy of ANN for DEC modeling. Kiran et al. [160] modeled IECs using three approaches (fuzzy inference system (FIS), ANN, and ANFIS) to predict outlet air temperature in primary channels and system energy efficiency. They found that the ANN model trained with the Levenberg–Marquardt (LM) algorithm performed best.

Zhu et al. [161] implemented a data-driven model based on an ANN to predict the performance of regenerative heat and mass exchangers (RHMX) in indirect evaporative cooling. Using this model, they identified an optimal air extraction ratio range of 0.3–0.36 under varying inlet air conditions, which is critical for optimizing RHMX design. Asfahan et al. [162] applied DL algorithms to evaluate the heat transfer performance of evaporative-assisted air conditioning (E-AC) systems. By extracting the most influential input features through correlation analysis, their model achieved fast and accurate predictions for direct, indirect, and M-cycle evaporative cooling systems, demonstrating strong generalization capabilities.

While ANNs have been widely used for performance prediction in evaporative coolers, their application in dew-point indirect evaporative coolers (DPIECs) remains limited. Longo et al. [163] developed an ANN model incorporating plate geometry, operating conditions, and refrigerant properties to predict condensation heat transfer performance in herringbone-type BPHEs, outperforming existing analytical models. Zhu et al. [164] proposed a machine learning-based ANN model to predict flow boiling/condensation heat transfer in serrated-fin microchannels. Sun et al. [165] established a BP ANN model using MATLAB 2018 to predict the performance of DPIECs. Experimental results showed acceptable errors in predicted evaporative efficiency. Further refinement using particle swarm optimization (PSO) led to a PSO-BP ANN model with improved accuracy.

Recent advances in multiphase thermal flow research have successfully combined visual data with machine learning and DL for two-phase heat transfer prediction. Inspired by this, researchers have integrated CV and AI algorithms to measure external surface condensation heat transfer using visual data. By tracking and analyzing dynamic droplet features, such as nucleation density, droplet growth rate, and average droplet diameter, they built datasets and trained AI models to link visual characteristics with heat transfer performance.

Damoulakis et al. [166] constructed a glass steam condensation chamber, using high-speed cameras to record condensation phenomena on copper surfaces. Image processing techniques extracted visual features such as droplet count, average diameter, coverage, and maximum droplet count under varying conditions. These data were used to train machine learning models, coupling droplet features with heat transfer coefficients measured by physical instruments to predict correlations between droplet characteristics and condensation heat transfer. However, this experiment had limitations: short recording times, fixed frontal camera angles, and analysis limited to 16% of the total area, making it difficult to represent overall condensation behavior. Additionally, the models focused on static features like droplet size and count while neglecting dynamic features such as growth rate and detachment/coalescence frequency, leading to suboptimal model performance. Suh et al. [167] developed a CV-based DL framework, capturing high-resolution condensation images with high-speed cameras. The framework included modules for object detection, tracking, and data processing to extract droplet distribution, growth rate, and heat flux density. At micro-scale resolutions (200 ms temporal, 300 nm spatial), the study revealed that overall heat transfer performance is collectively determined by heat transfer rates of individual droplets and droplet number density (weighted by droplet size). Khodakarami et al. [168] designed a CV and DL-based framework to measure external tube condensation heat transfer. A deep CNN extracted features from condensation images to track droplet detachment count and diameter, estimating detachment frequency and heat flux density. Results showed 97% droplet detection accuracy and less than 10% uncertainty in heat flux density. This image-only approach eliminated measurement errors from tube dimensions, coolant flow rates, and temperature sensors, enabling faster and simpler heat transfer measurement. The method also proved robust for localized pipe heat transfer analysis.

Combining CV and AI algorithms for condensation heat transfer measurement, particularly for DWC, offers a reliable and cost-effective solution. By reducing reliance on sensors and systematic errors, these techniques enable rapid performance estimation through droplet image analysis. However, research on external surface condensation remains limited, and further studies are needed to refine dynamic droplet feature extraction for broader applications.

7.4.5 Monitoring of Other Equipment

7.4.5.1 CV-Based Monitoring of Buildings and Infrastructure

The monitoring of buildings and infrastructure is critical for maintenance costs, improving operational efficiency, adapting to environmental changes, meeting regulatory requirements, enhancing residents' quality of life, and effectively responding to disasters. By providing real-time data support, this monitoring approach modernizes the management of buildings and infrastructure, ensures public safety, and promotes sustainable societal development. CV technology began its development in the 1960s, initially for extracting object shape information. As the technology advanced, CV started addressing more complex perceptual tasks, such as OCR and facial recognition. In recent years, the use of DL has driven significant advancements in CV, particularly in the recognition of image and video data [169].

The application of CV technology in building and infrastructure monitoring has become increasingly widespread. It enables automated detection and assessment of structural damage, such as cracks, corrosion, and spalling, thereby providing early warnings of potential structural issues. Additionally, CV systems can process large volumes of visual data to deliver detailed structural health reports, aiding engineers in making data-driven decisions. This non-contact monitoring method not only reduces safety risks during inspections but also lowers maintenance costs and prolongs infrastructure service life. Common damage detection methods include heuristic feature extraction, DL-based damage detection, and change detection [169].

Heuristic feature extraction methods involve manually designed filters to process images, combined with thresholding or machine learning classifiers to identify specific damage types. These methods are often used to detect concrete cracks, steel fatigue cracks, and similar defects, making them highly effective for identifying bridge cracks [170]. DL-based damage detection methods utilize DL, particularly CNNs, for image classification, object detection, and semantic segmentation to identify and localize structural damage. Change detection methods compare the structural state at different times using point cloud or image data to detect new damage or changes on the structure.

Structural component recognition is a process that employs CV technology to detect, locate, and classify structural elements, such as beams, columns, bridge decks, and walls, in images of buildings and infrastructure. It significantly enhances the automation and accuracy of structural health monitoring, enabling engineers to rapidly and accurately obtain status information on critical components, thereby facilitating data-driven maintenance and repair decisions. Additionally, this technology aids in the early detection of potential structural damage, preventing disasters, ensuring public safety, and extending infrastructure service life, all of which are vital for modernizing building and infrastructure management. As early as 2010, researchers used line segment pairs to identify reinforced concrete (RC) columns in images [171]. This method first detects long vertical lines in images/video frames via edge detection and Hough transforms, then constructs bounding rectangles for each line pair. If a rectangle resembles a column's shape and the color/texture of the line pair matches concrete samples in a knowledge base, the surface of a concrete column is assumed to be identified. However, this approach may fail to accurately detect surfaces of partially occluded or distant columns. Xiong et al. [172] proposed a method to automatically convert raw 3D point data (collected by laser scanners at multiple facility locations) into compact, semantically rich information models. The method identifies and models major visible structural components (walls, floors, ceilings, windows, and doors) in indoor environments, even under significant clutter and occlusion. While effective for rectangular openings, the algorithm may require adjustments for non-rectangular or complex-shaped openings. Perez-Perez et al. [173] utilized high-dimensional features (semantic features (193-dimensional) and geometric features (553-dimensional)) to recognize structural and non-structural components in indoor spaces. Their method employs a Markov random field (MRF) optimization framework to label subsets into semantic (walls, ceilings, floors, pipes) and geometric (horizontal, vertical, cylindrical) categories.

Machine learning-based structural component recognition methods enhance identification accuracy and reliability by automating and efficiently processing large volumes of data. These methods adapt to diverse building environments, effectively handle complex scenarios (e.g., mixed structural elements and occlusions), and improve recognition capabilities through multi-feature and contextual information. Additionally, they support data-driven decision-making, integrate with existing technologies for real-time monitoring, and reduce long-term operational costs, driving widespread research in this field. Yeum et al. [174] developed a robust image classification technique based on CNNs to locate and categorize regions of interest in visual assessments. Gao et al. [175] employed CNNs to classify input images into appropriate structural components and damage categories. Using weakly supervised learning, they inferred coarse localization of target objects based on outputs of the final convolutional layer.

With continuous technological advancements, CV-based detection technologies are progressively achieving higher levels of intelligence and automation. This field will integrate more advanced image recognition algorithms, such as DL and CNNs, to enhance the accuracy and speed of structural damage detection. Real-time monitoring systems, enabled by IoT technology, will deliver instant damage detection and early warnings, allowing preventive measures to be taken before disasters occur. The application of drones and robotics will facilitate regular visual inspections and data collection in hard-to-reach areas. Furthermore, AR and VR technologies will enable more intuitive visualization of structural models and monitoring results, improving engineers' comprehension of structural conditions. Cloud computing and big data analytics will streamline data storage, management, and analysis, while multimodal sensing technologies will provide comprehensive structural health insights.

7.4.5.2 CV-Based Monitoring in the Energy Industry

A stable energy supply is essential for sustaining all aspects of modern life, from heating and lighting to supporting global communication networks and data transmission. Additionally, as environmental challenges grow more pressing, the energy industry plays an increasingly critical role in advancing renewable energy and clean technologies, which are vital for addressing climate change and achieving sustainable development. The application of CV technology in the energy sector spans a wide range of areas, including oil and gas transportation and production, power system maintenance and monitoring, and the oversight of renewable energy systems. CV improves the efficiency and accuracy of energy monitoring, reduces human errors, lowers operational costs, and enhances safety protocols.

Solar energy holds a pivotal position among renewable energy sources. However, ensuring the effective operation of solar power plants requires high-quality installation, regular maintenance, and detailed inspections. Common failures in solar plants include electrical and non-electrical faults, such as process defects, mechanical damage, contamination, and humidity-related issues. These failures can lead to significant reductions in energy output and, in some cases, pose safety risks. Visible-light cameras are primarily used to detect visually apparent defects, such as cracked or broken cells, bird-dropping contamination, or material delamination, all of which severely impact photovoltaic (PV) plant performance [176–178]. For comprehensive FDD, however, IR and electroluminescence (EL) imaging techniques must also be integrated. These technologies reveal internal microcracks and other subtle issues within solar panels, providing deeper insights for maintenance and performance optimization.

The oil and gas industry plays a vital role in the global energy landscape, supplying a significant portion of the world's energy needs. However, the industry faces major challenges, most notably corrosion. The cumulative length of global oil and gas transmission pipelines (spanning from reservoirs, refineries, and storage facilities to end users) now exceeds 3 million kilometers [179]. Corrosion severely impacts the global oil and gas sector, with key contributing factors including temperature, carbon dioxide (CO_2) concentration, moisture content, pH, chloride ion (Cl^-) concentration, oxygen and salinity levels, hydrogen sulfide (H_2S) presence, metal cations, flow rates, and bacterial activity. Corrosion results in substantial maintenance costs and productivity losses. Traditional corrosion monitoring methods, such as corrosion coupons, electrical resistance probes, and ultrasonic testing, have been used for decades. However, these techniques often lack accuracy and timeliness in providing corrosion data, posing challenges for implementing proactive corrosion mitigation measures [180].

Machine vision offers a solution by detecting material or structural surfaces to identify and assess corrosion-related damage, such as rust, discoloration, pitting, cracking, or other surface irregularities [181]. Liu et al. employed image-capturing methods to detect rust spots, involving the development of image acquisition devices and strategies using varying natural light intensities. Idris et al. [182] utilized image filters to enhance corrosion data for accurate analysis through object recognition technologies.

In energy monitoring, challenges such as environmental interference, unstable lighting conditions, dynamic scene processing, and demands for high-volume, high-speed data handling impact machine vision performance. To address these issues, researchers are exploring advanced image processing algorithms, improving system adaptability and robustness, developing efficient data processing techniques, and pursuing cost-effective solutions. With advancements in DL, machine learning, and edge computing, these challenges are expected to be progressively resolved, enabling broader and more efficient applications of CV technology in energy monitoring. Future developments may include fully automated, intelligent monitoring systems capable of real-time complex task execution, delivering precise data analysis and predictive insights to enhance safety, efficiency, and sustainability in the energy industry.

7.5 CONCLUSIONS AND FUTURE DIRECTIONS

7.5.1 Integration of Infrared Imaging Technology

The integration of IR imaging technology with other technologies can enhance detection, monitoring, or analysis accuracy and efficiency, expand functionality, or broaden application scopes. Common integration approaches include:

1. Combination with Computer Vision

   Integrating IR imaging with CV algorithms enables automated object recognition and tracking in IR images. This not only improves image processing efficiency but also leverages DL to enhance IR image resolution, boosting performance in complex environments. With advancements in DL, researchers focus on improving IR image resolution, which is critical for CV and image processing tasks [183].
2. Combination with Multispectral/Hyperspectral Imaging

   By combining data from different spectral ranges, IR imaging provides richer information, widely applied in agriculture, geological surveys, and other fields. Hyperspectral imaging enables non-invasive imaging across multiple spectral bands, generating data cubes with spatial and spectral information. Mid-IR hyperspectral imaging has been extensively used in environmental monitoring, biomedicine, and materials science, facilitating in-depth analysis of material properties and environmental changes.

3. Integration with Machine Learning Algorithms

    Machine learning algorithms analyze IR images to automatically identify abnormal patterns for industrial inspection or safety monitoring. The rapid development of DL has made IR machine vision increasingly intelligent and automated, significantly expanding its applications [184].

4. Combination with Ultrasonic Imaging

    In healthcare, the integration of IR and ultrasonic imaging forms a multimodal approach, providing comprehensive views and information about tissues or objects. This multimodal imaging synthesizes data from different modalities, offering detailed biological insights to improve diagnosis and treatment outcomes.

    Through these integrations, IR imaging delivers comprehensive and in-depth analysis across diverse fields and complex scenarios, demonstrating vast application potential.

7.5.2 Enhancing Automated Frost/Dew Meteorological Observation

1. Mechanistic Study of Frost/Dew Formation

    Current understanding of frost/dew formation mechanisms remains incomplete. Machine learning and DL frameworks can extract new features from large image datasets, reflecting textural and spatial structural changes during frost/dew formation, aiding pattern discrimination and mechanistic analysis.

2. Automated Quantitative Measurement

    Automated quantification of frost/dew accumulation requires further exploration. Current methods are limited to indirectly estimating condensation levels via dew coverage rates.

7.5.3 Refining Intelligent Defrosting Control Strategies

1. Adaptive Defrosting Strategies

    Combine multiple frost suppression, delay strategies, and defrosting methods tailored to local conditions, optimizing control logic, and determining precise defrosting timings.

2. Energy Efficiency and Stability

    Minimize energy consumption during defrosting while ensuring indoor thermal comfort, maintaining stable system operation, and reducing defrosting frequency.

3. Image-Based Defrosting Techniques

    Advanced image processing-driven defrosting methods and intelligent defrosting by extracting new features from raw images to quantify frost severity.

7.5.4 Advancing Intelligent Condensation Heat Transfer Measurement

1. AI-Driven Measurement Models
   Integrate CV, machine learning, and DL to develop universal condensation heat transfer performance prediction models, advancing intelligent measurement technologies.
2. Experimental Optimization
   Reduce experimental errors by extending video recording durations and using multi-angle imaging of condensation surfaces. Develop new CV algorithms incorporating dynamic droplet features (growth rate, detachment frequency, coalescence count) to build robust visual datasets. Combining CV, image processing, and AI algorithms with frost/dew detection enables real-time automated analysis of frost/dew images, supporting meteorological observation, intelligent defrosting control, and heat transfer measurement. This approach saves labor and resources, improves efficiency, and promotes energy conservation. Furthermore, it provides novel insights into the heat and mass transfer performance of heat exchangers and extends to other nucleation-based processes in thermal fluid science, offering vast application prospects and significant research value.

REFERENCES

1. M. Vollmer, Infrared Thermal Imaging, Computer Vision: A Reference Guide, Springer, 2020: pp. 1–4.
2. E.F.J. Ring, History of thermology and thermography: Pioneers and progress, Thermology International 22 (2012) 3–8.
3. FLIR, Thermal Imaging Guidebook for Building and Renewable Energy Applications, FLIR, Wilsonville, OR, 2011.
4. H. Kaplan, Practical Applications of Infrared Thermal Sensing and Imaging Equipment, SPIE Press, 2007.
5. K. Chatterjee, S. Tuli, S.G. Pickering, et al., A comparison of the pulsed, lock-in and frequency modulated thermography nondestructive evaluation techniques, NDT & E International 44 (2011) 655–667.
6. S. Lu, L. Yang, C. Fan, et al., Numerical simulation and analysis of infrared thermal characteristics of aging power cable, Journal of Engineering Thermophysics 34 (2013) 332–335.
7. X. Maldague, Theory and Practice of Infrared Technology for Nondestructive Testing, Wiley-Interscience, Hoboken, NJ, 2001.
8. W.Z. Dai Jingmin, Infrared thermography non-destructive testing technology and its applications, Techniques of Automation and Application 26 (2007) 1–7.
9. H. Carslaw, J. Jaeger, Conduction of Heat in Solids, Clarendon, Oxford, 1959.
10. P.F. Maldague, Many-body corrections to the polarizability of the two-dimensional electron gas, Surface Science 73 (1978) 296–302.
11. V. Vavilov, V. Shiryaev, E. Grinzato, Detection of hidden corrosion in metals by using transient infrared thermography, Insight: Non-Destructive Testing and Condition Monitoring 40 (1998) 408–410.
12. D.P. Almond, S. Lau, Edge effects and a method of defect sizing for transient thermography, Applied Physics Letters 62 (1993) 3369–3371.
13. L. Favro, X. Han, P. Kuo, et al., Defect depth determination by thermal-wave imaging, Progress in Natural Science 6 (1996) S139–S141.

14. Z. Zhao, X. Wang, Y. Peng, et al., Development and application of medium- and long-wave dual-band infrared imaging technology abroad (in Chinese), Infrared Technology 42 (2020) 312–319.
15. M. Zhang, F. Han, J. Kang, et al., Application of infrared thermal imaging technology in civil fields (in Chinese), Infrared 40 (2019) 35–43.
16. R. Wang, Y. Chen, Application of infrared thermal imaging technology in thermal comfort of clothing (in Chinese), Jiangsu Textile 11 (2006) 41–43, 46.
17. M. Salim Al Jumaili, Digital image processing and applications, ResearchGate. https://www.researchgate.net/publication/364928386_Digital_image_processing_and_applications (2022).
18. C. Solomon, T. Breckon, Fundamentals of Digital Image Processing: A Practical Approach with Examples in MATLAB, John Wiley & Sons, 2011.
19. K. Yang, W. Xiang, Z. Chen, et al., A review on infrared and visible image fusion algorithms based on neural networks, Journal of Visual Communication and Image Representation 101 (2024) 104179.
20. Z. Lv, J. Li, X. Li, et al., Two adaptive enhancement algorithms for high gray-scale RAW infrared images based on multi-scale fusion and chromatographic remapping, Infrared Physics & Technology 133 (2023) 104774.
21. R. Soundrapandiyan, S.C. Satapathy, C.M. P.V.S.S.R., et al., A comprehensive survey on image enhancement techniques with special emphasis on infrared images, Multimedia Tools and Applications 81 (2022) 9045–9077.
22. M. Gao, Y. Bai, H. Liao, et al., Contrast enhancement algorithm for infrared images based on atmospheric scattering model, Computers and Electrical Engineering 118 (2024) 109318.
23. Y. Wang, X. Sui, Y. Wang, et al., Raw infrared image enhancement via an inverted framework based on infrared basic prior, Expert Systems with Applications 253 (2024) 124314.
24. W. Wild, Application of infrared thermography in civil engineering, Proceedings of The Estonian Academy of Sciences 13 (2007) 436–444.
25. V. Antonopoulos, Water movement and heat transfer simulations in a soil under ryegrass, Biosystems Engineering 95 (2006) 127–138.
26. J. Al-Karawi, J. Schmidt, Application of infrared thermography to the analysis of welding processes, Quantitative InfraRed Thermography (2004).
27. M.S. Jadin, S. Taib, Recent progress in diagnosing the reliability of electrical equipment by using infrared thermography, Infrared Physics & Technology 55 (2012) 236–245.
28. E. Johnson, P. Hyer, P. Culotta, et al., Evaluation of infrared thermography as a diagnostic tool in CVD applications, Journal of Crystal Growth 187 (1998) 463–473.
29. M.J. Hittel, R. Bingham, M.K. Sanders, NFPA 70B Recommended Practice for Electrical Equipment Maintenance 2002 Edition, IEEE, 2003.
30. S. Bagavathiappan, B.B. Lahiri, T. Saravanan, et al., Infrared thermography for condition monitoring – A review, Infrared Physics & Technology 60 (2013) 35–55.
31. R.A. Epperly, G.E. Heberlein, L.G. Eads, A Tool for Reliability and Safety: Predict and Prevent Equipment Failures with Thermography, IEEE, 1997.
32. Y. Li, X. Du, F. Wan, et al., Rotating machinery fault diagnosis based on convolutional neural network and infrared thermal imaging, Chinese Journal of Aeronautics 33 (2020) 427–438.
33. O. Janssens, R. Schulz, V. Slavkovikj, et al., Thermal image based fault diagnosis for rotating machinery, Infrared Physics & Technology 73 (2015) 78–87.
34. O. Janssens, M. Loccufier, R. Van de Walle, et al., Data-driven imbalance and hard particle detection in rotating machinery using infrared thermal imaging, Infrared Physics & Technology 82 (2017) 28–39.
35. X. Li, Y. Li, K. Yan, et al., Intelligent fault diagnosis of bevel gearboxes using semi-supervised probability support matrix machine and infrared imaging, Reliability Engineering & System Safety 230 (2023) 108921.

36. B. Li, X. Zhu, S. Zhao, et al., HV Power Equipment Diagnosis Based on Infrared Imaging Analyzing, IEEE, 2006.
37. J.H.E. de Oliveira, W.F. Lages, Robotized Inspection of Power Lines with Infrared Vision, IEEE, 2010.
38. Y. Cao, X. Gu, Q. Jin, Infrared Technology in the Fault Diagnosis of Substation Equipment, IEEE, 2008.
39. C.A. Balaras, A. Argiriou, Infrared thermography for building diagnostics, Energy and Buildings 34 (2002) 171–183.
40. M.A. Kregg, Benefits of using infrared thermography in utility substations, SPIE Defense + Commercial Sensing, 2004, pp. 249–257.
41. L. Dos Santos, E. Bortoni, L. Souza, et al., Infrared thermography applied for outdoor power substations, Proceedings of SPIE - The International Society for Optical Engineering, 2008.
42. Z. Azmat, D. Turner, Infrared Thermography and Its Role in Rural Utility Environment, IEEE, 2005.
43. J. Martínez, R. Lagioia, S. Edenor, Experience performing infrared thermography in the maintenance of a distribution utility, 19th International Conference on Electricity Distribution, 2007.
44. A.N. Huda, S. Taib, Application of infrared thermography for predictive/preventive maintenance of thermal defect in electrical equipment, Applied Thermal Engineering 61 (2013) 220–227.
45. Y. Lin, W. Zhang, H. Zhang, et al., An intelligent infrared image fault diagnosis for electrical equipment, 2020 5th Asia Conference on Power and Electrical Engineering (ACPEE), 2020, pp. 1829–1833.
46. G. Singh, T.A. Kumar, V. Naikan, Fault Diagnosis of Induction Motor Cooling System Using Infrared Thermography, IEEE, 2016.
47. F. Jeffali, B. Kihel, A. Nougaoui, et al., Monitoring and diagnostic misalignment of asynchronous machines by infrared thermography, Journal of Materials and Environmental Science 6 (2015) 1192–1199.
48. D. Chaturvedi, M.S. Iqbal, M.P. Singh, Intelligent Health Monitoring System for Three Phase Induction Motor Using Infrared Thermal Image, IEEE, 2015.
49. H. Luan, W. Tao, G. Zhu, et al., Review of the development of fully welded plate heat exchangers (in Chinese), Science China Technological Sciences 43 (2013) 1020–1033.
50. Y. Li, Research on Temperature Measurement Technology and Its Application Based on Infrared Thermal Imager (in Chinese), Harbin Institute of Technology, 2010.
51. J. Berce, M. Zupančič, M. Može, et al., A review of crystallization fouling in heat exchangers, Processes 9 (2021) 1356.
52. H. Ezuber, S. Zakir Hossain, A review of corrosion failures in shell and tube heat exchangers: Roots and advanced counteractive, Heat and Mass Transfer 59 (2023) 971–987.
53. A. Patel, Heat exchangers in industrial applications: Efficiency and optimization strategies, International Journal of Engineering Research & Technology (IJERT) 12 (2023).
54. J. Berce, M. Zupančič, M. Može, et al., Infrared thermography observations of crystallization fouling in a plate heat exchanger, Applied Thermal Engineering 224 (2023) 120116.
55. Y. Zhao, L. Nian, Y. Wang, et al., Coating defect detection based on electromagnetic induction infrared thermal imaging technology, Journal of Chinese Society for Corrosion and Protection 44 (2024) 1641–1648.
56. Y. Liu, C. Mei, Q. Tang, et al., Research status and development trend of infrared thermal imaging detection technology (in Chinese), Machinery Design & Manufacture 6 (2015) 260–262+266.
57. J. Huang, J. Ren, S. Su, et al., Current status and development trend of China's heat exchanger industry (in Chinese), Petro & Chemical Equipment 13 (2010) 5–8.

58. Y. Wang, Y. Xia, S. Lu, Research progress on fluid maldistribution in microchannel plate heat exchangers (in Chinese), Fluid Machinery 43 (2015) 83–87.
59. S. Zhang, Y. Li, Y. Xu, et al., A new method for estimating the refrigerant distribution in plate evaporator based on infrared thermal imaging, International Journal of Refrigeration 126 (2021) 57–65.
60. W. Li, P. Hrnjak, Quantification of two-phase refrigerant distribution in brazed plate heat exchangers using infrared thermography, International Journal of Refrigeration 131 (2021) 348–358.
61. M. Bobič, B. Gjerek, I. Golobič, et al., Dynamic behaviour of a plate heat exchanger: Influence of temperature disturbances and flow configurations, International Journal of Heat and Mass Transfer 163 (2020) 120439.
62. S. Dudzik, Investigations of a heat exchanger using infrared thermography and artificial neural networks, Sensors and Actuators A: Physical 166 (2011) 149–156.
63. X. Du, L. Feng, Y. Yang, et al., Experimental study on heat transfer enhancement of wavy finned flat tube with longitudinal vortex generators, Applied Thermal Engineering 50 (2013) 55–62.
64. K. Sarraf, S. Launay, G. El Achkar, et al., Local vs global heat transfer and flow analysis of hydrocarbon complete condensation in plate heat exchanger based on infrared thermography, International Journal of Heat and Mass Transfer 90 (2015) 878–893.
65. Z. Ge, X. Du, L. Yang, et al., Performance monitoring of direct air-cooled power generating unit with infrared thermography, Applied Thermal Engineering 31 (2011) 418–424.
66. K. Sarraf, S. Launay, L. Tadrist, Analysis of enhanced vapor desuperheating during condensation inside a plate heat exchanger, International Journal of Thermal Sciences 105 (2016) 96–108.
67. B. Kanargi, J.M.S. Tan, P.S. Lee, et al., A tapered inlet/outlet flow manifold for planar, air-cooled oblique-finned heat sink, Applied Thermal Engineering 174 (2020) 115250.
68. H.-Y. Li, M.-H. Chiang, Effects of shield on thermal-fluid performance of vapor chamber heat sink, International Journal of Heat and Mass Transfer 54 (2011) 1410–1419.
69. H.-Y. Li, S.-M. Chao, G.-L. Tsai, Thermal performance measurement of heat sinks with confined impinging jet by infrared thermography, International Journal of Heat and Mass Transfer 48 (2005) 5386–5394.
70. C. Xu, L. Yang, L. Li, et al., Experimental study on heat transfer performance improvement of wavy finned flat tube, Applied Thermal Engineering 85 (2015) 80–88.
71. H.-Y. Li, M.-H. Chiang, C.-I. Lee, et al., Thermal performance of plate-fin vapor chamber heat sinks, International Communications in Heat and Mass Transfer 37 (2010) 731–738.
72. R. He, P. Xu, Z. Chen, et al., A non-intrusive approach for fault detection and diagnosis of water distribution systems based on image sensors, audio sensors and an inspection robot, Energy and Buildings 243 (2021) 110967.
73. P. Zhang, P.S. Hrnjak, Air-side performance evaluation of three types of heat exchangers in dry, wet and periodic frosting conditions, International Journal of Refrigeration 32 (2009) 911–921.
74. E. Schmidt, W. Schurig, W. Sellschopp, Versuche uber die Kondensation von Wasserdampf in Film-und Tropfenform, Technische Mechanik und Thermodynamik 1 (1930) 53–63.
75. T. Kim, S.J. Kim, Vertical-length effect on dropwise condensation heat transfer at low heat flux – Part I: Experimental study, International Journal of Heat and Mass Transfer 224 (2024) 125329.
76. M. Bucci, Experimental and computational analysis of condensation phenomena for the thermal-hydraulic analysis of LWRs containments, Doctoral dissertation, Ph.D. thesis, University of Pisa, 2009.

77. S.G. Lim, Experimental study of external condensation and flashing-induced instability for passive containment cooling system, Doctoral dissertation, Ph.D. thesis, Korea Advanced Institute of Science and Technology, 2020.
78. X.-H. Ma, X.-D. Zhou, Z. Lan, et al., Condensation heat transfer enhancement in the presence of non-condensable gas using the interfacial effect of dropwise condensation, International Journal of Heat and Mass Transfer 51 (2008) 1728–1737.
79. Y. Hayashi, A. Aoki, S. Adachi, et al., Study of frost properties correlating with frost formation types, Journal of Heat Transfer 99 (1977) 239.
80. M. Jakob, Heat transfer in evaporation and condensation II, Mechanical Engineering 58 (1936) 729–740.
81. G. Tammann, W. Boehme, Die Zahl der Wassertröpfchen bei der Kondensation auf verschiedenen festen Stoffen, Annalen der Physik 414 (1935) 77–80.
82. Y. Song, D. Xu, J. Lin, et al., A study on the mechanism of dropwise condensation, International Journal of Heat and Mass Transfer 34 (1991) 2827–2831.
83. Y. Shen, S. Wang, Condensation frosting detection and characterization using a capacitance sensing approach, International Journal of Heat and Mass Transfer 147 (2020) 118968.
84. J. Weremczuk, Dew/frost point recognition with impedance matrix of fingerprint sensor, IEEE Transactions on Instrumentation and Measurement 57 (2008) 1791–1795.
85. W. Nie, X. Lin, G. Rao, et al., Development and validation of a laser-absorption-based method for in-situ and high-precision water vapor dew point temperature measurement, Infrared Physics & Technology 123 (2022) 104137.
86. R. Kozak, K. Khorsand, T. Zarifi, et al., Patch antenna sensor for wireless ice and frost detection, Scientific Reports 11 (2021) 13707.
87. X. Li, S. Ma, K. Wu, et al., A new automated dewgauge and experiment (in Chinese), Meteorol 38 (2012) 501–507.
88. L. Zhu, Z. Cao, W. Zhuo, et al., A new dew and frost detection sensor based on computer vision, Journal of Atmospheric and Oceanic Technology 31 (2014) 2692–2712.
89. L. Zhu, Z. Cao, Y. Xiao, et al., Automated surface frost detection based on manifold learning, Journal of Computer Applications 35 (2015) 854.
90. L. Zhu, X. Zhang, J. Wu, Surface dew and frost phenomena detection based on small-sample learning (in Chinese), Modern Electronics Technique 42 (2019) 130–135.
91. J. Wang, F. Deng, N. Li, et al., Research on recognition method of ground dewbased on machine vision (in Chinese), Chinese Journal of Sensors and Actuators 33 (2020) 1734–1741.
92. R. Wachal, J.-S. Stoezel, M. Peckover, et al., A computer vision early-warning ice detection system for the Smart Grid, PES T&D 2012, IEEE, 2012, pp. 1–6.
93. Z. Zhou, J. Wu, L. Zhu, Dew and frost image classification based on multi-path feature fusion and deep learning (in Chinese), Computer Applications and Software 35 (2018) 205–210+231.
94. J.M. Maldonado, A. de Gracia, G. Zsembinszki, et al., Frost detection method on evaporator in vapour compression systems, International Journal of Refrigeration 110 (2020) 75–82.
95. K.-S. Lee, W.-S. Kim, T.-H. Lee, A one-dimensional model for frost formation on a cold flat surface, International Journal of Heat and Mass Transfer 40 (1997) 4359–4365.
96. M. Kandula, Frost growth and densification on a flat surface in laminar flow with variable humidity, International Communications in Heat and Mass Transfer 39 (2012) 1030–1034.
97. K. Matsumoto, M. Koshizuka, M. Honda, et al., Measurement on nano scale by scanning probe microscope for obtaining real ice adhesion force, International Journal of Refrigeration 41 (2014) 181–189.
98. K. Qu, S. Komori, Y. Jiang, Local variation of frost layer thickness and morphology, International Journal of Thermal Sciences 45 (2006) 116–123.

99. J. Xiao, W. Wang, Q. Guo, et al., An experimental study of the correlation for predicting the frost height in applying the photoelectric technology, International Journal of Refrigeration 33 (2010) 1006–1014.
100. W. Wang, J. Xiao, Y. Feng, et al., Characteristics of an air source heat pump with novel photoelectric sensors during periodic frost–defrost cycles, Applied Thermal Engineering 50 (2013) 177–186.
101. Y. Ge, Y. Sun, W. Wang, et al., Field test study of a novel defrosting control method for air-source heat pumps by applying tube encircled photoelectric sensors, International Journal of Refrigeration 66 (2016) 133–144.
102. X. Bai, W. Wang, Y. Sun, et al., Operating performances of an ASHP unit operated in a mild and humid region using tube-encircled photoelectric sensor based defrosting initiation strategy, Energy and Buildings 177 (2018) 140–153.
103. B. Mandelbrot, The Fractal Geometry of Nature, WH Freeman and Co, New York, 1982.
104. Y. Hao, I. Jose, X. Yong, Experimental study of initial state of frost formation on flat surface, Journal of Southeast University 35 (2005) 149–153.
105. P. Hou, L. Cai, W. Yu, Experimental study and fractal analysis of ice crystal structure at initial period of frost formation, Journal of Applied Sciences 25 (2007) 193–197.
106. Y. Liu, Z. Liu, L. Huang, et al., Fractal model for simulation of frost formation and growth, Science China Technological Sciences 53 (2010) 807–812.
107. Y. Liu, Z. Liu, L. Huang, Experimental and theoretical investigations of the fractal characteristics of frost crystals during frost formation process, Experimental Thermal and Fluid Science 36 (2012) 217–223.
108. S. Chen, S. Yao, F. Xie, et al., Thermal conductivity of frost layer on finned-tube vaporizer based on fractal theory (In Chinese), CIESC Journal 63 (2012) 3855–3860.
109. S. Yao, S. Chen, Y. Liu, et al., Heat transfer analysis of frosting under cryogenic condition, Cryogenics 6 (2013) 15–20.
110. S. Negrelli, R.P. Cardoso, C.J. Hermes, A finite-volume diffusion-limited aggregation model for predicting the effective thermal conductivity of frost, International Journal of Heat and Mass Transfer 101 (2016) 1263–1272.
111. A.R. Tahavvor, Modeling of frost crystal growth over a flat plate using artificial neural networks and fractal geometries, Heat and Mass Transfer 53 (2017) 813–823.
112. C.-H. Cheng, C.-C. Shiu, Frost formation and frost crystal growth on a cold plate in atmospheric air flow, International Journal of Heat and Mass Transfer 45 (2002) 4289–4303.
113. J. Wu, G. Ouyang, P. Hou, et al., Experimental investigation of frost formation on a parallel flow evaporator, Applied Energy 88 (2011) 1549–1556.
114. A.N. Malik, S.A. Khan, I. Lazoglu, A novel demand-actuated defrost approach based on the real-time thickness of frost for the energy conservation of a refrigerator, International Journal of Refrigeration 131 (2021) 168–177.
115. L. Cai, R. Wang, P. Hou, et al., Study on restraining frost growth at initial stage by hydrophobic coating and hygroscopic coating, Energy and Buildings 43 (2011) 1159–1163.
116. A.D. Sommers, N.L. Truster, A.C. Napora, et al., Densification of frost on hydrophilic and hydrophobic substrates – Examining the effect of surface wettability, Experimental Thermal and Fluid Science 75 (2016) 25–34.
117. J.W. Yoo, Y. Chung, G.T. Kim, et al., Determination of defrosting start time in an air-to-air heat pump system by frost volume calculation method, International Journal of Refrigeration 96 (2018) 169–178.
118. X. Zheng, R. Shi, S. You, et al., Experimental study of defrosting control method based on image processing technology for air source heat pumps, Sustainable Cities and Society 51 (2019) 101667.

119. H. Miao, X. Yang, D. Yin, et al., A novel defrosting control strategy with image processing technique and fractal theory, International Journal of Refrigeration 138 (2022) 259–269.
120. Z. Li, W. Wang, Y. Sun, et al., Applying image recognition to frost built-up detection in air source heat pumps, Energy 233 (2021) 121004.
121. S. Wang, W. Wang, Y. Sun, et al., Research on image recognition frost measurement technology for air-source heat pumps based on light adaptation (in Chinese), Heating Ventilating & Air Conditioning 52 (2022) 113–117+68.
122. A.R. Tahavvor, M. Yaghoubi, Prediction of frost deposition on a horizontal circular cylinder under natural convection using artificial neural networks, International Journal of Refrigeration 34 (2011) 560–566.
123. A. Zendehboudi, B. Wang, X. Li, Application of smart models for prediction of the frost layer thickness on vertical cryogenic surfaces under natural convection, Applied Thermal Engineering 115 (2017) 1128–1136.
124. D. Datta, S. Tassou, D. Marriott, Frost prediction on evaporator coils of supermarket display cabinets using artificial neural networks, Proceedings of Clima 2000 Conference, 1997.
125. B. Zhao, R. Yi, D. Gao, et al., Frosting weight and refrigerating capacity prediction of fin evaporator based on random finite element method and ridgelet neural network, International Journal of Refrigeration 99 (2019) 37–46.
126. J. Tang, C. Herman, G. Gong, A novel self-adaptive control strategy of frost prevention and retardation for air source heat pumps in winter conditions, Applied Mathematical Modelling 83 (2020) 284–300.
127. Y.H. Eom, Y. Chung, M. Park, et al., Deep learning-based prediction method on performance change of air source heat pump system under frosting conditions, Energy 228 (2021) 120542.
128. K.F. Rabbi, K.S. Boyina, W. Su, et al., Wettability-defined frosting dynamics between plane fins in quiescent air, International Journal of Heat and Mass Transfer 164 (2021) 120563.
129. W. Wang, Q. Zhou, G. Tian, et al., A novel defrosting initiation strategy based on convolutional neural network for air-source heat pump, International Journal of Refrigeration 128 (2021) 95–103.
130. S.A. Iranagh, A.R. Tahavvor, M. Yaghoubi, et al., Experimental and artificial neural network evaluation of frost formation on square finned tube under natural convection, Heat Transfer Engineering 44 (2023) 368–389.
131. J. Chen, J. Wu, J. He, et al., A novel defrosting initiating strategy for automotive air conditioner heat pumps based on frost thickness growth prediction, International Journal of Refrigeration 134 (2022) 242–252.
132. J. Rose, Dropwise condensation theory, International Journal of Heat and Mass Transfer 24 (1981) 191–194.
133. C. Extrand, A. Gent, Retention of liquid drops by solid surfaces, Journal of Colloid and Interface Science 138 (1990) 431–442.
134. C. Dietz, K. Rykaczewski, A. Fedorov, et al., Visualization of droplet departure on a superhydrophobic surface and implications to heat transfer enhancement during dropwise condensation, Applied Physics Letters 97 (2010).
135. N. Miljkovic, R. Enright, E.N. Wang, Effect of droplet morphology on growth dynamics and heat transfer during condensation on superhydrophobic nanostructured surfaces, ACS Nano 6 (2012) 1776–1785.
136. S. Kulju, L. Riegger, P. Koltay, et al., Fluid flow simulations meet high-speed video: Computer vision comparison of droplet dynamics, Journal of Colloid and Interface Science 522 (2018) 48–56.
137. A. Mudugamuwa, S. Hettiarachchi, G. Melroy, et al., Vision-based performance analysis of an active microfluidic droplet generation system using droplet images, Sensors 22 (2022) 6900.

138. S. Hu, J. Yan, The research on drop radius distribution of Marangoni dropwise condensation (in Chinese), Journal of Engineering Thermophysics 32 (2011) 230–232.
139. P. Guan, L. Jia, D. Wang, Visualization research of condensation over horizontal fin tube (in Chinese), Journal of Thermal Science and Technology 10 (2011) 12–17.
140. Z.Z. Chong, S.B. Tor, A.M. Gañán-Calvo, et al., Automated droplet measurement (ADM): An enhanced video processing software for rapid droplet measurements, Microfluidics and Nanofluidics 20 (2016) 1–14.
141. W. Xu, Z. Lan, Q. Liu, et al., Droplet size distributions in dropwise condensation heat transfer: Consideration of droplet overlapping and multiple re-nucleation, International Journal of Heat and Mass Transfer 127 (2018) 44–54.
142. B. El Fil, G. Kini, S. Garimella, A review of dropwise condensation: Theory, modeling, experiments, and applications, International Journal of Heat and Mass Transfer 160 (2020) 120172.
143. B. Zhou, H. Yang, W. Feng, et al., Self-propelled and size distribution of condensate droplets on superhydrophobic surfaces (in Chinese), Surface Technology 49 (2020) 170–176+190.
144. G. Bravo, E. González, Thermal comfort in naturally ventilated spaces and under indirect evaporative passive cooling conditions in hot–humid climate, Energy and Buildings 63 (2013) 79–86.
145. S.T. Smith, V.I. Hanby, C. Harpham, A probabilistic analysis of the future potential of evaporative cooling systems in a temperate climate, Energy and Buildings 43 (2011) 507–516.
146. H. Campaniço, P.M. Soares, P. Hollmuller, et al., Climatic cooling potential and building cooling demand savings: High resolution spatiotemporal analysis of direct ventilation and evaporative cooling for the Iberian Peninsula, Renewable Energy 85 (2016) 766–776.
147. A. Ahmad, S. Rehman, L.M. Al-Hadhrami, Performance evaluation of an indirect evaporative cooler under controlled environmental conditions, Energy and Buildings 62 (2013) 278–285.
148. S. De Antonellis, C.M. Joppolo, P. Liberati, et al., Experimental analysis of a cross flow indirect evaporative cooling system, Energy and Buildings 121 (2016) 130–138.
149. H.-J. Kim, S.-W. Ham, D.-S. Yoon, et al., Cooling performance measurement of two cross-flow indirect evaporative coolers in general and regenerative operation modes, Applied Energy 195 (2017) 268–277.
150. Y. Wan, A. Soh, Y. Shao, et al., Numerical study and correlations for heat and mass transfer coefficients in indirect evaporative coolers with condensation based on orthogonal test and CFD approach, International Journal of Heat and Mass Transfer 153 (2020) 119580.
151. W. Shi, Y. Min, Y. Chen, et al., Development of a three-dimensional numerical model of indirect evaporative cooler incorporating with air dehumidification, International Journal of Heat and Mass Transfer 185 (2022) 122316.
152. Y. Chen, Y. Luo, H. Yang, A simplified analytical model for indirect evaporative cooling considering condensation from fresh air: Development and application, Energy and Buildings 108 (2015) 387–400.
153. Y. Chen, H. Yang, Y. Luo, Indirect evaporative cooler considering condensation from primary air: Model development and parameter analysis, Building and Environment 95 (2016) 330–345.
154. Y. Min, Y. Chen, H. Yang, et al., Characteristics of primary air condensation in indirect evaporative cooler: Theoretical analysis and visualized validation, Building and Environment 174 (2020) 106783.
155. D. Meng, J. Lv, Y. Chen, et al., Visualized experimental investigation on cross-flow indirect evaporative cooler with condensation, Applied Thermal Engineering 145 (2018) 165–173.

156. Y. You, G. Wang, B. Yang, et al., Study on heat transfer characteristics of indirect evaporative cooling system based on secondary side hydrophilic, Energy and Buildings 257 (2022) 111704.
157. Y. Min, W. Shi, B. Shen, et al., Enhancing the cooling and dehumidification performance of indirect evaporative cooler by hydrophobic-coated primary air channels, International Journal of Heat and Mass Transfer 179 (2021) 121733.
158. A. Pacheco-Vega, M. Sen, K. Yang, et al., Neural network analysis of fin-tube refrigerating heat exchanger with limited experimental data, International Journal of Heat and Mass Transfer 44 (2001) 763–770.
159. M. Hosoz, H. Ertunc, A. Ozguc, Modelling of a direct evaporative air cooler using artificial neural network, International Journal of Energy Research 32 (2008) 83–89.
160. T.R. Kiran, S.P. Rajput, An effectiveness model for an indirect evaporative cooling (IEC) system: Comparison of artificial neural networks (ANN), adaptive neuro-fuzzy inference system (ANFIS) and fuzzy inference system (FIS) approach, Applied Soft Computing 11 (2011) 3525–3533.
161. G. Zhu, T.-T. Chow, C. Lee, Performance analysis of counter-flow regenerative heat and mass exchanger for indirect evaporative cooling based on data-driven model, Energy and Buildings 155 (2017) 503–512.
162. H.M. Asfahan, U. Sajjad, M. Sultan, et al., Artificial intelligence for the prediction of the thermal performance of evaporative cooling systems, Energies 14 (2021) 3946.
163. G.A. Longo, G. Righetti, C. Zilio, et al., Application of an artificial neural network (ANN) for predicting low-GWP refrigerant condensation heat transfer inside herringbone-type brazed plate heat exchangers (BPHE), International Journal of Heat and Mass Transfer 156 (2020) 119824.
164. G. Zhu, T. Wen, D. Zhang, Machine learning based approach for the prediction of flow boiling/condensation heat transfer performance in mini channels with serrated fins, International Journal of Heat and Mass Transfer 166 (2021) 120783.
165. T. Sun, X. Huang, C. Liang, et al., Prediction and analysis of dew point indirect evaporative cooler performance by artificial neural network method, Energies 15 (2022) 4673.
166. G. Damoulakis, M.J. Gukeh, S. Moitra, et al., Quantifying steam dropwise condensation heat transfer via experiment, computer vision and machine learning algorithms, 2021 20th IEEE Intersociety Conference on Thermal and Thermomechanical Phenomena in Electronic Systems (iTherm), IEEE, 2021, pp. 1015–1023.
167. Y. Suh, J. Lee, P. Simadiris, et al., A deep learning perspective on dropwise condensation, Advanced Science 8 (2021) 2101794.
168. S. Khodakarami, K.F. Rabbi, Y. Suh, et al., Machine learning enabled condensation heat transfer measurement, International Journal of Heat and Mass Transfer 194 (2022) 123016.
169. B.F. Spencer Jr, V. Hoskere, Y. Narazaki, Advances in computer vision-based civil infrastructure inspection and monitoring, Engineering 5 (2019) 199–222.
170. C.M. Yeum, S.J. Dyke, Vision-based automated crack detection for bridge inspection, Computer-Aided Civil and Infrastructure Engineering 30 (2015) 759–770.
171. Z. Zhu, I. Brilakis, Concrete column recognition in images and videos, Journal of Computing in Civil Engineering 24 (2010) 478–487.
172. X. Xiong, A. Adan, B. Akinci, et al., Automatic creation of semantically rich 3D building models from laser scanner data, Automation in Construction 31 (2013) 325–337.
173. Y. Perez-Perez, M. Golparvar-Fard, K. El-Rayes, Semantic and geometric labeling for enhanced 3D point cloud segmentation, Construction Research Congress 2016, 2016, pp. 2542–2552.
174. C.M. Yeum, J. Choi, S.J. Dyke, Automated region-of-interest localization and classification for vision-based visual assessment of civil infrastructure, Structural Health Monitoring 18 (2019) 675–689.

175. Y. Gao, K.M. Mosalam, Deep transfer learning for image-based structural damage recognition, Computer-Aided Civil and Infrastructure Engineering 33 (2018) 748–768.
176. S. Leva, M. Aghaei, F. Grimaccia, PV power plant inspection by UAS: Correlation between altitude and detection of defects on PV modules, 2015 IEEE 15th International Conference on Environment and Electrical Engineering (EEEIC), IEEE, 2015, pp. 1921–1926.
177. A.M. Sizkouhi, M. Aghaei, S.M. Esmailifar, A deep convolutional encoder-decoder architecture for autonomous fault detection of PV plants using multi-copters, Solar Energy 223 (2021) 217–228.
178. M. Meribout, V.K. Tiwari, J.P.P. Herrera, et al., Solar panel inspection techniques and prospects, Measurement 209 (2023) 112466.
179. A.A. Mazraeh, F.B.I. Alnaimi, Multi-diameter pipeline inspection gauge for lang distance industrial application, International Journal of Scientific Engineering and Research 6 (2015) 646–651.
180. M. Oki, A.A. Adediran, P. Anawe, Corrosion monitoring in the oil pipeline industry, Journal of Multidisciplinary Engineering Science and Technology (JMEST) 2 (2013).
181. A.H. Khalaf, Y. Xiao, N. Xu, et al., Emerging AI technologies for corrosion monitoring in oil and gas industry: A comprehensive review, Engineering Failure Analysis 155 (2024) 107735.
182. S.A. Idris, F.A. Jafar, Image Enhancement Based on Software Filter Optimization for Corrosion Inspection, IEEE, 2014.
183. Y. Huang, T. Miyazaki, X. Liu, et al., Infrared image super-resolution: A systematic review and future trends, IEEE Journal of Selected Topics in Applied Earth Observations and Remote Sensing 18 (2025) 25439–25463.
184. Y. He, B. Deng, H. Wang, et al., Infrared machine vision and infrared thermography with deep learning: A review, Infrared Physics & Technology 116 (2021) 103754.

8 Summary and Outlook

Zeyu Zhao, Xinrou Han, Shuang Yang, and Bin Yang

Recent advancements in artificial intelligence (AI) have significantly accelerated the adoption of non-contact measurement technologies across three critical domains: indoor thermal environment monitoring and regulation, outdoor environmental monitoring, and heating, ventilation, and air conditioning (HVAC) equipment diagnostics. These innovations demonstrate remarkable potential for enhancing precision, efficiency, and adaptability in environmental management systems.

8.1 INDOOR THERMAL ENVIRONMENTAL MONITORING AND REGULATION

Non-contact measurement methods based on infrared and visible light images, which detect personnel information related to the thermal environment by analyzing human skin temperature and posture, have been proven to be efficient and effective.

1. Contact and semi-contact methods often produce a sense of foreign body, causing discomfort to the subjects. In contrast, non-contact methods monitor through sensors and video image processing technology, avoiding interference with the subjects.
2. Combined with computer vision and deep learning technologies, non-contact methods can efficiently predict the thermal state of personnel, thereby supporting the intelligent control of HVAC systems.
3. Future detection technologies should enhance accuracy and robustness to respond dynamically to changes in human thermal states and distribution. A well-integrated system combining multiple monitoring techniques with multimodal data collection is essential for comprehensive thermal state assessment.

With economic growth, technological advancements, and an increasing focus on sleep quality and overall health, continuous monitoring of sleep conditions has garnered significant attention. Various measurement devices and methods are being utilized for long-term sleep quality assessment.

1. Portable and wearable monitoring devices can continuously monitor sleep quality and multiple physiological indicators, helping users to understand their health status in real time. However, the sense of wearing produced by contact monitoring devices can have a certain impact on sleep. Therefore, non-contact sleep monitoring methods have emerged.

 DOI: 10.1201/9781003611455-8

2. With the rapid development of computer vision technology, its application in sleep monitoring has been significantly enhanced, effectively reducing the discomfort caused by traditional contact measurement methods.
3. The currently widely used non-contact measurement methods mainly include two types: one is the detection method based on infrared sensors (such as infrared thermography and micro-sensor technology); the other is the detection method based on ordinary cameras (such as Eulerian video magnification technology and skeleton keypoint detection technology).
4. Non-contact measurement approaches must meet several criteria: (a) non-intrusiveness to preserve natural sleep conditions, (b) real-time physiological parameter collection and monitoring, and (c) high measurement accuracy and reliability. These methods enable precise thermal perception monitoring and physiological data collection.
5. Current non-contact measurement research primarily focuses on adjusting the environment based on individual sleep thermal comfort. However, HVAC control based on the varying thermal states of multiple occupants in shared spaces remains unrealized. Future research should aim to balance individualized comfort preferences while optimizing energy efficiency.

In recent years, occupant-centric control (OCC) strategies have been proposed to reduce system energy consumption while enhancing thermal comfort. The OCC strategies generally utilize occupancy presence, count, and location data to identify indoor occupied areas. Additionally, they predict occupant thermal states based on behavioral indicators (e.g., window operation, human–machine interaction) to optimize HVAC system operation.

1. Chapter 4 provides an in-depth discussion of occupant behavioral characteristics and their impact on HVAC system performance. It explores methods for adjusting operation parameters based on occupancy data and predicting thermal comfort using adaptive behavior models to guide HVAC regulation.
2. This book presents comprehensive discussions on various ventilation strategies based on occupant demand, including full ventilation, personalized ventilation, personal comfort systems, and natural ventilation.
3. Future thermal comfort prediction models should consider geographical and environmental diversity as well as population characteristics. Additionally, quantifying control system performance across different regions and building types is crucial for optimizing regulation effectiveness.
4. Future research should integrate psychological factors with other thermal comfort determinants to improve prediction accuracy and enhance HVAC system control efficiency.

8.2 OUTDOOR ENVIRONMENTAL MONITORING

Outdoor thermal environments significantly influence urban residents' quality of life and health, making environmental monitoring a critical aspect of urban planning and management. The study of urban microclimates, particularly pedestrian thermal comfort, has gained increasing attention. Remote sensing imagery and street-view imagery (SVI) exhibit distinct advantages among various monitoring methods, providing large-scale, multi-temporal land surface temperature data and rich urban infrastructure information.

1. Urban microclimate parameters significantly affect pedestrian thermal comfort and are closely linked to urban heat islands, pollution, and energy consumption. Monitoring methods primarily include field sensors and numerical simulations. Future research should emphasize quantitative assessments and refined urban design to address climatic and urbanization challenges.
2. Remote sensing imaging utilizes satellites, aircraft, or unmanned aerial vehicles (UAVs) equipped with electromagnetic signal sensors to acquire land surface information. Widely employed in GIS and environmental monitoring, this technique plays a vital role in urban thermal environment assessment. During extreme weather events, it supports emergency response and early warning systems, ensuring public safety and contributing to sustainable urban development. Future advancements should focus on enhancing monitoring precision to facilitate livable city planning.
3. SVIs provide micro-scale, detailed 3D visual information, aiding in urban street thermal environment assessment and pedestrian thermal comfort evaluation. It quantifies urban geometry, greenery, and sky view factors. However, challenges such as temporal variability, dynamic changes in pedestrians and vehicles, and inconsistencies in image capture times persist. Privacy protection measures may also affect research accuracy. The evolution of deep learning and big data analytics offers promising avenues for improving urban environmental assessments.

For outdoor environmental monitoring, particularly in construction site applications, high-quality image datasets are crucial. Emerging technologies, including UAVs, behavior recognition, real-time positioning, spatial planning, wearable devices, EEG, eye tracking, and virtual reality, have been extensively applied in construction safety monitoring.

1. UAV technology, integrated with computer vision and deep learning algorithms (e.g., YOLO), overcomes limitations of traditional construction safety inspections. It enables automated recognition of construction elements, enhancing monitoring efficiency. The integration of advanced data analytics and automation further modernizes site management.
2. Computer vision combined with LiDAR and radar scanning facilitates data collection, allowing machine learning algorithms to accurately detect worker and machinery positions and postures. Behavior recognition advancements contribute to the automation and intelligence of construction safety management.
3. Real-time positioning technologies (e.g., ultra wide band (UWB), global positioning system (GPS)) help prevent collisions by tracking workers and

machinery locations. Spatial planning, particularly building information modeling (BIM) integration, enhances visualization-based safety management. Worker postures and physiological data can be monitored by wearable devices to detect unsafe behaviors and falls. Safety training efficiency can be improved by eye-tracking and VR technologies. The convergence of these technologies enhances safety management precision and promotes the construction industry's advancement.

4. Future research should focus on technology integration and the application of emerging AI and machine learning techniques to expand safety monitoring capabilities.

8.3 HVAC EQUIPMENT MONITORING

HVAC systems comprise multiple parts, including refrigeration, electrical, and air distribution systems. Their complexity and interdependent failure mechanisms pose challenges for fault diagnosis. Integrating unstructured data (e.g., images and audio signals) with structured data collected from building automation systems and inspection robots can enable real-time automated fault detection.

1. Infrared thermal imaging identifies heat exchanger inefficiencies and fouling accumulation. Additionally, it enables real-time temperature anomaly monitoring, preventing safety hazards such as overheating and leakage. Accurate thermal data facilitates well-informed maintenance strategies, enhancing equipment reliability and cost-effectiveness.
2. Video- and audio-based inspection robots aid in diagnosing chiller room faults. Equipped with infrared cameras, standard cameras, and microphones, these robots capture image and audio data for fault detection. While currently used for diagnosing pump and gauge malfunctions, future applications may extend to other HVAC components.
3. Multi-strategy frost suppression and defrosting techniques should be optimized to enhance defrost timing precision. Effective defrosting minimizes energy consumption while ensuring stable operation and thermal comfort.
4. Accurate quantification of frost/dew accumulation requires further exploration. Advancements in computer vision-based intelligent defrosting strategies should focus on deriving new feature parameters from raw images to quantify frost accumulation levels.
5. Experimental visualization accuracy can be improved by extending video recording durations and employing multi-angle imaging of condensation surfaces. Novel computer vision algorithms incorporating droplet growth rates, detachment frequencies, and merging patterns should be developed to construct reliable condensation datasets.
6. AI-driven models combining computer vision and deep learning should be developed to create generalizable and precise predictive frameworks for condensation heat transfer performance.

Index

A

B

C

D

E

F

G

H